***New York Times* Extended Bestseller • *Los Angeles Times* Bestseller • Amazon Best Food Lit 2010 • *Good Morning America* Top Cookbooks/Food Books 2010 • IACP Award in Literary Food Writing • "Th[illegible]**
***(Entertainment Weekly)* • "The Best Books [illegible] by [illegible]**
***Journal)* • "The Holiday's Most Sumptuous [illegible]**

PRAISE FOR

As Always, Julia

"This book feels like chick lit . . . of an exalted order . . . about all kinds of things, from shallots, beurre blanc, and the misery of dried herbs to politics, aging, and sex . . . Funny and forthright opinions about food and life."
—***New York Times***

"A testament to the fortitude that went into creating a classic." —***Saveur***

"Behind-the-scenes nuts-and-bolts on [a] culinary landmark . . . The letters take on the resonance of a plainsong kind of poetry."
—***Los Angeles Times***

"Alternately meaty and frothy, appropriate for two women who loved to cook." —***Wall Street Journal***

"Delightful . . . Expertly edited by the culinary historian Joan Reardon."
—***Bloomberg News***

"A pleasure both culinary and literary . . . A lesson in how to become an American original." —***Whole Living***

"An important piece of culinary history through the never-before-seen letters between these two witty women." —***New York Post***

"Brilliantly edited and with comments by culinary historian, cookbook author, and biographer Joan Reardon, the book is a feast for any Julia fan."
—***Boston Globe***

"A reminder of the power of persistence, of a dream or a friendship, and of the lost art of writing a letter." —***Oregonian***

"Fascinating and engrossing." —***Dallas Morning News***

Avis DeVoto, Julia Child, and Chef Max Bugnard cooking together in Paris in 1956.

As Always, Julia

The Letters of Julia Child and Avis DeVoto

FOOD, FRIENDSHIP, AND THE MAKING OF A MASTERPIECE

Selected and Edited by Joan Reardon

MARINER BOOKS
HOUGHTON MIFFLIN HARCOURT
Boston New York

First Mariner Books edition 2012

www.hmhbooks.com

Library of Congress Cataloging-in-Publication Data
As always, Julia : the letters of Julia Child and Avis DeVoto : food, friendship, and the making of a masterpiece / selected and edited by Joan Reardon.
p. cm.
Includes index.
ISBN 978-0-547-41771-4 ISBN 978-0-547-57748-7 (pbk.)
1. Child, Julia—Correspondence. 2. De Voto, Avis—Correspondence.
3. Cooks—United States—Correspondence. I. Reardon, Joan, date.
II. Child, Julia. III. De Voto, Avis.
TX649.C47A4 2010
641.5092—dc22 2010025840

Book design by Melissa Lotfy

Printed in the United States of America
DOC 10 9 8 7 6 5 4 3 2 1

To the trustees, administrators, and staff of the Schlesinger Library at Harvard University, whose vision made available the letters and special collections that have made all of my publications possible, including *As Always, Julia*.

And to William Truslow, the trusted lawyer, friend, and custodian of Julia Child's legacy, and Mark DeVoto, the beloved and accomplished son of Avis DeVoto.

Julia views Paris from the balcony of the Hôtel Pont Royale after her arrival in 1948.

Contents

Introduction

A letter truly belongs to the sender and the recipient, and to encroach on their privacy, in effect to eavesdrop over their shoulders, seems to me an intrusion of their personal exchange. But when a group of letters tells a terrific story—creates a complex narrative of professional growth, deep curiosity, and culinary awakening—it's important, yes, even imperative, to open those envelopes again and look at what the correspondence reveals.

The letters included in this collection begin when Julia Child was living in Paris with her husband, Paul, and wrote a fan letter to Avis DeVoto's husband, Bernard, which Avis answered. They continue until 1961, when volume 1 of *Mastering the Art of French Cooking* was published. Beyond telling the warm and human story of a friendship between two dynamic women, they document the evolution of the book from its inception to its publication. They cover a wide range of other topics as well, including the various foods that were available in the United States at the time and their quality; other cookbooks and magazine recipes; the pitfalls and successes of a career in the foreign service; the national temperament and character of the French, Germans, Norwegians, and English; even the Kinsey Reports on human sexuality—all juxtaposed against the hysteria of the McCarthy era and the general political climate of the 1950s and 1960s, mirroring to a surprising degree some of the issues of the current American scene.

Julia, of course, remains a culinary icon, and her rise to fame is familiar. But the woman everybody claims to know is not the one we see here. Julia's letters to Avis reveal the ups and downs of her early efforts to "translate the genuine taste of French cooking into American." She was comfortable writing to Avis about her developing professionalism and the personal toll that Paul's multiple postings took on her cookbook project. Alternately self-assured and self-doubting, she is always genuine, witty, and highly opinionated. We see her during moments of frustration and fulfillment, disillusion and delight.

Avis was Julia's confidante, the one to whom she could write, "For a while, what with all this moving and settling I was becoming frantic, schizophrenic, and even, perhaps, necrophilic, combined with anguish, frustration and ill-temper. But, finally getting into the book-work again, I find I work awfully slowly. There is so much that has been written, by people so much more professional than I, that I wonder what in the hell I am presuming to do, anyway." Avis was also the person whom Julia admired, cajoled, leaned on, complained to, identified with, and considered a partner in her culinary efforts. When Julia exchanged photos with Avis early in their correspondence, she wrote that Avis's face was the one that "I always try to wear when I am in New York, with no success."

One of Bernard DeVoto's students at Harvard described Avis as "very good looking and very sexy-seeming and the only faculty wife who might have said 'horseshit' even to [Harvard] President Lowell." She was the mainstay of and secretary to her husband, the mother of two sons, a reviewer of whodunits for the *Boston Globe,* an editor, an accomplished cook, and a Bread Loaf Writers' Conference regular. When she answered Julia's first letter to Bernard, it was the beginning of what would perhaps be the most important chapter in her life as the inspiration and soul mate of writers and the ultimate facilitator.

After Bernard died in November 1955, Avis scouted for Alfred A. Knopf, read manuscripts for Houghton Mifflin, was house secretary at Harvard's Lowell House, and held a position in the dean's office at Radcliffe College. In 1970, Knopf asked her to proofread volume 2 of *Mastering the Art of*

French Cooking. A year later, she joined the staff of WGBH, Boston's public TV station and home to Julia's cooking show, *The French Chef*. As a member of Julia's television team, she organized Julia's fan letters and assisted in the production of three more shows after the series ended. She was Julia's Cambridge connection when Julia resided at her home in the South of France or traveled around the United States.

Early on, both Paul and Julia recognized that their friendship with Avis offered an opportunity to be part of the Cambridge social scene. At their legendary Sunday afternoon cocktail parties, the DeVotos associated with the Kenneth Galbraiths, the Arthur Schlesingers, and many other members of the Harvard intelligentsia. Paul had lived in Boston when he was young, and Julia felt a kinship with Massachusetts as her mother's birthplace and because of her own years at Smith College. To live in Cambridge when Paul retired from the State Department became a goal that the Childs enthusiastically embraced, and Avis made it happen. After the Childs took up residence there in 1961, telephone conversations and close personal contact largely replaced Julia's and Avis's frequent letters.

What emerges from this collection are portraits of two notable women and their ambient worlds, which amazingly capture the cultural landscape of America in the 1950s and 1960s. Read from beginning to end, these letters also provide insights into the expanding role of women, who were just beginning to achieve recognition and pursue meaningful careers apart from their husbands' positions or fame. It is a compelling story.

Editor's Note

The year was 2006, and lilacs bloomed along Brattle Street. Every day for a week, I made my way to the welcoming Schlesinger Library in Cambridge, Massachusetts, to read Julia Child's letters to Avis DeVoto. Archived in the Avis DeVoto Papers, they had been sealed for thirty years. Now, inventoried and cataloged, they were open to readers.

During the course of many years of research at the library, I had collected most of Avis's letters to Julia in the Julia Child Papers, but I had not been able to read those of Julia to Avis. My anticipation was high as I opened the boxes, took out the chronologically arranged folders, and started to read the letters, typed single-spaced on onionskin, sent from Paris, Marseille, Plittersdorf, Washington, D.C., Oslo, Plascassier, France, and elsewhere to Avis's home on Berkeley Street in Cambridge. I already knew that Avis's letters were intelligent, witty, and engaging, and I soon discovered that Julia's were equally so.

The entire correspondence extended from Julia's initial fan letter to Bernard DeVoto on March 8, 1952, to letters written only a year before Avis's death in 1989—more than four hundred in all. For this book, I have selected those that trace the story of volume 1 of *Mastering the Art of French Cooking*. Because more of Avis's letters have survived than Julia's, I have omitted some of Avis's to achieve a balance.

I have made minor adjustments in the letters for readability, adding italics where appropriate and accents to French words. I have also occasionally supplied punctuation for clarity and silently corrected obvious typing errors, but I have preserved most of the original misspellings, as well as the capital letters and underlining in the originals. If a letter's date appears in brackets, I have supplied it using reasonable conjecture, and I have standardized the form of the dates in the letters. Although I have arranged the letters in sequence, references to subjects therein are not always perfectly chronological because of delays in mail delivery.

Julia and Avis were cosmopolitan women who traveled widely, read extensively, and moved in diverse political, intellectual, and social circles. They had many acquaintances in common and frequently referred to public figures and friends by their first names or initials. Where necessary, I have inserted full names in brackets. I have also added identifying notes for people with whom a broad audience may not be familiar, explaining their significance at the time the letters were written. If a name was misspelled in a letter, I have corrected it in a footnote on the first appearance.

In a few instances, I have deleted references in Avis's letters to personal family matters in the interest of preserving privacy, and I have sometimes omitted references to Cambridge residents and issues because they required more supporting explanation than seemed warranted. All such deletions are marked with an ellipsis centered on its own line. Otherwise, I have made every effort to reproduce the letters exactly as they were written.

—JOAN REARDON, Chicago, 2010

An enterprising cook, Avis steered *Mastering* first to Houghton Mifflin and then to its eventual home at Knopf.

In Paris, Julia thought seriously about learning to cook and enrolled in the professional course at the Cordon Bleu in the fall of 1949.

PART ONE — 1952–1954

All from One Kitchen Knife

The Initial Exchange of Letters Between Julia Child and Avis DeVoto, the Contract with Houghton Mifflin, and a Developing Friendship

Pictures arrived this morning . . . I am so damn happy to have them and thank you both from the heart. All from one kitchen knife. It was a miracle, wasn't it? To think that we might easily have gone through life not knowing each other, missing all this free flow of love and ideas and warmth and sharing . . . We share really almost everything.

AVIS DEVOTO to JULIA CHILD, September 1, 1956

Julia in her Paris kitchen at 81 Rue de l'Université, where she assembled an impressive *batterie de cuisine*.

PREVIOUS PAGE: Julia at the Cordon Bleu with one of her favorite instructors, Chef Max Bugnard.

Almost three and a half years after arriving in Paris with her husband, Paul, Julia Child read an article in a 1951 issue of *Harper's* written by the historian and prolific journalist Bernard DeVoto. In "Crusade Resumed," he revisited what he considered "the only mission I have ever set myself, that of trying to get for the American housewife a kitchen knife she can cut something with." DeVoto criticized American-made stainless steel knives for their inability to hold an edge, and he detailed his continuing search for a carbon steel paring knife. As a cook who had already acquired a substantial *batterie de cuisine,* Julia sent him one, and Avis, who answered most of her husband's letters, acknowledged the gift. Soon "Dear Mrs. Child" and "Dear Mrs. DeVoto" became "Dear Julia" and "Dear Avis."

As an employee of the State Department's U.S. Information Service (USIS), which operated as a sort of propaganda agency, Paul set up photo and other art exhibits that would present the United States in a favorable light. Meanwhile, Julia explored the markets and dined in the small restaurants of Paris. Blessed with a hearty appetite, she had never been particularly interested in food until she began working with Paul in the Office of Strategic Services (OSS) during World War II in Ceylon (now Sri Lanka) and China, where she enjoyed exotic meals. But in the City of Light, she experienced a culinary epiphany, and she had all the fervor of a religious convert regarding French food, wine, and cooking utensils. Above all, she valued the importance the French placed on *métier* (skill). Propelled by her enthusiasm, she began a formal culinary education at Le Cordon Bleu in 1949 and sought out friends who thought about food the way she did. She met Simone "Simca" Beck, a Parisian who was well versed in the cuisines of Normandy, Alsace, and Provence, at a party for embassy personnel. The two clicked immediately and began to discuss food and "how to make a valid professional project out of it." When Simca and her friend Louisette Bertholle urged Julia to join Le Cercle des Gourmettes, an exclusive women's club started in the late 1920s, Julia was delighted. The three women often arrived hours before the scheduled Gourmettes' luncheon to assist the chef of the day in what they considered a private cooking lesson.

Cooking in tandem became a heady experience. So in January 1952, when a handful of Julia's American friends who either lived in Paris or were on holiday asked her to teach them something about French cooking, she persuaded Simca and Louisette to join her in organizing classes in a venture that came to be known as L'École des Trois Gourmandes. The three women taught twice a week. Julia organized the lesson plans and typed the recipes. The format included two hours of instruction and hands-on cooking, after which everyone sat down to a leisurely meal in the Childs' dining room, with Paul selecting the wines.

Teaching cooking classes together soon led to writing a cookbook together. A few years earlier, after Louisette returned to France from a visit to the United States, she and Simca had submitted about six hundred recipes for a book to be published by the New York publishing house Ives Washburn. The head of the house, Sumner Putnam, had hired a translator/writer named Helmut Ripperger to prepare some of the recipes for a small spiral-bound book called *What's Cooking in France*, which was published early in 1952 but not widely promoted. Ripperger was not interested in working on a much larger cookbook, no contract was negotiated, and the project stalled.

Simca and Louisette turned to Julia to help them realize their plan to publish their big book. Although reluctant at first, Julia soon recognized the benefits of the project, which presented an opportunity to test and refine recipes and "translate the genuine taste of French cooking into American." Julia also knew that simply being an American might give her an advantage in dealing with U.S. publishers, and she began to question the wisdom of publishing the book with Ives Washburn. In the fall of 1952, she requested that all of the material already sent to Putnam be returned for translation and editing. Then she, Simca, and Louisette began to outline the tentatively titled "French Cooking for All" and discuss its scope. They told Ives Washburn that they wanted the manuscript published in a sequence of five individual volumes.

Meanwhile, Julia sent the first draft of the book to Avis, who immediately saw its potential and communicated her enthusiasm to Julia, along with the advice to extricate the project from Ives Washburn. Avis was an enterprising cook who knew how rare it was to find shallots even in the specialty shops of Cambridge and understood that there were limited places where an American cook could find a variety of herbs. She quickly became Julia's stateside adviser on ingredients, utensils, and the preferences of American cooks, as well as a valuable sounding board for Julia's staunch liberalism, ambition, and occasional insecurity.

Because Houghton Mifflin was her husband's publisher, Avis knew the staff at the venerable Boston publishing house, and she contacted her friend Dorothy de Santillana, senior editor there. De Santillana instinctively knew that this technique-focused cookbook was unlike the Americanized French recipes that were being offered in women's magazines and upscale cookbooks. She was interested.

In the following early letters between Julia and Avis, the latest victims of Senator Joseph McCarthy's Communist witch hunt get equal time with experiments on beurre blanc. Avis's accounts of forays into the West and excitement over Democrat Adlai Stevenson's 1952 presidential campaign cross paths with Julia's descriptions of the catch at the Old Port in Marseille, where Paul was posted in 1953. ("What luck for us," Julia says. "It could have been AbbisAbababababa.") Family members are introduced, and recipes are discussed at length.

Based on de Santillana's assessment of the manuscript and Avis's enthusiasm, Houghton Mifflin began discussing a contract in late 1953. It was signed on June 1, 1954. One-third of the $750 advance was forwarded to Julia as the representative of the Trois Gourmandes, with instructions to submit the manuscript of "French Cooking in the American Kitchen" (as it was now called) as soon as possible. Avis signed on as Julia's unofficial line editor. It was the perfect arrangement all around.

81 Rue de L'Université, Paris, 7

MARCH 8, 1952

Dear Mr. de Voto:

Your able diatribe against the beautiful-beautiful-rust-proof-edge-proof American kitchen knife so went to my heart that I cannot refrain from sending you this nice little French model as a token of my appreciation.

For the past three years here, I've had the good fortune to be able to spend my life studying French cooking and have amassed a most satisfyingly professional *batterie de cuisine,* including a gamut of excellent French knives. When we were in the USA last summer I picked up four beautiful-beautiful American stainless steel housewives knives, of different makes, to try them out. But I have been quite unable to sharpen them satisfactorily. I am therefore wondering if the average American housewife really wants a sharp knife in the kitchen, as many of my compatriots accuse me resentfully: "But your knives are so sharp! They're dangerous!"

If you are in need of some good professional knives, I would be very pleased to get some for you, and the prices are modest:

This one is about 70¢ (280 francs)
8-inch blade, about $2.40
7-inch flexible fish filleter, about $1.00

Mailed from here Fourth Class, one or two at a time, there seems to be no duty to pay at your end.

We do enjoy you in *Harper's*!

Most Sincerely,
Julia Child
(Mrs. Paul Child)

8 Berkeley Street
Cambridge 38, Massachusetts

APRIL 3, 1952

Dear Mrs. Child:

I hope you won't mind hearing from me instead of from my husband. He is trying to clear the decks before leaving on a five weeks' trip to the Coast and is swamped with work, though I assure you most appreciative of your delightful letter and the fine little knife. Everything I say you may take as coming straight from him—on the subject of cutlery we are in entire agreement.

This is the first knife which has come from France—we have had one from Spain and one from Germany and a great many from all over the States. I am particularly happy to have it because I have known all along that no French cook would be caught dead with a stainless steel knife. I have an enormous collection of knives—all sent since the *Harpers'* piece and an earlier one in *Fortune* last spring—but the prizes of my collection remain two ten-inch chef's knives made by a firm called Pouzet which I have owned for twenty-five years and which I plan to hand down to my children. I was in Paris for a couple of weeks in 1950 but was rather too busy eating to think about hunting for knives, a temptation I am sure you will understand.

An aside on eating—I am green with envy at your chance to study French cooking. There are two dishes served at Bossu on the Quai Bourbon that I remember in my dreams, and if by any possible chance you know how to make them I would be forever in your debt if you would let me know. One is a mixture of eggs, cream and fresh tarragon, done in a saucepan. I probably can't get the right kind of cream for it but I live in hopes—I have tried it a dozen times with little success. The other is their veal in cream with tarragon. I am fairly successful with this—using sometimes a dry white wine or even vermouth—but Bossu probably has access to younger veal than I can get here.

You are quite right in thinking that stainless steel knives can't be sharpened properly. If they are cheap, the steel is too soft to take an edge. The very expensive ones will hold an edge for a long time but can only be re-sharpened at the factory—I am thinking of the brand called Frozen-Heat, made by Robeson at Perry, New York. These people have sent us two five piece sets—retailing at $19.95—and I must admit they are good. But who wants to send knives to the factory to be reconditioned?

We have been immeasurably heartened, however, to find that there are a number of small manufacturers who are quietly turning out excellent carbon steel knifes at reasonable prices, though where they are sold I cannot tell, as the American housewife wants stainless steel and that's what the retailers give her. A butchers' supply house seems to be the answer, and one of these houses in Kansas City sent us a boning knife, which sounds like your flexible fish filleter—a wonderful knife. The best value of all is a set made by the Goodell Company in Antrim, New Hampshire—long slicer, short chef's knife, boning knife, paring knife, pot-roast fork, plus a magna-grip to hold them on the wall—it retails for $7.95. This set restored our faith in American industry, and my hardware man at Harvard Square now stocks them and they sell like hotcakes.

Since you are on your way to becoming an expert in these matters, I wish you would tell me how you sharpen your knives. Steel, hone, or what? I haven't mastered the hone, though if I ever get time I mean to practice until I have learned to do it correctly. I do use a steel frequently, but I find that once a year or so my knives need a trip to a really good professional who grinds them very gently and then finishes with a hone. *The New Yorker* last week spoke highly of a sharpener called the Emmons Rifle which has been made in Connecticut for a hundred and fifty years and sells for all of ninety-five cents. It's a fourteen inch paddle covered with some special Turkish emery—I have sent for one but it hasn't come yet. Most patent sharpeners are death on knifes, I find—especially the kind that give a sharp edge but remove ribbons of steel in the process. Of course you are right in thinking that lots of women are terrified of sharp knives—but once having used a good one I don't see how they can go back to hacking and mashing.

We have learned a tremendous lot this last year on the cutlery question, and sometime DeVoto is going to write another piece on the subject. We want to visit some of the manufacturers first—at Antrim, and at Southbridge and Ayer, Massachusetts, where Dexter and Murphy make fine knives. We are even working our way through a six-volume affair called *La Coutellerie Depuis l'Origine Jusqu'à Nos Jours,* by a gent named Pagé, but since it was written before the invention of stainless steel, it won't be much help except to show off our learning to the enemy, who probably won't be impressed.

Thanks again for the knife, which is a little gem. My husband, I regret to say, has snitched it for his own use—cutting the lemon peel the proper thinness for the six o'clock Martini—but it will be mine while he is in California. We are both pleased that you like the Easy Chair,* and we both enjoyed hearing from you.

Sincerely—
Avis W. DeVoto

81 Rue de L'Université, Paris, 7

MAY 5, 1952

Dear Mrs. De Voto:
I have been planning to write you a long and well-organized letter in response to yours. But, with the tourist season and the sudden inauguration of "Mrs. Child's Cooking School, Paris Branch," there has been too little time. As to the tourists, we have been to the Follies Bergères, the Lido, the Ritz Bar, the Meurice Bar, the American Express, the Coq Hardi, the Tour d'Argent, the Table du Roi, and are now taking quite a bit of bicarbonate of soda. I don't ever want to go to any of them again except the American Express. And what is worse, a great many of these people are for Taft† and think McCarthy is doing a fine thing, which makes digestion even more difficult. What is the country coming to, I sometimes wonder.

The cooking school, "L'École des Trois Gourmandes" is a joint enterprise with two French women. Tuesdays and Wednesdays from 10 until 2, including lunch, maximum of 5 pupils. Object, to teach the fundamentals of French cooking to Americans. So far we've given 22 lessons, and it is working out well, and a tremendous amount of work for us professors, as each of us has slightly different methods, and we want everything to be as perfect and complete as possible. I'm enjoying it immensely, as I've finally found a real and satisfying profession which will keep me busy well into the year 2,000. But I wish I had started in when I was 14 yrs. old.

I was fascinated with your account of the KNIFE. You must have a kitchen full by now. And particularly pleased to hear that there are some good ones being manufactured in the USA. The "Frozen Heat" types sound interesting, but who, as you say, wants to send them back to be re-sharpened. You'd have to have two sets. But, having seen lately, how few women have any idea how to use a knife (including French women), and remembering our American ideal of "glittering beauty," it will take a great deal of awakening to make the average housewife knife-conscious.

* Bernard DeVoto's column in *Harper's*.
† Robert A. Taft, a conservative Republican senator from Ohio, who was vying for the GOP nomination for president in 1952 but lost to Dwight D. Eisenhower.

I suppose one method will be the slow but sure one of the home economics courses, such as *Life* wrote up several weeks ago.

As to knife sharpening, I am no expert. My husband, who is a one-man art factory, goes over them once in a while with a soap stone (carborundum with oil); and I keep an edge on them with a big steel *fusil.* I have once this year taken them all to a professional, who did a beautiful job, but I did feel he took off quite a bit of steel. However, I don't expect them to last a life time, I just want them to cut well. I don't know anything about hones. The next time I have the "man" sharpen them, I will hover over him, and report what he does. I think he has a mechanical wheel, and also a hand stone. Would be interested to hear how you like the Emmons Rifle, and what it is.

You mentioned French cream. I think the difference between theirs and ours is how the separator is set. In theirs, the separator is set so that what comes out of the "other" spout is practically water, which is given to the pigs or thrown away. So it, the cream, is a most concentrated thick mass. And what you get is "whole cream." Some of it they make butter out of, the rest is sold as "crème fraîche." And they don't put it under refrigeration, so it gets that slightly sour taste. I wonder if one took US cream and put it through a separator one wouldn't get the same result. You could always get a richer flavor to your sauces by "buttering them up," i.e., after the sauce is off the heat, beat in bits of butter, a bit at a time; but never reheat to even under the boil again or the sauce will thin out or the butter will release itself. And you could get a slight sour by a few drops of lemon juice.

Have been only once to Le Bossu, and had neither of the two dishes you mentioned. Scrambled Eggs, French, they do not scramble them as long as we do; the eggs remain in a soft broken custard, rather than in flakes. Done over very low heat, almost like a Hollandaise, stirred constantly until they "custard." A bit of butter beaten in at the end will add richness. Tarragon could be chopped up with the butter, and so impregnate itself.

Veau à la Crème, à l'Estragon (a method). Heat butter in pan until it has foamed and is just turning blonde. Sauté the veal, about 2 minutes on each side. Salt and pepper it. Stir in minced shallots. Pour in some brandy, set it aflame, let burn ½ minute and cover pan. Pour in a bit of white wine (dry) and some good reduced meat stock. Sprinkle in ½ your tarragon (chopped). Cover pan and let stew about 5 minutes. Uncover pan, turn up heat, let sauce reduce. Add heavy cream, and let that reduce. Sprinkle in fresh herbs before serving. The veal shouldn't be overcooked, and all of this shouldn't take more than 15 minutes.

Mr. DeVoto's article on the hopelessly antediluvian monsters that Eisenhower will face in Congress* is a most sobering thought piece. But I want the Republicans in anyway; they need to "grow up" to their responsibilities. I have faith that the nation is strong enough to withstand them and to teach them, though my faith is not without dreadful qualms.

Mon mari se joint à moi, chère madame, en vous priant d'accepter, vous et votre mari, l'expression de nos meilleurs voeux,†

Julia Child
(Mrs. Paul Child)

8 Berkeley Street
Cambridge 38, Massachusetts

MAY 30, 1952

Dear Mrs. Child:

The knife has come and it is a dream—I am somewhat overwhelmed by your kindness. I have not used it yet but I expect it will perform nobly—I want to try it first on a fish. In about two weeks my twelve-year-old son and I move down to the shore for the summer, and if he can't catch what passes for sole in those waters (flounder), he can provide me with a mackerel. But I think the knife will also be very good for vegetables—it really is a joy to use knives like the two you sent me. They are as easy as possible to sharpen with a steel. Some of my best American knives are hell to sharpen, especially the hollow-ground ones. Hollow-ground is a hollow fraud—it may be fine for razors but in the kitchen it is perfectly useless.

Of course I know the Flint knives whose pretty picture you sent me, and I have a fine specimen. As I think I told you, these stainless steel knives, the expensive ones, hold an edge for quite a long time compared with a carbon steel knife, but it is practically impossible to sharpen them at home as the steel is too hard and too brittle. I have been faithfully using the Frozen Heat knives sent me by Emerson Case,‡ because I promised him I would, but they weren't as sharp as your two knives to begin with. Funny thing happened with one of those Robeson knives. I had an extra

* Eisenhower hoped to win over the isolationist Republicans in Congress to an internationalist approach.

† "My husband joins me, dear lady, in hoping that you and your husband will accept the expression of our best wishes."

‡ Case, president of Robeson Cutlery Company in Rochester, New York, developed a process for heat-treating stainless steel blades to make them harder that he called Frozen Heat.

one, roast carving size, sent me by the then president of Macy's last summer, but when Mr. Case sent me two of his sets I gave this odd knife to my part-time cook. Mary never abuses a knife; she has been too well indoctrinated by the DeVotos. But she was slicing an onion on a board one night at home and the Robeson knife was suddenly in two neat pieces. I sent it back to Mr. Case in great glee, and he replaced it at once since those knives are guaranteed for a lifetime, no less, but he has still not explained why or how it could possibly have happened. In all my days I have never had a knife come apart in my hands, even when digging dandelions out of the lawn.

Of course there is nothing I would like better than to come and look at French knife factories—except possibly go to England which is the great love of my life. But, fat chance. Or rather, there is about one chance in a million, because there is a vague deal on foot which would bring both the DeVoto's to France for a month or more to take a long look at French industry and the plan Schuman.* I really doubt if this will come off, and DeVoto takes a dim view of it anyway, since he has always been oriented West like a homing pigeon and indeed has just returned from a six week's trip to California, Oregon, Utah, Colorado, as you will duly read in *Harpers*. However, I can dream, and the powers that are talking about this trip can easily deal with such minor matters as transportation in a summer when all space is at a premium. I am very superstitious about this possibility however, so don't mention it to anyone at all.

The Emmons Rifle is not anywhere as good as Sheila Hibben† says it is, or perhaps I have not mastered its use. I am mastering the hone or carborundum stone, but I use my steel every day. I have an absolute beauty, a French import sent to me by a butcher's supply house in Kansas City.

Recipes noted, and many, many thanks. I shall try your method for Veau à la Crème, minus shallots which I can never find in the market. It is very like my own method, except that I do not use brandy, hence no flaming, and that I use olive oil instead of butter. I like to do vegetables in olive oil too—mixtures such as celery, green peppers, onions—plus sometimes eggplant and tomatoes. I like to do young carrots and scallions that way too, very little oil and let them cook in their own steam.

---

* The 1950 Schuman Plan, a precursor of the European Coal and Steel Community (1952) and eventually the European Union, called for establishing a single organization to pool European coal and steel resources.

† A food writer who had consulted on the White House menus for Eleanor Roosevelt and was a columnist for *The New Yorker*.

I still hope you can eat the eggs with tarragon at Le Bossu, because there is some element there that I have not hit on. There is one more never to be forgotten egg dish that I hope you can help me with. This is called pipérade and is I think of Basque origin—it is very soft scrambled eggs with tomatoes and various kinds of peppers, so that it is full of little hot pockets. The best place to eat it is a tiny little hole in the wall between the Rôtisserie Périgourdine and a little nightclub named L'Ecluse—I have forgotten the name of the restaurant which has two or three tables on the sidewalk and a few more inside and is very inconspicuous and cheap. At least it was in 1950. I really cannot think of any dish more suited to a quivering stomach after too much wine the night before! Not that I ate it as a hangover cure; I have not had a hangover in fifteen years, thank you.

I am anxious to get to the beach—I'm tired, dispirited after a hard winter, and nothing smoothes me out like sea air and long afternoons on a hot beach. My son Mark is great fun to live with and we gourmandize a lot, since Mary remains in town to look after my husband and elder son. Address, incidentally, 21 Leonard Street, Annisquam (Gloucester), Mass. This is on Ipswich Bay and rather heavenly. Cape Ann also provides what I stubbornly maintain are the world's best lobsters. I also stubbornly maintain that the only real way to cook lobsters is in three or four inches of sea water, in a covered kettle, for about twelve minutes (pound and a quarter lobsters being the ideal size). You then drape these dazzling creatures over the rocks until they cool off a bit, tear them apart with the bare hands, dip each piece in melted butter, and guzzle. There should be from two to six lobsters per person. While the lobsters cook and cool off, two dry Martinis à la DeVoto should be served. Nothing whatever else should be served—we are eating all the lobster we want, we are not fooling around with salad or strawberry shortcake or even coffee. All you need are the martinis, plenty of lobsters, millions of paper napkins, and a view. This is one of the culinary matters where the Americans have it all over on the French, as I hope a loyal American like you will agree. Ditto strawberry shortcake, season for same being upon us when the New Jersey berries come in next week. A strawberry which has traveled from Florida or California has no taste at all. I am not saying that the French lobster in all its sizes and varieties is not a fine thing, but when you have plenty of lobsters right out of the ocean I think it is a crime to obscure that heavenly flavor with any sauces.

We will be in Annisquam until September 10th or so, DeVoto commuting week-ends as it is only 45 miles.

I am fascinated by your account of your cooking school—what a wonderful job to tackle. Do you plan to come back to America eventually and set up shop here? You will have no competition except Mrs. Lucas,* who at present is sweeping all before her, on radio, television, and in person. She comes to Boston every spring under the auspices of the Smith College Club to make six or eight public appearances, which are mobbed. I must admit she is a very good showman, and I believe she is making a fortune. She also hits the road and has appeared as far west as Seattle. Plenty of room in the field for you, and a very good living. I would give anything to take such a course as I really do love to cook and there is so much I have to learn. But there just isn't time. I do a weekly newspaper column on detective stories—very mere, badly paid, and I do all my husband's secretarial stuff, all the typing, take dictation from a tape recorder, and handle a lot of his routine mail on my own. The house is big and the children need me a lot. Plus all this I am now correcting proofs on the new book, which will appear late in the fall, and it's a hell of a job, the footnotes and bibliography especially which have some horrifying items in illiterate, 16th–17th century French, Spanish and so on. I am losing my mind.

Well, it is very relaxing writing to you, but there are other fish to fry. I am keeping your letters in my fat file on kitchen knives—we still have hopes of another article on the subject but we haven't got round to visiting these local foundries yet. Maybe next fall, if things calm down a bit. If you ever hear of a stainless steel knife made in France, I wish you would let me know—I cannot believe that French manufacturers will ever fall for that fraud, but in the interest of expanding markets they may have to. Let us hope they find chrome too hard to come by.

Now I must bully my eldest into washing the car, a proper job for Memorial Day when any cautious American stays off the roads. I love my new knife, and so do all the visitors who have seen it. Thank you a thousand times.

Yours,
Avis DeVoto

---

* Dione Lucas, an English chef, cooking teacher, and cookbook author; an early champion of French cooking; and the first woman to have her own cooking show in the United States.

8 Berkeley Street
Cambridge 38, Massachusetts

OCTOBER 3, 1952

Dear Mrs. Child:

I have been terribly delayed in writing to thank you for the little cookbook. I have had a stubborn virus for three weeks now, my maid only comes three days a week, a compromise between what I need and what her husband wants, and of course everybody in the United States is perfectly hysterical about the campaign. Including me. I hope to God you two have absentee ballots in good working order. The spectacle of Eisenhower's complete capitulation to Taft is something to see, and l'affaire Nixon* is and has been quite terrifying. I was completely confident until Nixon's soap opera was so successful and can only pray that serious Republicans return to their senses.

Anyhow, during all this turmoil I have managed to make eggs pipérade three times with complete success except that I found there was a little too much fat in it—going a little light on the oil and butter takes care of that and I am overjoyed to have the recipe. Thank you very, very much. My family thanks you too. It is all so simple but I never would have thought of three kinds of fat in one recipe—it works like a charm.

If we live through the election perhaps I can get back to some serious cooking, but now life is far too full of telephones ringing. What with the boss going out to Springfield and helping with other speeches here, I have to swing quite a lot of his affairs myself. After a blistering hot summer during which my weight fell to 112 pounds, I could do with some peace and the opportunity to putter around a stove. And I have told DeVoto that if he ever gets back to the knife question, if he tells the truth he will have to accept my verdict that French knives are better than American or any others. My hand automatically reaches for the two you sent me and I vastly prefer them to any others. They are far easier to sharpen, too.

I hope all goes well with you and with your school, but I am sorry you have to miss the campaign doings. They are to be sure quite exhausting, but I have never seen the population so excited—nobody talks of

* During Eisenhower's campaign for president, the vice presidential candidate Richard Nixon of California was accused of having appropriated campaign donations for his own use. In a nationally televised speech, he defended himself and said the only political gift he intended to keep was the family cocker spaniel, Checkers.

anything else. And you are not seeing Adlai [Stevenson] on Television. (I never looked at TV until the conventions but now I see why it was invented.) Please do not think me too immoderate when I say that this is a completely new kind of man in politics. I think he is going to be bigger than Roosevelt—but then, I didn't always admire FDR though I voted for him four times. I hope the French press is covering all this adequately—the English papers and weeklies have been very satisfactory. The *Economist* certainly isn't missing anything.

Sorry to sound so frenetic—I am in a state, what with my throat and everything. Now I must dress to go to the Ritz to have dinner with Alfred Knopf, that great gourmet [and publisher]—I'm perfectly certain he has been harrying the chef all afternoon. I hope there are no Republicans present as I would like to relax with the vintages Alfred probably ordered weeks in advance.

Blessings. And Vote.
Avis DeVoto

81 Rue de L'Université, Paris, 7

DECEMBER 15, 1952

Dear Mrs. De Voto:
I have not yet answered your very nice letter of way back before the election, but I have thought of you both with affection and interest. Was delighted to see that Mr. De Voto was in on the Stevenson campaign. It must have been a bitter disappointment to all of you, after the work you put into the campaign. Stevenson appears to be a marvelous fellow, from all points of view, and the more he talked, the better I liked him. He made a tremendous hit here with the French, which probably did him no good at home. But I do think he would have had one awful time, with more and louder McCarthyism. I am, of course, "persona traitoria" to my family in California, who are Old Guard Republicans of the blackest and most violently Neanderthal stripe. My nice step-mother wrote me the other day saying please not to write anything more to my father about either politics or Charlie Chaplin,* as it upsets him too much. And my brother, from Pittsfield, Mass., writes to inform me I have no right to foster my ideas on other people, and that furthermore the only real red-blooded Americans

* The comic film star was forced to leave the United States by the FBI in 1952 because of his leftist political views.

are the Republicans. YE GODS. Well, I hope tempers will subside and a bit of mature reason shall reign.

I am sending you with this letter a part of the Sauce Chapter from our forthcoming book FRENCH HOME COOKING. We are so steeped in it, we cannot look at it objectively, and need some intelligent American opinion.

We seem to be sewed up morally (not legally) with Ives Washburn Company. Do you know anything about them as a publishing house? I immodestly think that this could become a classic on French cooking, as it is a complete re-studying of classical methods and recipes, in view of making them easier to do, and of bringing them up-to-date, and of making them understandable to the novice, interesting for the practiced cook. I do not think, moreover, that we are at all in competition with the *Gourmet* publications, or any of the other French cook books we've seen, as we are not trying to be an encyclopedia, only a good book on the fundamentals . . . with the object of making cooking make sense.

Whether or not it makes sense to you, also, interests me very much, so if you have the time, I shall be deeply appreciative of any comments you can make. And please be frank and brutal.

WISHING YOU BOTH A VERY MERRY CHRISTMAS!

Julia Child

PS: Please do not show this ms. to anyone. I think cooking recipes and methods are too easily stolen, and, as quite a bit of this is new stuff . . .

8 Berkeley Street
Cambridge 38, Massachusetts

CHRISTMAS DAY 1952

Dear Julia:

This must be very fast. Christmas day, I'm cooking a dinner single-handed, roasting a turkey and I baste constantly, guests coming, general confusion. But your manuscript came yesterday and I am wildly excited. I haven't had time to do more than dip here and there, and my report will have to come when I have gone over it in detail. But I most heartily approve the general scheme and am absolutely convinced that you really have got something here that could be a classic and make your fortune and go on selling forever. Every place I dipped I came up with something

I hadn't known or only got to know by experimenting and making mistakes. Such as aluminum tends to make egg yolks greenish.

Now look. Ives Washburn is a perfectly good publisher but small, poor, not known well. I want to grab this book for Houghton Mifflin. They have published B. [Bernard DeVoto] since *Across the Wide Missouri* and he is in every way deeply satisfied with them. They would give you a great deal more promotion—advertising, generous advances if you want them. And they are absolutely wonderful to work with, on the editorial end as well as all others. And they publish a lot of cookbooks—I've just found half a dozen on my shelf, including June Platt.* Little, Brown publishes Dione Lucas and I know HM would love to wipe their eye with you.

Lovell Thompson who is Vice-Pres. and head editor is dropping in here today as he and his family always do, bearing bourbon, and I'm going to talk to him, preparing the way. He knows nothing about cooking so it won't be at all difficult to keep from showing him your manuscript. But I do want your permission to show it to Dorothy de Santillana, who knows more about food than any woman I know. She is their top editor. She used to be married to Robert Hillyer.† She is now married to Giorgio de S., who is an Italian marquis and teaches history of philosophy at MIT and is a darling. They go abroad every summer—well, almost—and might very well see you in Paris this year. You'll die when you meet Dorothy because she is very beautiful and enormously fat—I think this is really one of the rare glandular cases—it makes no difference because she is a great natural force and men gravitate to her like flies. I'm quite sure she'd give her eye-teeth to get this particular book.

Because there isn't any cookbook like it. I've got half a dozen French ones and you know only too well what they are like. They take it for granted that you've spent five years mastering all the basic techniques. Lucas came closer to what Americans want but then again she doesn't explain anything much.

So this is being fired off air-mail to ask if I can show the ms. to Dorothy. If you want me to I will stay with her while she reads it and bring it home with me and not leave it in the office, though you are perfectly safe as HM Co. is a monument of integrity and frequently loses money by re-

* A well-known author of American regional cookbooks during the 1940s, '50s, and '60s.

† A Pulitzer Prize–winning poet and novelist, formerly a Harvard University professor.

fusing to descend to the sharp practices of some other publishers I could mention.

And don't forget advances, dear. Just between us, they advanced DeVoto twelve thousand dollars so he could finish *Course of Empire* (which thank God has sold enough in the first two editions to very nearly wipe out that advance).

So please write me at once air-mail giving me a green light. B. says there is no such thing as a moral obligation to a publisher. He also says there is no conceivable doubt that HM would do much, much better by you. Must stop now as the kitchen calls. Let me know at once. And I'll report at length on the ms. when I have time to read it.

Blessings on you both and Merry Christmas,
Avis DeV.

81 Rue de L'Université, Paris, 7

DECEMBER 30, 1952

Dear Avis:
We've just returned from a week in Angleterre and found your letter. How nice you are to take such an interest, and to write in the midst of turkey time. Your letter has thrilled us all (would say excited, which is my real reaction, but am learning not to use that word because of its more carnal implications in French!). Yours is the first word we've had, and as my dear Paul is really thrilled when he receives an evaluation from another painter or photographer, your reactions, coming from a knowing source, fill me with a great flood of happiness. *Je ne saurais comment vous dire, chère amie, à quell point j'en suis touchée.** So without more ado, but with brimming heart, I shall plunge in as follows:

Your comments on Ives Washburn confirm what I have managed to pick up about them. Before going into this, I asked Paul's nephew, Paul Sheeline, of Sullivan and Cromwell, to act as our legal advisor. (He is receiving copies of all correspondence we receive and send.) As you can probably gather, we don't know beans about the publishing business, but want to avoid as many stupidities as possible. He is a fine, hard-headed young man, Harvard graduate, in his early thirties. Paul Sheeline wrote us that from what he had been able to fathom, Ives Washburn did not have too good a reputation, particularly among book dealers, as they do

* "I don't know how to tell you, dear friend, how touched I am."

no proper advertising and are slipshod in their accounts. However, Sheeline had talked to his father-in-law, Donald Moffat,* and Moffat felt that almost any deal which can be made by a budding writer with a publisher was a good one, in that the publisher is taking considerable risk. Although the news about Ives Washburn was discouraging, I have never been at all impressed with the fact that we are unknown writers . . . as I am convinced that if we can get the book into the hands of someone who knows about cooking, it will sell itself. So, although Ives Washburn appears to think they've got us in the bag, we are not committed to them in any legal way.

The moral obligation was contracted by Louisette Bertholle (as you will see from the attached "History"). Neither Simone Beck nor I have had any personal contact. Sumner Putnam, of Ives Washburn, who became a real friend of Louisette's (she is a charming and sweet person), introduced her around, entertained her, etc. etc. etc. And his great friend and legal counsel, Peter Trafford, is the uncle of her son-in-law, etc. etc. etc. Putnam sounds like a very nice man indeed, but I have the impression he is a fellow with the mazuma who has taken up publishing as a hobby; and that he would like to publish a French cookbook because it would be sort of the chi-chi thing to do. (Maybe I'm wrong.)

Anyway, besides their evidently miserable reputation, my two other grave hesitations about Ives Washburn are:

1) As we are planning to do the book in sections, suppose they fold up in the middle of the series.
2) Neither Putnam nor his present staff knows anything about cooking (glaring example of which is *What's Cooking in France*).

We like everything you say about Houghton Mifflin. Firstly, they publish DeVoto, which is both endearing and completely reassuring. Secondly, they know about cook books and have, from your description, a most able editor in your friend Dorothy de Santillana. This is of prime interest to me, for she sounds like just the woman we need, who can evaluate a ms. professionally and give competent criticism.

So, we Trois Gourmandes have unanimously agreed that we would like very much for you to show her the ms., and have attached the rest of the chapter (including the TOP SECRET recipes, which we'd love to

* A longtime author and regular contributor to *The New Yorker.*

have you try out and report on). Also, for you to use at your discretion, copies of the correspondence we've had with Putnam, and a history of our relations with him. They may feel the story is too involved for them to enter, but, if they happen to like what they see, I hope not. In any case, you have here the whole story, and are just as au courant as we are.

We have as yet heard nothing from Putnam, though the ms. was sent to him on Dec. 10th. We shall do nothing until we hear from you, except to begin getting our soups into shape.

You mention *Course of Empire,* is that the new book of which the Lewis and Clark episode in *Harpers* was a part? In any case, I'm sending for it right away, I guess we missed the notices. It's a period which fascinates Paul, and we both chewed into the *Harpers* bit with great satisfaction.

With much love to you both, and all our
most succulent wishes for 1953,
Julia Child

Les Trois Gourmandes (*left to right*): Louisette Bertholle, Simca Beck, and Julia prepare to teach their cooking class.

Enclosures:

1) History of our relations with Ives Washburn
2) Copy of letter to Putnam, Dec. 10, 1952
3) Copy of letter, Putnam to Child, Dec. 9, 1952
4) Copy of letter to Putnam, Nov. 30, 1952
5) Table of Contents for book, by sections
6) Balance of Sauce Chapter, pages 28 to 58 (censored)
7) Censored section, pages 36, 43, 54

To A. De Voto, Dec. 30.

ENCLOSURE I.

History of our relations with Ives Washburn

Louisette Bertholle and Simone Beck started writing their big book on French cooking about 5 years ago, it covered the field in the form of a collection of recipes, regional specialties, wines, cheeses, etc. It was in French and had an English translation which needed re-writing. Louisette Bertholle took the ms. to New York in 1951 and it interested Sumner Putnam of Ives Washburn. Although Louisette suggested an editor, Putnam insisted on having one Helmut Ripperger, a "food advisor," evidently quite well-known in the New York trade. They then decided to put out a little teaser for advance publicity, called *What's Cooking in France* (I sent you a copy). This was almost entirely the work of Ripperger, who chose the recipes, wrote them up, but never happened to send a proof back to Beck and Bertholle in Paris. They never saw anything of the book until it appeared in print. (!) It is an attractive-looking little book, and the introductory and bridge passages are charmingly written, the recipes themselves are not very professional. Even so, selling at $1.25, one might expect it to go over, but it was given only about one page of publicity, and has not done well at all, or if it has, Ives Washburn has not let on. A contract was signed for this little book. In the meantime, Ripperger was to put the big book into shape, and Putnam hired him at something like $50 a week. But no contract or advances were made between Putnam, Bertholle and Beck. Sometime last summer, Ripperger, having produced very little, finally threw in the sponge or had the sponge thrown at him. Bertholle and Beck suggested that I take over the editing job, to which Putnam agreed in a letter to Bertholle. However, no agreement of any kind was gone into, and I never heard from him directly until he sent a letter to me on Nov. 20, saying he was returning the Beck/Bertholle ms. via our embassy pouch. Nor did I communicate with him until Nov. 31,

1952, when we wrote him a joint letter telling him the sauce chapter was about to be sent, and that the whole book was changed. (In the meantime, we had solicited the services of a lawyer, Paul Sheeline, of Sullivan & Cromwell, New York.)

The title FRENCH HOME COOKING, is the title chosen by Putnam, and belongs to him. However, the book as it is now being written has no relation whatsoever to the former book, and the various papers attached will explain all about it.

On Dec. 10, 1952, we sent the sauce chapter to Putnam, via Paul Sheeline, with the covering letter which is attached. We have as yet heard nothing from him.

8 Berkeley Street
Cambridge 38, Massachusetts

JANUARY 2, 1953

Dear Julia:

This is just to report that your second installment arrived this morning and I have just finished reading it through. I must say I am in a state of slight stupefaction. I am so keen about this proposed book that I am also feeling it can't possibly be as good as I think it is. And knowing the publishing business, I am in a state of despair at the time it is going to take to have Houghton Mifflin make up their minds—I am nothing to them except wife of one of their authors, friend of most of the executives, and occasional reader of ms. and consultant. I am now trying to get Dorothy de S. on the telephone and she is still out to lunch and also it is that horrible week right after holidays and she may not be back this afternoon. I want to take the manuscript in to her house tomorrow afternoon and spend a couple of hours with her, showing her correspondence and so on. I know she will take fire as I have. She has had private lessons with Dione Lucas in N.Y. and she will know far more than I do about your methods.

If she is as excited as I think she will be, then we have got to work our way through Lovell Thompson and Paul Brooks.* I have sounded out Lovell already. But only in broadest outline. We were at the champagne stage of our Christmas dinner when the Thompson's arrived with their three staggeringly handsome children so I was in no particular condition to talk business. But Lovell is most amiably disposed to cookbooks and

* The editor in chief of Houghton Mifflin.

bent an attentive ear to all I poured forth. I had just written you for permission to show Dorothy what you had, and he agrees she's the right gal. What he will say about this Ives Washburn entanglement I don't know. What I am afraid of is that Dorothy Fisher* may try to snag it for somebody else. But please, you three, possess yourselves in patience until I have a professional opinion or two. These things take time—and it is maddening to wait.

But I will keep a finger on the pulse and am in a position to go in and scream at excessive delays. And first I must see Dorothy [de Santillana] and that will be as soon as humanly possible.

You wanted criticisms and I haven't enough continuity in my life these days to give them in detail. But one or two things occurred to me as I read the second installment today. (Anyway I think I don't want to criticize in detail because that is the job of the editor when the book has been contracted for, and if it's HM as I pray, Dorothy is more than qualified to do the job you need.)

But a few small items. I don't like your word "recipient." Think you should in each instance specify what you mean—container, receptacle, bowl, saucepan or what not. Recipient isn't American in that sense at all.

Are you going to say anything about different kinds of olive oil? Suppose you use regularly a light French oil—but lots of us here use the rougher, more strongly flavored Spanish or Greek oils or Italian. I cook a lot with olive oil—particularly mixtures of vegetables—and the flavors do vary considerably. Then of course lots of Americans use the vegetable cooking oils like Wesson and Mazola.

Stale bread—in your recipe for Aïoli. I'm afraid you've got to say something about American bread here—you at least, Julia, are familiar with the various mixtures that are called bread, and are sliced and packaged in waxed paper. We never have it in the house, but most Americans do. Until Pepperidge Farm bread appeared on the market I made all my own bread—now I only bake when we want brioche or something of the sort. You can't get real French bread hereabouts—our "French" bread looks the same as you get in France, but is much crustier and drier. So what do you mean when you say a slice of stale bread? This will give you a fine headache.

---

* Dorothy Canfield Fisher, to whom Simca and Louisette had dedicated their previous small book of recipes. As a best-selling author and member of the editorial board of the Book-of-the-Month Club, Fisher knew many publishers, and Avis is suggesting that she might try to interest other publishers in the book.

I'm afraid I'm inclined to agree with the reactions of your English critics and Mr. Sheeline on the French pronunciations, and God knows I am not bi-lingual. I never dare to speak French because I am so timid about it. But I can read it, and I have a peculiar passion for hearing it sung—Fauré,* Debussy, etc.—that turns me into a jelly. But I do feel that anybody who will be interested enough in French cooking even to open the covers of this book will also know enough high-school French pronunciation not to have too much difficulty with the terms. I am rather torn. I do think your phonetic pronunciations are amusing, even helpful—I also think if you are going to have them at all they ought to be in a demure little section in the back of the book. Because if this book works out as I believe and hope it will, it <u>is</u> going to be a classic—a basic and profound book even if it is written readably. I like the style enormously—it is just right—informal, warm, occasionally amusing. But it is basically quite a serious book and I feel that the phonetics detract a bit from its dignity. However, this certainly is a question for the editor to decide, and I would leave it to her.

Also, and again this is a question for the publisher and editor to decide (and I'm just talking off the top of my head and could be so wrong)—I am against having it published in parts. I would much prefer one big book and I don't give a damn how big or how expensive. And if you get the right publisher it won't matter how long it takes to test and try out and write and edit. If the publisher is interested enough, he will wait until you are finished with your work. But the cook wants it all between the same covers, and I'll bet a hat the publisher does too. I think Dorothy will say the same.

Alright—I have just got Dorothy on the telephone—and I am going in to her house tomorrow at four (Saturday) with the ms. which I hate to let out of my clutches. She is excited. We can have a couple of hours over it all alone and then we have to collect our men and go to a cocktail party which can't be avoided. I will send this off, and will write after I have talked to her. And she is going abroad this summer. We will now join in a moment of silent prayer. I'll keep in touch.

Lord—another postscript—I don't seem to be thinking very consecutively. I just do not agree at all with Donald Moffat's notion that an apprentice writer had better be thankful to get any kind of contract

* Gabriel Fauré, a late-nineteenth-century and early-twentieth-century French composer of chamber music and songs.

from a publisher. Much as I admire Donald Moffat, I can only say phooey to that. Certainly a border-line publisher may take advantage of a new writer, which is why you must stay out of the hands of any publisher who isn't long established and absolutely first-rate. But there are twenty firms who rate that way. And a good publisher like HM or Harpers or Knopf or Little Brown and so on will give you the standard undeviating contract, pay you what advances are necessary, advertise as much as they can afford, even gamble on advertising appropriations if they believe in the book enough. Moffat simply couldn't have had any idea what kind of a property you have here. Did he see any of it? My God, he knows enough about French cooking to recognize what kind of book this can be.

But you must resign yourselves to TIME. I don't know how much of this you have written down, but the editing job alone is going to take months and months and months. Here I am talking as if HM had already signed a contract. And that will take time too. Don't for the love of heaven let anybody rush you into anything. One step at a time. And the next step is tomorrow when I go into a huddle with Dorothy.

Blessings,
Avis

B. gave me Chamberlain's *Bouquet de France* for Christmas—lovely, lovely book. Practically nothing explained about how to do things. Sells for $10 and selling <u>well</u>.

8 Berkeley Street
Cambridge 38, Massachusetts

JANUARY 4, 1953

Dear Julia:
Short bulletin. I saw Dorothy yesterday—we spent better part of two hours over the ms. and the correspondence. She is just as excited over this book as I am, though she has not read it word for word—she is doing that now. She read enough to judge its quality, however. She sees it as I do—as a property which if properly handled would sell—perhaps not spectacularly but very solidly—for many years and bring you in a steady if not large income.

This is what she is going to do now. Tomorrow she will put it up to Lovell Thompson who is executive vice-President of HM. He knows nothing about cooking. However, he always knows the exact questions to ask about a book—and he will depend heavily on Dorothy's opinion and on mine. We see no point in having anybody else at HM even look at the book—this for your protection mainly. Tuesday Dorothy has to go to New York on business and will make every effort to see Mr. Sheeline and get his opinion as to the exact legal status of the property. Lovell will not take a step in the matter until he is convinced that there is no slightest chance that HM will be accused of pirating, or getting around another publisher. So that will be entirely up to Sheeline. Dorothy and I do not feel that you are bound in any way to Ives Washburn, but Lovell is a maniac on the subject. Lovell also moves very, very slowly, so even if Sheeline gives us a green light, it may be some time before Lovell gets around to taking an option or even sending you a letter to bind the matter. I hope to God Sheeline is clear about all the ramifications of this matter.

Incidentally Dorothy exploded, just as I did, when she got to what Donald Moffat said about publishers and new authors. He just is so wrong, and he should know better. No established publishing house <u>ever</u> takes advantage of a budding author. There is, after all, the Author's League which exists mainly to regularize contracts and all that sort of thing. Any publisher who takes advantage of any kind of author is on very shaky ground indeed. The legal contract is on a sliding scale, ten percent, twelve and a half, fifteen after so many thousands sale, or was when I looked last. And any author who <u>pays</u> to get anything published is a mug and deserves what is coming to him—no reputable house ever engages in anything of the sort. Don't dream of questioning any contract you get.

Dorothy also says you must entirely give up the notion of having this published in parts. No publisher would touch it that way—it greatly increases the expense of publishing. It's got to be one big book, if it takes you the rest of your lives. But if you sign with HM they will help you with advances, and in every other way, and not push you to hurry it up. (So, of course, will any of the other big houses.)

Dorothy will get in touch with me as soon as she has seen Sheeline, and will keep in touch with me right along. And if all goes well, and you sign with HM as I devoutly hope, I will then bow out of the picture and leave everything in her very capable hands. So watch and pray. I will fire off a note to you as soon as I hear anything at all. I turned over everything—ms. and all correspondence, to her, and she will guard it with her life. I will be on edge until I hear from her one way or another. Oh—D. also said, when we were discussing your idea of publishing in parts, that you certainly could have sections published in *Gourmet* or another magazine, for extra money, and it would not hurt the sale of the book in the least, in fact it would help. But let's get the title clear first.

Hastily—
Avis

PS. D. wants to know if either of your associates has ever published a cookbook in France. And we both want to know how much of this book you have got written—you mentioned getting your soups in shape—are any other sections written or roughed out? We're willing to wait, years if necessary, but it would help to know how much of the actual work is done.

This on a separate sheet, TK* your remarks about the French, so you won't have to pass it on to your associates. Your remarks about the undue influence of French gastronomes fill me with joy. I had always suspected as much. It's by way of protecting their monopoly, and easy enough to understand. But what really makes me want to stand up and cheer, and at the same time fills me with a wild new hope, is what you say about the servant problem. Nobody can cook as she wants to, lovingly, while coping with children, housecleaning and all that. Food is what suffers first—and that's why we depend more and more on sliced bread, frozen foods, pressure cookers. I can't wait to see what you do about casseroles. Dorothy and I discussed that subject with great heat. There isn't one casserole in a hundred that is fit to eat. But because everybody wants to sit around and drink cocktails in a convivial fashion, and because the cook doesn't want to miss the fun, nine times out of ten we wind up with a casserole. (Not me, God forbid—I find that a rump roast of beef can wait just as well.) Dorothy says she ducks invitations to dine with young married people because she can't, at her age, take the casserole any more—she described one composed of pork chops and canned Bing cherries, after which she came home and was sick. Nobody has more than a part time servant any more, except the rich. I pay Mary eight dollars a day for three days a week—she comes at ten, cleans the house, serves us dinner at six-thirty—a perfectly horrible hour—because naturally she wants to get home to her husband, and he is more and more demanding, so that she is

* Publishing term meaning "to come." Avis wrote the following personal remarks on a separate sheet of paper.

torn between us. I keep her because her disposition is divine, she cooks well and she suits our crazy household. I might have to struggle with a dozen live-in maids at forty a week before I found one I could bear to live with. DeVoto fumes at the necessity of eating early, but he just doesn't understand that no maids want to serve dinner later than that, except a very few old-timers who survive in proper Bostonian families.

Because you have persisted in living in an American way in Paris, you are going to be able to cast a great deal of light on the revolution in living standards that has changed our lives so much. It will be absolutely fascinating when this servant shortage hits the French. Not wishing them any hard luck, but won't it be something to watch? And I wonder how long it will take them to find out that lots of labor-saving devices, so called, are a snare and a delusion. The things that show up! Dishwashers, for instance. I haven't got one. As long as I have Mary, even part time, I don't see the point. And I don't mind washing dishes. And when I cook I wash up as I go along. But my friends all have electric dishwashers. You have to rinse everything that goes into them. A dishwasher won't take the milk off a glass—it has to be washed. Pots and pans have to be done by hand. The washer will only hold dishes enough for four or five, and it takes about half an hour to do the job from start to finish—after you've rinsed everything and packed them in the dishwasher, a Chinese puzzle. So if you have a big party you have to do it over and over. And now my friends are discovering that the detergents used in a dishwasher have a nasty facility for wearing the finish off dishes—if there is gold on your dessert service, it won't be there long. Do you know that if you live in the country and have a cesspool, and use detergents in the kitchen, all the chemical activity in your cesspool stops, so that you periodically have to start up the breaking down of your sewage with yeast? Smart, aren't we? Pressure cookers, stainless steel knives, deep freezers, frozen foods, detergents—we'd probably be better off without them.

Oh how I hope you've got some good casseroles. I don't want to eat out of a casserole very often, but when I do it's got to be something as honest and satisfying as cassoulet.

I will be very happy to get an omelet pan—Dione Lucas peddles them in her courses, but I never got round to buying one. Bless you. And now I've got to get this off the dining room table.

Affectionately—
Avis

A Pulitzer Prize–winning writer, historian, conservationist, and social commentator, Bernard DeVoto set in motion a long friendship between his wife, Avis, and Julia when he published an article in *Harper's* criticizing American knives.

81 Rue de L'Université, Paris, 7

JANUARY 5, 1953

Dear Avis:

Just a short note, to acknowledge yours which came in this morning. I'm about to rush out to do my marketing, and then settle down to my soups. I can't tell you my emotions of love and gratitude for all your interest and hard work in the behalf of our book; you display the true marks of a Great Gourmande . . . which always includes the warmest and most generous of natures . . . and is why people who love to eat are always the best people.

We shall do nothing until we hear from you. No word as yet from Putnam, so he can just wait, and we shall continue to work.

A few remarks:

Pronunciations: If you feel the same way, I think they should be out, and we'll just forget about them. If anyone is offended, they're no good.

Recipient, and other indefinite words, will correct these, and be exact in the future.

Olive oil, type of bread, etc. We plan to have a section in the beginning, or perhaps even along with the pink pages, on ingredients and substitutions. For instance, my sister writes that shallots are very hard to find, and very expensive when she does find them. Her young married friends want to know when margarine can be used, etc. etc. Concerning strong olive oils, we love them too, especially when they are rather green and very fruity. But I have discovered (so understand why they are not used so much here) they are too strong to be used when you are drinking a fine wine, as they are oily and fruity in the mouth, and kill the taste of any of the great wines. But fine for *le gros rouge* which is a drink rather than an experience. Have also learned why US French bread is not like the French, it is the processed flour made from hard wheat. The French "French type" crusty bread is made out of cheap soft-wheat flour.

Book in Sections. Can do the whole thing, but wonder if it would not be possible to have the desserts and cakes in a separate volume. This is such a special business, and we need quite a bit of study and research. We're working with a wonderful old chef now, but have to pick his brains and methods, then work out our own. How lucky we are to have these professionals to work with: He has that wonderful old-timey "art for art's sake" approach, and nothing short of perfection satisfies him at all. It's an inspiration to work with such a man . . . much the same gleam of wonder and flame for beauty and succulence that shines out of Carême's* writings.

Corrections. We have in one report, from my sister, noting that for Mirepoix in the Waring mixer [blender] veg. should be cut ¾" instead of 1½". (That was a slip on my part, and shows to me how careful one must be.) I can also see, from what she writes, that things will have to be explained even in more detail, as she followed the directions exactly. Should mention why ham was not included in the mirepoix on a large scale (because sometimes you don't want it, for fish, etc.); should also say that all veg. will not be chopped evenly, but some will be a puree, some larger lumps, but to keep putting back too-large lumps until the largest is only about 1/16 inch. Etc. Etc. Also, thought that in SHELLFISH BUTTER,

* Marie-Antoine Carême, an early-nineteenth-century French chef and author of voluminous culinary books; considered the father of haute cuisine.

Electric Blender Method, should mention a bit more about how butter will turn into a pomade quickly, and that is OK (so people will know what to expect). Etc. Etc. There will be plenty of bugs to straighten out, and maybe we will need more guinea pigs to test things out.

Wine: The wine section will be small, as there are several good books out about it. I think the most useful things to know, besides a general guide of what goes with what, are:

1) Guide to buying (know how the merchant stores his wine, as it can be ruined by him).
2) How to keep it yourself, lying bottles on side, and what temperature. Letting it rest after a journey & why.
3) Why baskets or decanters (not for chi chi).
4) Temperature to serve, and when to uncork and *chambrer.**
5) How to pour from a bottle.
6) Wine glasses.

Etc.

I find a lot of this stuff is taken for granted, and it took me a long while to find out. And all of it doesn't seem to be written down anywhere.

Herbs and spices: Will go into these, not at great length, but enough to show substitutions, and dry for fresh, and also will do some experiments on freezing. Am sure one can blanche chives, tarragon, etc, and freeze in small packets . . . which would be wonderful for winter use. Am also going to send for some from Spice Island, or other, to test them out.

Frozen Foods: We have for the moment in our Embassy Commissary frozen peas, limas, beans, broccoli. I've run a few experiments on these, and find the peas *à la Française* (½ cup water, bit of hashed shallot or onion, 5 or 6 leaves shredded lettuce, salt & pepper and 1 TB [tablespoon] butter, pot covered, high heat . . . breaking apart pea-lump, then continue with high heat and cover until all water evaporates) . . . works out dandy. Have tried beans in big pot of boiling water, and then refreshed in cold water, the French system and a life-saver for doing veg. ahead . . . don't work, no flavor. Will continue experiments . . . My main interest, besides having all possible flavor, is to be able to get them done ahead of time so you don't have any last minute worries. Have found the frozen chicken absolutely awful and tasteless and stringy . . . What have you found about it. Perhaps

* "To bring to room temperature."

ours has been frozen and unfrozen in the voyage. French chickens are so good, the competition is pretty one-sided. But shall try some more, my feeling at present is that frozen is like sawdust, and to get a decent meal you'd have to marinate it, and then you might as well not use it at all.

Canned Soup: Though not French, they're so useful, we'll put in some hints and ideas for taking away the taste of the can, etc.

Pressure Cooker. Will mention pressure cooker, particularly for stews and soups. I think it is fine for some things, but food should be finished off in open pot or in oven, for release of "pressure gas" taste, and also for reduction of sauce. My sister says she will use any time saver available, but will also do anything to give taste to her food. I think one has to be very careful using potatoes in a pressure cooker, as they get a most awful taste if cooked too long. Etc.

This is very random.

I sent for the Chamberlain *Bouquet de France*. (One of the expenses of this profession!) It is a wonderful and beautiful book. What fun it would be to travel with, and what fun they must have had doing it together. I just regret, too, that the recipes are not more professionally done. If only she had had time to study a bit here. They are fascinating recipes, and good ideas. But for the novice, there is not enough explanation, and even some pitfalls. Just glancing through, I haven't had time to study it up.

p. 219 (1) Langouste à l'Américaine: No mention of reducing the cooking liquid after stewing for 30 minutes (which is too long). No indication of how to "finish the sauce" by blending flour, butter and coral. Off heat? Boil afterwards? How to mix, etc. (Actually, butter and flour paste shouldn't be boiled . . . and not enough butter used, I would think, to "butter" . . .)

p. 259 (2) Escalope de Veal, etc. "Cook them slowly in butter until the slices are tender and a fine golden-brown color on both sides." They won't brown if cooked slowly, butter has got to be just turning blond before meat is added, then browned quickly, then heat lowered. Etc.

It's too bad. If the book had been properly edited, these little slips would not have occurred. It's a shame, in an otherwise so wonderful a piece of work. It surprises me that *Gourmet* didn't catch those things. I only conclude they didn't have a professional going over the recipes. Diat*

* Louis Diat, the French-born chef at the Ritz-Carlton in New York City, a cookbook author, and the first head of *Gourmet*'s test kitchen.

should have looked at it. The tragedy is, young brides will try out the recipes and conclude that only a genius can cook.

Au revoir, ma chère, and thanks so much for your good letter. Much love to you both, and my Paul sends his warmest also (he is just as fascinated as I am by this true life drama).

Julia

JANUARY 9, 1953

Dear Julia:

I haven't heard from Dorothy [de Santillana] yet so this letter bears no news. She may be held up in New York by the terrific ice-storm which grips the coast, or she may not have been able to reach Sheeline, or he may not have been able to give her definite word yes or no. At any rate she promised to phone me when she had any news, and I am not going to bother her. This is just a partial reply to your last, so after Monday I dunno when or how I can write. The painters are moving in, to do two living rooms and my bedroom and bathroom, and where I will set my typewriter for a week I do not know. But I will keep this and send it along when I have heard from Dorothy. I am chewing my nails off to the elbow, like Noel Coward's lady.*

Don't entirely give up the idea of pronunciations. Wait for editorial opinion. I rather like the idea of having them in a separate section, for the timid. Having desserts and cakes in a separate volume is also up to your editor. None of these things have to be decided now.

I can see that your sister is working a lot harder at this job than I am. Truth is, I am only a part-time cook. I have a maid now only three days a week (why will they get married and move out?) who does all the cleaning and cooks our dinners on the days she is here. She has worked for me a long time and I have taught her to be quite a good cook, if I do say so. The rest I swing myself, and it is great fun when I have time, which I seldom do. I do a lot of manuscript copying for the boss, answer a great deal of his mail (which is how I came into your life!), read proof and so on. I also have a minute but well-paying job reviewing detective stories for the *Boston Globe*—a fairly loathsome task but I love the check coming in and put it all into clothes. My house is big and people come in a lot. My old-

* The female characters in Noel Coward's plays were often jittery. The specific reference here is not clear.

est son is in the army now but his brother, who will be thirteen tomorrow (a birthday he shares with his father) is around a lot. So I don't spend as much time as I would like in the kitchen.

Then the horrid further truth is that DeVoto is pretty much a barbarian as far as food goes. He likes things beautifully cooked but his range of likes is limited. Roast beef, steak (don't we all?) but lamb in no form (because he spent a year on a sheep ranch in Idaho where they fed the help on fresh killed mutton twice a day). He loves Boston baked beans (of no appeal to me) and a range of western and Mexican foods very highly spiced—chili, enchiladas and the like. He likes nearly anything in a red hot sauce and practically nothing in a bland sauce, though he will eat them politely. He likes fish, but he dislikes asparagus, artichokes, or even the commoner vegetables, and will eat large quantities of mixed green salad twice a day. Fortunately we all adore the onion family, in any way shape or form.

My anemia also comes into this picture. I've had it for twenty-five years (it's secondary and well under control) and it means that I practically live on red meat and fish and eggs. Keeps me nice and thin, too. And if I start worrying about anything, and I have just as many worries as the next person, I stop eating. During the last few weeks of the campaign I practically lived on coffee and orange juice, and got down to 112 pounds. During these spells of what my psychoanalytical friends call "anorexia," I have no interest in food at all, and very little in cooking. So I am far from being an expert. Though I like to think that if my life had been different I could have become a first-class cook.

Well, I don't know what all this maundering is about. Alibi, probably.

Now down to answering questions. I'm hopelessly conservative about kitchen equipment. I haven't got a Waring mixer. I've got an electric mixer and I never use it. I practically never use a pressure cooker. I never had a cook who would use them either. You have to clean the damn things. You have to hover over a pressure cooker. So we wind up using the egg beater or a whisk, and cooking all the long-time mixtures in an iron Dutch oven or iron frying pan. I haven't got a deep freeze and never expect to have one. The freezing compartment in my refrigerator will hold half a dozen cans of orange juice (now that is one of the great discoveries and I will never squeeze another orange) and a couple of boxes of French fried potatoes (also extremely good). Most of my friends have deep freezes and it is just beginning to dawn on them how expensive it is to have one. These dear people buy half a steer and by the time they have had it cut up and processed it has cost them ninety cents a pound which to be sure gives them some fine steaks and roasts at thirty cents less than the market price, but also gives them thirty or forty pounds of hamburg at a great deal more than the market price. I just don't think it makes any sense unless you live on a farm and have a lot of vegetables and fruit you want to freeze. And if you live on a farm along comes a big electric storm and the power goes off and maybe you lose five hundred dollars worth of food at a whack. So I'll have none of it, and occasionally when one of our Western friends sends us a few elk roasts I can always park them in a friend's freezer until we need them. Elk can be frozen and thawed and refrozen without damage.

Frozen food is a very mixed business anyway. I think the beans are appalling, especially the Frenched or slivered ones. The cut beans are a little better. But both lack flavor and if overcooked even a minute have a repulsive texture. Granting that you can only get perfect beans when they are picked in an immature state from your own garden, still we get pretty good beans from California and Texas the year around. The frozen peas are too sweet for my taste, though I use them occasionally. I've never tried the broccoli. I only like broccoli when it's fresh and then with hollandaise—or cold vinaigrette. The baby limas are pretty good. So is the asparagus, remarkably enough. But on the whole, I prefer fresh vegetables, and I like them undercooked, Chinese fashion. The Chinese have a lot to teach us about cooking vegetables. Which is why I like to cook them in a bit of oil—a mixture of celery, onions, green peppers, for example, cooked about twenty minutes in a heavy covered pan with a couple of tablespoons of oil and no water. Or tomatoes, eggplant, onions, celery—what is the French name for that fine dish—ratatouille? Or young carrots and scallions. I'd love to hear more about the method of cooking vegetables ahead and freshening—but don't take time off, stick to your puddings. I like braised vegetables too—celery, endive and the like.

Frozen chicken must vary enormously according to brand. I agree absolutely about complete lack of flavor in the white meat—it isn't fit to eat. But at the beach last summer I tried frozen drumsticks and second joints and they were pretty good—I don't know what brand. But then the dark meat is juicier and has more flavor anyway. Wish to God our American poultry raisers would stop trying to develop turkeys and chickens with more and more white meat, which has few takers in my household. You are quite right that French chickens are better than the American variety, though some producers are beginning to develop better birds—and my God are they expensive. Domestic ducks are generally good, but it's all but impossible to buy a goose.

Oddly enough one meat they freeze quite successfully is veal cutlet. I am passionately fond of veal and cook it a dozen ways, and at the beach where I couldn't get fresh veal I tried the frozen. One brand was quite good—again I can't remember which brand. You have to open the package and take a good look—if it's dark veal don't buy it. A package has four small cutlets which is just right for two people. I used this most successfully with tarragon and white wine, finishing off with heavy cream and butter when the sauce cooked down. Fish, too, freezes quite well. The best is called fillets of ocean perch (actually rosefish, which was considered a trash fish here for years and is still used to bait lobster pots). Cheap—43 cents for enough to feed three people—and very good in half a dozen ways. I like it sautéed in crumbs, then sharp cheddar grated over it when it comes from the pan, with a Mouli grater. Have you got a Mouli grater? Because I will send you one if you haven't. It is perfection for cheese, nuts, chocolate, in small quantities, and is completely simple to wash. My God the gadgets I've bought and discarded because they are a nuisance to wash.

Speaking of gadgets, B. gave me an electric deep-fryer which is wonderful for fish, wiener schnitzel and the like—but which smells up the house so much that I seldom use it. Even with kitchen doors shut, windows open, couple of deodorizing gadgets going, you still get pockets of that deep frying smell for days after. I need one of those hoods with an electric fan in it, but short of tearing out a kitchen wall I don't know how I'm going to manage it.

Do you want me to send you some dried herbs? Herb Farm is good and I use their tarragon all the time—it is impossible to buy fresh tarragon and I've always killed every plant I've set out in the garden by picking the leaves too fast. Herb Farm has a new one that is delicious—grated orange peel, very fragrant. They make a fine dill salt and a fine garlic salt. Say the word.

I am again mixing French dressing in job lots. Stubbornness, probably. I like to mix it each time right in the wooden bowl, use a variety of vinegars and a variety of greens. One of the canned things I depend on is artichoke hearts which are fine in salads once in a while. I don't know what you mean by the taste of the can. Seems to me we have pretty well got away from that in this country. As with frozen foods, some are fine and some are not. I use S. S. Pierce canned chicken broth by the case, in risottos, gravies, sauces—it's strong, well-flavored, and not gunked up with rice. Most of the other brands are terrible. We aren't soup eaters, particularly. Stews and chowders, yes. I don't know anything better than a good fish chowder. A firm called Snow is canning minced clams and they are absolutely wonderful. (Send you a sample?) Add thin cream, butter, freshly ground black pepper and you have the most wonderful bisque, strong of clams. Fish cans pretty damn well, when you come right down to it. Though why, oh why, we can't can crab meat as well as the Japanese or the Russians I can't think. They have king crabs on the West coast, after all, but let the Americans can them and they are awful. Canned tuna is good, if you pay for the best, and canned shad roe is divine. The only good beet comes from a can, and what we would do without canned tomatoes I cannot think, as you can only buy good fresh tomatoes for about a month in the year.

I think the bride has simply got to do some market research—if she can't buy good fresh spinach, make do with the frozen second best. In general, frozen will be second best, no matter how much time it saves her. The only thing I <u>know</u> is best frozen (and there are a few who disagree) is orange juice. Which has nothing whatever to do with the French or English breakfast.

Shallots are hard to find, and certainly expensive. I can get them by ordering ahead. There must be an infinity of places in this country where they have never been heard of. I should think the nearest substitute would be the sweet red Italian onions which are pretty well all over the place now. Or scallions which we get nearly all year, from the West or South. They can be very hot, if overdeveloped. When in doubt, taste. Could leaks possibly be used as a substitute? You tell me—I haven't tried. We like leeks cooked as a vegetable.

You are so right on the over-cooking of the langouste. Why is it that most cookbooks give far too long a cooking time for shellfish? A pound or pound and a quarter lobster should have no more than fifteen minutes and twelve is better—I think I sounded off to you about lobsters before. Shrimp (even the big ones) no more than eight. And I hope you'll come down hard on the desirability of letting shrimp, lobsters, salmon cool in the broth they were cooked in, to retain their juiciness and flavor. Too many little mistakes in *Bouquet*—but just you wait. I have gone blind proof-reading DeVoto—and then picked up the bound book and found a mistake on the first page. On *The Hour* we swore that there wouldn't be a mistake on page one, and there wasn't, but there was a howler on one of the last few pages. It always happens. It will happen to you. If you proof it fifteen times you won't get every error. No more now. Thank you so much for your kind and loving remarks. I like you too.

I knew I'd forget the most important thing. Which is report on your Beurre blanc. I took a copy of that one before I gave the ms. to Dorothy, and I will not share it with anyone. I followed directions exactly and the result tasted quite heavenly, but I don't think I got the right thickness. You say "thick and creamy." Well, it was creamy all right, but very thin cream. Perhaps it should be like that, but I had got the idea from the recipe that it would be more like warm hollandaise in consistency. Not knowing how long the reducing process would take, I did it a little ahead of time, and the pan was quite cool when I started putting in the butter. Next time I will stay right with it, not doing anything ahead. The butter did not separate out, but it certainly didn't get what I would call thick. I used red wine vinegar plus lemon juice—3 TB and 2 TB. I must say it tastes like heaven. We had it on plain steamed scrod fillets, and it was eaten by four of us to the last flake and drop.

One great problem for the American cook who uses your book is going to be wine. We are not a nation of wine drinkers and most cooks are going to have to buy wine specially for cooking. If your sister lives in California maybe she would do a little research on the domestic wines that could be used for cooking. Many of these are quite respectable and quite cheap, but so many Americans just don't know the difference between a sweet and a dry white wine, and neither does the dealer at the grocery or the package store. What we need desperately is some half bottles or quarter bottles. Before the war I could buy quarter bottles of imported French wine—a perfectly good Médoc and a Graves, and a Pommard, for thirty-five cents a bottle, and I always had a couple of cases on hand for cooking. The bottle would be just enough for an ordinary recipe, or if too much, what was easier than to finish the dear little bottle yourself? Came the war, and no French wines coming in, and of course we drank them all up. I must check at S. S. Pierce and see if they can import them again, though S. S. P's wine department has fallen off horribly. I have tried some quarter bottles, red and white, bottled by the Almaden-Madrone Vineyards in California which, though recommended by Frank Schoonmaker,* are

* A wine writer and importer.

quite unspeakably awful. Maybe I better tackle Schoonmaker—I used to know him. I use a Roma California Sauterne at ninety cents a bottle which is pretty good. But if we have a really good wine we drink it all up.

I am not really a wine drinker. B. is, and has worked at it and is very good at the parlor game of identifying the vineyard and the year. My trouble is that I can't drink a Martini before dinner and then drink wine. When I was abroad I was perfectly happy with no Martini at all, and plenty of wine, but I do not think they mix at all well. B. is so wedded to his Martini that I will never be able to solve this problem to my own satisfaction.

I like the whites and rosés better than the reds, in general. Before the war we used to drink a *pétillant** Vouvray which was wonderful. I know you won't believe this because Vouvray isn't supposed to travel. I can only say that this one did and it was cheap. Gone forever. And when I was in Paris there were no Vouvrays, so I settled on blanc de blanc and was quite happy. We <u>must</u> get some quarter bottles of good wine for cooking, because the cocktail habit is too deeply ingrained for Americans to drink wine every day, regrettable as it is.

When you have time, can you send me a recipe for *oeufs en gelée*—the poached egg kind? I will have to omit truffles and tarragon leaves—can't get them. Wonderful lunch dish.

JANUARY 13, [1953]

[Dear Julia]:

Just talked to Dorothy on the phone—B. was talking to her about something else and I horned in. She did not see Sheeline while she was in New York because Lovell Thompson, after reading the correspondence, said that in his opinion nobody had any legal claims on you at all, that you are perfectly free to take any offer you like, and that he wants Dorothy to talk to Paul Brooks and if Paul is agreeable to go ahead and give you an option, contract and what have you. Dorothy, therefore, has been trying to catch Paul, who has been ill. She expects to get him soon, says that he and Lovell will want to talk to me, and then we are probably set. I warned you that these things go around in circles. So just sit tight and wait. I will keep you informed right along.

Paul Brooks is HM's chief editor, and one of the best in the busi-

* "Sparkling."

ness. B. says the editorial job Paul did on *Across the Wide Missouri* and *Empire* couldn't have been excelled by any living man. I wouldn't think Paul knows beans about cooking, and his very nice wife Susie not much more. They are outdoors people—bird-watchers, live in the country, camp out, ride horses and such. Lots of children. And two nicer people never lived. Paul is about 45—looks 30. He and Lovell really run the business. Henry Laughlin (Jones-Laughlin Steel Co.) is president but doesn't do much actively.

Dorothy has read every word of the ms. and is tickled pink with it. Agrees with me in every detail. Said she hadn't been doing much cooking for quite a while—too busy, ready to settle for a chop and some frozen peas—but this ms. made her itch to go into the kitchen and cook like mad. Interesting reaction. Same as me.

I am sending to her the part of your letter of the 8th which deals with background, change in plans of the book and so on. Will be very helpful to Lovell and Paul. About to be dispossessed from my desk, as the painters are walking in the front door (day and a half late) and will get this into the mail. Watch and pray. If Lovell thinks your title is clear, don't see how Paul can possibly disagree. But oh God, I will be glad to know that an option is on its way across the Atlantic. There are so many slips in this business.

Hastily,
Avis

81 Rue de L'Université, Paris, 7

JANUARY 19, 1953

Dear Avis:
We both enjoyed your nice long letter immensely, and this will be but a short one from me, as I've got to get to work. I find I'd much rather write to you than to work, and I must not indulge myself. However, a new item has come into our lives. We knew Paris could not last forever, as we've been here just over four years. We've been living on borrowed time, really. So the other day Paul is notified that he will in all probability be appointed Public Affairs Officer for the South of France, with GHQ* in Marseilles. We'll know definitely in a few weeks, and then move down on the first of March, if all goes through as it should. But, what luck for us. It could have been AbbisAbabababa, or something. And it will be an abso-

* General headquarters.

lutely fascinating job, as he'll have everything from the Italian border to the Pyrenees, going up almost as far as Grenoble, and all those wonderful places such as Albi, Rodez, Conques, etc. Public Affairs Officer means information and propaganda, including press, movies, professors, exchange of persons, culture (that un-American thing). Here we have been able to adjust as we please, and have been in the fortunate position of being just below the level of diplomatic brouhaha so we haven't had to get involved in any of that boring stuff. There, we shall have to entertain prefects of police, newspaper people, mayors, etc. etc.; and do a lot of traveling, and know everybody in as many villages, hamlets and *villes* as possible.

This all will mean it's going to be a bit slower on the book . . . and I'll have to grind to a stop for about a month so we can find ourselves a house and get ourselves moved in to it. However, it will mean a wonderful acquaintance with Provençale cooking . . . which is actually more to the taste of our wineless culture in the US than the more classical French cooking.

It means, of course, that we three G's can't work closely together, and the school will suffer, and from the point of view of cooking career, a real blow. I needed another 6 months here. But we can commune, we three, perfectly well, and we'll be coming to Paris, and they'll be coming south. And most of the main work we do separately anyway. It's the personal getting together and quick exchange, and experiments together that will not be always possible. *C'est la vie*.

Of course, there is always the possibility that the new Congress and administration will decide they don't want the information program at all. Or the possibility that Mrs. P. Child, yak-yaking away against the Old Guard Republicans, will be considered an undesirable person which will make her husband persona non grata.

We shall not know definitely for a bit, but consider it a 95% certainty.

To answer your letter specifically. Fascinating that DeVoto, though a great wine fellow, prefers the "barbarian" foods. But I think that is very American male. Paul loves that kind of food, also, and if there is anything he loathes, it is something that is white. However, he is becoming very partisan to the classical stuff when it comes with a good wine . . . but he adores highly spiced and garlicked food. And we are sorry about your anemia, but thank heaven it is well under control. Don't shots of something do any good, too? And I know very well what you mean about loathing the thought of food or cooking when you don't feel well. We both got amoebic dysentery when we were in the Far East during the war, and it was never discovered in the USA . . . But when we got over here, all this good rich food brought it on with violence. Luckily we ran into a wonderful French doctor who happened to specialize in it, and we are gradually getting rid of it. But we find we can't eat with the pure gourmandise we'd like, and that *les alcools* don't agree with us at all (no martinis). But wine seems to go down very well, if not overdone.

I am somewhat old-fashioned about kitchen equipment, too. But find I do love that Waring mixer, and use it all the time to make Mirepoix. That's about all I use it for, but can make enough for 2 weeks in half an hour, and it would take 3 times that long otherwise. Tried making mayonnaise in it, and it was such a mess to clean out! And I quite agree with you about most gadgets. If you have good knives you certainly don't need tomato slicers, parsley choppers, onion choppers, etc, nor potato slicers. The pressure cooker, as you say, you do have to watch like a hawk. But I am trying to use this as much as possible where it can be useful, because I feel for the book it has to be talked about. I just tried a Demi Glace in it, and it is quite useless, as the point of cooking the Demi Glace is that it is continually skimmed. After PCing it, I had to cook it 2 hrs. anyway. Deep frying I just hate, as it not only gets all over the house, but gets in my hair. Deep freezing I am in favor of for Mirepoix, Demi-Glace, orange juice, herbs, lima beans, mushroom Duxelles, etc. Or, useful ingredients; or emergency stuff that is better frozen than canned. I like French dressing mixed each time, too. But find, although we have some beautiful wooden bowls, that I am sure I can taste a rancid oil taste in them, so have switched to porcelain. And I've started the bowls out by never washing, etc. etc. (Super nasal, perhaps.) Now, that's too bad about shallots; as I feel they are essential. Here there are two types, the *Échalote Grise,* with a very hard skin, and a white interior. It is expensive. Then there is the *Échalote-Oignon,* which has a thin reddish skin, and is purple inside, and is the one most people use, and is inexpensive. This is sometimes called the *Échalote de Jersey*. Maybe it is like a small purple sweet Italian onion. What are scallions, little green onions? I think if shallots are not available, one could use finely minced onion, blanched for a minute in boiling water. (This would freeze, and one could make a lot in a Waring mixer, and put them up in little packets. I'll have to try it out.) Or it can just be left out altogether, or a bit of chives added to finish the sauce. (Then chives put up in some kind of oil that some house like Spice Island makes, perhaps.) Leeks might work, but have a very special taste.

Wine. Vermouth may be used, but is a pretty strong and herbal taste.

If wine is not used, especially in the fish sauces, it just won't taste French. The Wine Advisory Board of Calif. has kindly sent us a big bunch of excellent literature, giving names and types of wines, and say they will be glad to go over our copy, if we'd like them to. As for wines not traveling, Paul says a lot of that is nonsense. How would one ever have wines in Paris, if they don't travel. Of course, an old wine is like an old lady, and traveling can disturb her. We love Vouvray Mousseux, and always serve it instead of champagne . . . It is 280 francs a bot, as compared with 900 francs at least for a good champagne. And if champagne is not top notch, it is pretty dull, I think. Now, if you'd just come over here, I'm sure we could convince you to like the reds too. We ran into a beautiful Bordeaux 1929, that is just perfectly matured, and is everything one reads about that a wonderful Bordeaux should be, but one rarely tastes. It is really something to swoon over, that wonderful rich exciting bouquet, that excitement as it fills the mouth with its mellow (I can't go on, I keep thinking of that old cartoon in *The New Yorker,* "a little wine of no pretensions, etc."). Well, here this wonderful stuff is only 400 francs a bottle . . . My god, I never tasted anything like it. And we have some red Burgundies . . . Oh, I'm swooning over the typewriter just at the thought of them. I find I'd really never tasted a good wine until I came over here, and now I'm pretty well gone for good . . . and willingly give up my martini, because I can't mix them at all without getting one hell of a hang-over.

Casseroles. I even hate the name, as it always implies to me some god awful mess. The one you described that D. de S. [Dorothy de Santillana] ate of pork chops and Bing cherries! Yes indeed, we plan to have plenty, but hope we can get around the name. Most of the sautéed chicken dishes can be made way ahead, and the wonderful Boeuf Bourguignonnes, Blanquettes de veau, etc. etc. I do both when I'm entertaining, I mean, both "casseroles" and roasts. As you can get a roast almost roasted, then leave it in the warming oven and it will finish itself off. I always use a meat thermometer. Or you can roast ahead of time, then slice it up just before serving and put it in its sauce or juice and run it into a hot oven to heat up. And there are such good veal dishes as Veau Orloff, in which you roast it, then cover each slice with a rice and onion "soubise," sprinkle the top with grated cheese, and put it in the oven to heat and brown the top. All this chef-hostess stuff is my particular interest . . . but it does take practice and experience, so the stuff is really hot, but not over-done, etc. When we get into "recipes for dishes," we plan always to have "make ahead" notes for everything, including veg. (I also think the young hostess should be advised never to say anything about what she serves, in the way of "Oh, I don't know how to cook, and this may be awful," or "poor little me," or "this didn't turn out" . . . etc. etc. It is so dreadful to have to reassure one's hostess that everything is delicious, whether or not it is. I make it a rule, no matter what happens, never to say one word, though it kills me. Maybe the cat has fallen in the stew, or I have put the lettuce out the window and it has frozen, or the meat is not quite done . . . Grit one's teeth and smile.)

My, for a short note, this ain't. The Beurre Blanc. It should be like a not very thick hollandaise, thick enough to "nap" a fish. Pan should be warm when first butter goes in. Maybe your low heat when you were making it was too low. Maybe you let it sit over the heat when it was done, and it thinned out. Its OK to do the reduction ahead of time, but heat it up before using, maybe adding 1 tsp. more vinegar. The acid is extremely important, as it is that which starts off the precipitation and thickening. And, it can sit off the heat, but never in hot water. If your pan is too thick, it holds the heat too much, and that can start a thinning process. Those are the only things I can think of, but will do some more experiments on it. (We got rather bilious working it out, but I think I can stand some more of it now.) As the reduction takes only a minute, I think it is much better to do the whole thing just before serving. It's tricky, no doubt about it . . . but the only trickiness is really, if it has to wait. But if the fish is hot, it doesn't make much difference if the sauce is fairly cool.

Oeufs en gelée. I enclose a business about adding gelatin to bouillon.

1) Poach eggs and put them in icebox to get cold (to hasten the jellifying). Put the egg molds in the ice box, also, to chill.
2) Quick method: Put a thin slice of ham in the bottom of the mold, lay the egg on top, and pour in the jelly. Then set in the ice box to set.
3) Fancier method (which is fun if you have time). When jelly is at that syrupy stage so it is almost about to set, pour it into the cold mold and swish it all about so there is a thin film around bottom and sides of mold. Set mold in icebox until jelly hardens. Then you can decorate the bottom and sides of the mold any way you want.

Decorative elements: matchsticks of red pimento, Blanched leak green, cooked carrot, string beans, ham, tongue, black olives, etc. etc.

(Use black olives instead of truffle.) Dip them in the jelly, and place them around the sides and bottom of the mold. Cut a piece of ham the size of the bottom of the mold, and lay it in. Put in the egg, and pour around the rest of the jelly (or other shapes, besides matchsticks). To vary, you can put in a mousse of ham, or of liver, etc. Mix the pureed meat with a bit of stiffly whipped cream or thick Béchamel, and a bit of the jelly.

To un-mold, dip each mold in very hot water for a few seconds, quickly run around the sides of the mold with a small knife, and set it upside down on serving platter. If by chance you've let mold get too hot, set the mold in cracked ice to cool off. (What a nuisance, and why they should be un-molded ahead of time, in case of accidents.)

This isn't a very well written recipe, but I am just dashing it off. You can just do whatever you want. Important thing is to test out jelly consistency ahead of time, so you won't have any disappointments. I find I don't want too much jelly about the egg, as it just makes too much jelly, so I use as small a container as possible. But that is a matter of taste.

This is an appallingly long letter. But it is so much fun to write to you. And it is dreadfully typed. You type so beautifully. Practically never a mistake. I have been typing like mad for nigh onto 25 yrs, and it has never been any better than this.

Still no word from Putnam, thank heaven. But don't you think it a bit odd that they never even sent a postcard, saying "ms. received"? Maybe they never received it. What you say about Paul Brooks and Lovell Thompson is very encouraging, especially that they feel we are in no way involved with Ives Washburn. So we shall wait, but with little glimmers of *espérence*.* (Do you think a cliché is less of a cliché when it is in two languages?)

In the midst of this, I am making a split pea soup, and have been so happy to have my frozen mirepoix. If I hadn't had that it would have taken me at least 15 extra minutes (no, about 33 minutes) to cut up the veg, and then to stew them in butter. As it is, I just dump in everything and there she is. We have found some of the most wonderfully smoked bacons and hams . . . all done by a little charcuterie on my market street. That real old-fashioned smokey smoke. That is the wonderful thing here, all these little people who do little things themselves, and none of this poking a ham with a needle stuff. (Not yet, anyway, but I suppose it will come. But maybe not entirely, because the Frenchies love their food so much.) Ham

* *espérance*: "hope."

is terribly expensive, however, 1400 fr. the kilo. Which means $2.50 per pound. Ye gods. Never have bought a ham what am here, and never intend to at that price. But awfully good.

With much love, as always,
Julia

Could easily go on for pages. So much to say. US and French life are so terribly different. We noticed that 2½ yrs ago when we came back. The important thing here is that food is a great national sport, indulged in by all classes. One's best evenings are composed of a good dinner, and nothing else is necessary, and it takes the whole evening. And then, the woman is in the home & the husband comes home for lunch—that in itself is such a difference. (I love that!) Etc. etc. etc.

Julia writing notes in the South of France.

81 Rue de L'Université, Paris, 7

JANUARY 20, 1953

Dear Avis:

Enclosed are two "Guinea-pig responses" which Mrs. de Santillana might like to have. I expect we shall get about 6 or 7 more from Pasadena, and then a bunch from my sister-in-law from New Hope, Pa. Unfortunately, the Pasadena ones are all from good friends, but I think they are good enough friends to be cold-blooded.

Now I am beginning to think the book is not so much for a mass circulation. I keep forgetting everyone does not have a passion for French food, and unless people have been over here or have eaten it at home, it doesn't mean much to them. It is really probably more of an "upper middle brow and upper-brow" audience. Now, when I come to think of it, it takes most Americans a couple of months of living over here to develop a real appreciation . . . and some people we know never develop one at all, but stick to hamburger and ketchup and baked potatoes. Well.

Also enclosed is a recipe sheet on boiling and refreshing green veg.

In haste, but lovingly,

Julia

JANUARY 22, 1953

Julia, my pet!

We are all set, and a contract will be on its way to you very soon. It would have gone through even sooner—though Paul [Brooks] pointed out to me over the phone just now that this is very fast action for so deliberate a house as HM—only DeVoto gummed the works. *Course of Empire* has been chosen for the history award of the year by the National Book Award Committee—and Archie* MacLeish's collected works has been chosen for the poetry award, also a HM book, and as HM has never had a book tapped by this august committee they are in a perfect tailspin. It is all right to tell you now though it has been deeply secret, as the big doings take place Tuesday in New York. Everything else has ground to a stop at 2 Park Street, and Paul says he has been nothing but a social secretary.

But—Dorothy submitted a report, which I will copy later though I have no business to, Lovell was agreeable, Paul even more so, and the directors initialed the whole business this morning. I feel wonderful, though I had no real doubts that it would go through. I shouldn't think

* Archibald MacLeish, a poet, dramatist, Harvard professor, diplomat, and close friend of the DeVotos.

they would dally too long in getting off the contracts. Dorothy has sent all your material back to me, and you can either send her another copy of everything that is ready, or wait until you retrieve the stuff you sent to Ives Washburn and which seems to have vanished into a void. What a way to run a business! I think you might be fairly preemptory with them — and you ought to notify Sheeline too. I would like to see Donald Moffat's face when he hears the news. How an established writer can have such naive notions I do not know.

Your news about your transfer to the south is staggering. I didn't even know that Paul was in public service. Is it State Department? Of course I share your regrets about leaving Paris — but I am certain that you can work out the details of the cooking research, and as you point out, there's all that wonderful Provençale cooking. Frogs legs, Provençale — ah me. Until the old Lafayette Hotel in New York folded up, every New York trip took me straight to that ugly dining room to eat frogs' legs dripping in garlic and butter, and their gratinéed potatoes which were the best in the world.

I'm very rushed for time today — moving back to four rooms done by the painters and paperhangers was interrupted by Mary's absence for several days, and now she is back there are floors to be washed and waxed before we can be completely settled, and the books oh my God to be put back on shelves. I am going to New York with B. on Tuesday to stay three days. The award presentations will be a big do with many dignitaries — Archie is flying up from Antigua, where the Dean Achesons* are going to spend their much needed vacation with the MacLeishes and who knows maybe Acheson will be on hand Tuesday. Archie has very grand friends. Elmer Davis† is coming up from Washington — one of our oldest and dearest friends. Francis Biddle‡ whom I also love was on the committee. Of course the books chosen are very fine books, that goes without saying — but I have a hunch that the committee is also saying, this is what we think of the guys who were attacked by that louse McCarthy. The presentation is attended by lots of publishers — this is their choice — it was established four years ago because the publishers thought the Pulitzer Prize didn't mean much any more. Rachel Carson's *The Sea Around Us* got it last year. No money — unfortunately — but a placque and lots of ku-

---

* Dean Acheson, the secretary of state in the Truman administration.

† A well-known radio news commentator.

‡ A lawyer, judge, and appointee to many important government agencies, Biddle was the attorney general during World War II and chairman of Americans for Democratic Action (ADA).

dos and publicity and very real help in sales. Then there's a reception for all the bigwigs. Then HM throwing sanity to the winds is having thirty people to dinner at 21.* So it will be very gay. B. also has to do a couple of radio and TV appearances which he hates but this time he has to be graceful about it. So I am a little out of my mind at the moment.

This isn't a reply to yours — maybe next week when the heat is off.

Right now I have only time to copy Dorothy's report and don't ever tell her I did. Tain't ethical. She's got some minor matters wrong, but of no account.

> Julia Child is an American woman who has been living for the last four or five years with her painter husband in France. She is, I believe, middle aged (are you?), and has all her life maintained a lively interest in cooking.
>
> In response to one of Benny DeVoto's articles — about the poorness of American knives — she sent him a set of French cooking knives, and this led to a long and intimate correspondence with Avis. It turns out that Mrs. Child, a graduate of the French Cordon Bleu School, has, for the last four years run a cooking school in Paris with two Frenchwomen, a cooking school for Americans. All this time they have been making notes, discussing, and planning a basic cookbook of French home cooking for Americans.
>
> It would be as different from the two Chamberlain books *Clementine* and *Bouquet de France* as the *Joy of Cooking* is from the host of chi chi special books which flood the market.
>
> One chapter is done, the chapter on sauces, and has been sent to Avis who has allowed me to read and study it. I am profoundly impressed by the explicitness of the directions, the meticulous detail, which could enable a novice to turn out a perfect product. The usual cookbook slurs over the minute techniques — this explains them very, very clearly, telling you why you do certain things. There are 3 receipts here labeled "top secret" which really are that — I have never seen them before — for hollandaise, mayonnaise, and vinegar-butter sauce. The tone of the book is also most attractive; not condescending, not chi chi, not coy, not gourmet club, but wise and friendly advice from one cook to another.
>
> I have read Dione Lucas's original English cookbook, have been to her cooking school in New York before she was so well known, and have the Little Brown "Americanized" version of her book.

* 21 Club, a celebrated New York City restaurant.

> This is much better, much less complicated in both choice of food and directions. It is also much more French. I repeat, this is a book for housewives—not for hostesses. It is a book for people who will do the cooking themselves. Mrs. Child, unlike most Americans who seize on the opportunity of cheap service in Europe, does all her own work and serves as well as cooks. She has also full American equipment of electrical aids, beaters, mixers, pressure cookers, freezers etc. and will make full use of them in this book.
>
> Boiled down, this is *How to Cook French* in your American kitchen. Several publishers are interested, and Mrs. Child, through family connections and through her French Associates, who once did a little primer with Ripperger, has been in correspondence with Ives Washburn. But they don't want to publish with Ives Washburn because this is a big, good thing and Ives is not a big, good publisher.
>
> I think we should feel most grateful to Avis for steering it our way, and at this moment I should like to clear the picture and clear Ives Washburn out of Mrs. Child's mind by offering a contract with an advance of $250.00 on signing, $250 on receipt of half a satisfactory manuscript and $250 on receipt of a complete and satisfactory one.
>
> This is not a seasonal book but should have excellent continuing sales. Avis DeVoto is a cookbook expert. She has collected and studied them for twenty years. She thinks this is an essential book. I do too. For my money, it's TK.

So there you have it. I can't resist quoting Paul [Brooks] over the telephone—I asked if he'd read the ms. and he said he had, and went on—"I always tell Susie I don't care what I eat, so long as it's a little good plain French cooking." He said he thought it was a book that would make a great difference in his own home life. This from Paul is something—I always figured he and Susie had minds above such matters.

I will write later, when there is a free hour, if there ever is. Right now I feel swamped. But I am <u>so</u> overjoyed at the way this has turned out. How I wish I could come across next summer and meet you both. But we are definitely going West—Pa, Ma, and darling Mark whom you would love. He is really the most enchanting child. And I will be eating very badly all summer, except in such places as Butte and Helena, or when we settle down for a few weeks and I do it myself. Dammit I've got to stop. I am as pleased as if I had a new baby.

Bless you both,

Avis

81 Rue de L'Université, Paris, 7

JANUARY 28, 1953

Dear Avis:

What a letter from you, bursting with wonderful news at every seam. You must feel you have given birth to not just one baby, but a whole litter.

Paul and I are so happy and so proud that DeVoto won that award. It is heartwarming and encouraging and just as it should be. He is certainly some one whom we have admired immensely for years, and for a "Real American" (how I hate that term when used by a politician) DeVoto's it, for our money. He's a courageous, vigorous, outspoken historian and political observer, and his blood and guts and learning blaze through everything he writes. My God, he's the kind of man that makes America the home of the brave and the land of the free, and it's most fitting that he should be recognized and honored. We both send you our most fond and happy congratulations. Wish we could see him on TV, or even a photo facsimile of same. We're searching the *NY Times* for an account of the doings.

Course of Empire arrived in the mail yesterday, and if we finish *Boswell in Holland** tonight, we'll be starting right in on it. (Paul reads to me in the kitchen while I'm getting dinner.) It looks awfully good. Only criticism Paul has is that the end paper maps would be nicer if 4 feet in size.

And the HM news of our cook book. HOORAY. Naturally we know nothing is in the hand until we hear from them officially, but we know you would not have written if it were not pretty well sewed up. How marvelous for us to work with a good, serious, reputable, capable, honorable and honored house. The book will be dedicated to you, my dear, and to *La Belle France*. With all the things you have to do in your extremely busy life, that you have taken all this time and devotion and energy to promote something by people you only know through two pieces of cutlery, rustable at that . . . But how nice it is that one can come to know someone just through correspondence, and become really passionate friends. And may I say, for us both, that we are yours to command now and forever more.

It occurs to me I have never told you a thing about my darling Paul. He is 52 yrs. old (I am almost 42, *au commencement d'un certain age*). Brought up in Boston with his twin-brother, Charlie. Spent 5 or 6 years in France

* The journal of the Scottish lawyer and writer James Boswell, who spent 1763–1764 in Holland.

in the 20's, where he was a schoolmaster in the Dordogne, etcher, wood-carver, painter and photographer, and judo expert. Then spent the 30's as a schoolmaster at the Avon School in Avon, Conn. When the war came he joined the OSS, was sent to the China, Burma, India area to make war-rooms for Wedemeyer, Mountbatten, etc.* We met in Ceylon, where I was also with OSS (not as a spy, unfortunately, but in charge of all the files with the euphemistic title of Registrar). After the war he went to the State Dept. and we came over here in November 1948, where he has been in the US Information Service. He is an able painter, a most talented photographer (Edward Steichen has just chosen 5 of his photos for the Museum of Modern Art's permanent collection). He can really do just about anything, including making a French type omelette. Carpenter, cabin-builder, intellectual, wine-bibber, wrestler. A most interesting man, and a lovely husband. You'd like him very much, I know.

Now, I must get this off. We are writing Putnam today, as even if there is a slip between us and HM, I want to break off relations with that no-good firm, and have nothing to do with them. And want to get the letter off right away, in case he decided to write within the next month. Will, of course, send copy to Sheeline, whom we have always kept informed, including exchanges between you and me. And will send copy of Putnam letter for you. Hope he doesn't make any trouble. And there is no telling how many people he has shown the ms. to by now . . . but *tant pis* . . . our form can be swiped, but not our methods and experience!

Much, much love to you both . . . How rosy life is for us all!

Julia

PS: Will send biographies of Simone and Louisette, and a nice photo of a cold decorated fish.

81 Rue de L'Université, Paris, 7

JANUARY 30, 1953

Dear Avis:

We sent off the enclosed letter, breaking off relations with Putnam, yesterday via Sheeline. He might, for some reason, unknown to our illegalized minds, not wish to transmit it. But we shall know in a few days what

* Albert Coady Wedemeyer was the chief of staff to Lord Louis Mountbatten, Supreme Allied Commander South East Asia, during and immediately after World War II.

he has to say, and whether or not he did or didn't. Whether or not a contract arrives from HM, Simone and I are anxious to have nothing more at all to do with Ives Washburn, and right now. This has all killed poor Louisette, who is one of those sweet and darling people who can't bear ill feelings, and it is all somewhat involved with her son-in-law, etc. But, as Simone says, she is such a romantic, and when I think how she must have been pushed around by this ego-centric pansy, Ripperger, lending him money, making him delicious little meals, trying to get him to work. And I know quite well how she feels. It's lucky Simone and I have no personal tie-up at all in the affair, and I can pose, with our nephew, as a real tough egg.

Also enclose two more guinea pig responses, which Mrs. de Santillana might like to have. We shall have a synthesis when they are all finally in.

I have just served my poor husband the most miserable lunch of frozen haddock Duglere,* frozen "French" string beans and "minute" rice. It is just no fun to eat that stuff, no matter how many French touches and methods you put to it. It ain't French, it ain't good, and the hell with it. (I do realize that we may get poor quality way over here, and it may be un-frozen and re-frozen, so I will reserve my judgment like a good girl. But I am not going to fool around with that nasty, tasteless, depressing, A&P garboozova.) My, I get so depressed after a poor meal; that's why I can never stay in England for more than a week. The sun is out, and I'm going to do a bit of "footing" about the Rue de Seine area, which always puts me immediately and romantically in a wonderful mood.

Just found a tiny squib about the MacLeish/DeVoto prize-giving in the *NY Times* international edition for Thurs. But they didn't say enough, and then spent three paragraphs talking about Ellison's novel,† instead of about the fellow who interests us. No pictures, either. Hope you will write us all about it.

We are all infuriated about the article about France in the new *Life*. I think us Americans are much more upset than the French . . . who bolster us up, saying, "Well, you know Luce;‡ don't be so upset." I simply cannot understand such utter lack of responsibility in a publication that knows how important it is in forming mass opinion. It cannot be excused

---

* *Dugléré,* a sauce of shallots, mushrooms, white wine, and fish stock.

† Ralph Ellison won the National Book Award in fiction for *Invisible Man* in 1953.

‡ Henry R. Luce, the cofounder of *Time* and founder of *Life,* and an influential member of the Republican Party.

as mere naivety on their part. They are always speaking lightly and slightingly of France in *Time,* which enrages me. I have always heard Luce was an old SOB, but he is beginning to sound more & more like Col. McCormick.*

Much Love,

J.

8 Berkeley Street
Cambridge 38, Massachusetts

JANUARY 30, 1953

Dear Julia:

God knows when I'll have time for a coherent letter. We're just back from New York where we were wined and dined to exhaustion—enormous party at 21 with fabulous food and drink and my darling Elmer Davis coming up from Washington specially and shedding great light on the political picture and many other near and dear ones contributing to the fun. I got into such a euphoric state, quite beyond my usual meek behavior, that I was able to tell Bergdorf Goodman that their evening clothes simply didn't suit me at all. And then thank God I found the perfect dress elsewhere—hyacinth blue chiffon, very sleek and smart as befits my advanced years. I wear evening clothes about twice a year.

Anyway this is just to say thank you thank you for the wire whip which awaited my return, and for the cookbook which came Monday. It looks great fun and I appreciate the implied compliment to my French. This will take time to read as I do not bother with a dictionary—I would rather plunge in and read away until I get the hang of it as I do sooner or later. I am slightly awed by the arrival of the whip—telepathy or something. I had one which I used happily for years—took it to the beach this summer where it vanished. Our cat vanished too, same day—maybe he took it with him. I meant to get another in New York at Bazaar Français but didn't get that far downtown in the rush. All very odd.

When the omelette pan arrives I will practice—your directions are completely clear. I love eggs, love to cook eggs, understand the general principles, never had a real pan. Also by these directions hope to finish the reclamation of two iron frying pans ruined by my men folks when they were alone last summer. Think they left them on high heat and forgot them. Speaking of utensils, I bought myself one of those folding French

* Robert R. "Colonel" McCormick, the owner of the *Chicago Tribune* and a conservative Republican.

steamers which I adore—wonderful for peas (French style) and beans, fish and the like. On the other hand, I don't like the wire dryer for salad greens—it's decorative, but I can get greens crisper and have them stay fresh longer by wrapping them in a towel.

What I really wanted to tell you was that I got a smuggled copy of Stevenson's speech to the Gridiron Club,* a priceless document. It begins—"A funny thing happened to me the other day on my way to the White House—" What a man. What a man. I hope to God you've been able to catch up with the finer nuances of the Wilson affair†—we Democrats have been in a state of uncontrolled glee. It couldn't have been more perfect if we'd set it up ourselves.

Small world division. I went down to New York with Anne Barrett—she works for Houghton Mifflin, is divorced from a psychiatrist, was a great friend of the Kennedys,‡ and remembers Paul. She didn't make the connection until I told her about the cookbook. Does Paul remember her? She is about six feet tall, very thin, now grey-haired and quite distinguished looking, and a great friend of ours.

Now after all this upheaval about the Nat. Book Awards, let us hope HM will return to the workaday world and get a contract off to you if it has not already done so. They really lost their senses over this binge—put us up in a suite at the Ambassador and paid all expenses. I hope Henry Laughlin picked up the tab at 21—in addition to endless drinks for thirty, the food was staggering. A cold ham and a cold turkey. The biggest hot roast of beef I ever saw. Curried chicken, a wonderful hot seafood thing with plenty of lobster and shrimp. Plus salads and everything else and more drinks far on into the night. I kept thinking Lawd a' Mercy on me, how can this be I? So here I am peeling potatoes again. Queen for a day. And came back and found that Mary hadn't been here all week, being smitten with a bug, and my breakfast dishes still in the sink and the beds unmade. Worth it, though.

Blessings and I will make more sense next time.

Avis

PS: I did not abandon my son. He was staying at his school in West Newton.

* A prestigious club for journalists in Washington, D.C.

† Charles E. Wilson, Eisenhower's first secretary of defense and the past president of General Motors. A controversy arose during his Senate confirmation hearings over his relationship with GM.

‡ The family of Edith Kennedy, Paul Child's former love, who died of cancer before Paul joined the OSS.

81 Rue de L'Université, Paris, 7

FEBRUARY 3, 1953

Dear Avis:

Just a short note (I think), to acknowledge yours of Jan 30th. My, the doings sounded wonderful in NY and at 21. Where did you wear the hyacinth blue chiffon? And how nice all your friends came up, as they should have. Louisette called me up, as we were in the midst of giving a "young peoples" cocktail party, to say DeVoto was on the Voice of America, but we couldn't break up the party (which was to introduce a darling young American spending her junior year abroad, to 3 attractive young French boys . . . and it worked like a charm, they pounced on her and took her out to a party, and was old Auntie J. pleased as a cat) so we didn't hear it. But I am delighted Louisette did, which will confirm and re-enforce the DeVoto name in her troubled mind.

Am glad the whip and book arrived. Funny you needed a whip. The book is not of vast importance, but, as I know you collect them, you are probably like me . . . any cook book is fun, and you can always pick up something. Speaking of French books, do you have the *Larousse Gastronomique,* which weighs about 10 pounds, and is an encyclopedia of cooking history, cooking terms, and every recipe anyone ever heard of in the French cuisine? It is great fun. What French books do you have, maybe I could help you fill in.

You asked me if there was anything I wanted. And thank you, but not at present. I do have a Mouli grater. (They were invented over here, I think, but I got mine in the States.) They are wonderful. I just have an awful lot of gadgets, because people are always sending them. When we finally get settled in Marseille, I might want a typical US onion and a scallion. I have just sent you three shallots air mail, and hope the girl at the post office didn't mash them as she hit them awfully hard all over with her rubber stamp. But I didn't dare say anything to her, as I was afraid there might be some rule about sending veg. into the US, or something.

Speaking of small world Dept., this morning chez Louisette I met a girl (young matron) who said she had met <u>you</u> and knows Helen Everitt,* and Dorothy de S. Her name is Helen Cummings Cook,† and her husband makes wires in Williamstown. And Paul has been saying off and on, Anne Barrett . . . Well, what do you know. He says of course he remem-

* Bernard DeVoto's literary agent, who later became the first director of the Radcliffe Publishing Course.
† The third wife of William H. Vanderbilt III, the former governor of Rhode Island. The reference to "wires" is obscure.

bers her, and with great pleasure. She was a great friend of Edith Kennedy's, and he always thought she was a nifty gal with a lovely subtle mind. He sends her his love. (And I hope you have a subtle mind, too. Myself, I am infinitely subtle and extremely worldly.)

Now it looks as though we really are going to Marseille, though I never quite believe anything in this business until I see an official paper. We go down next week for a week's look around, to meet the consul, and see the territory. Then we come back here for a week. Then we go down there for 2 weeks and wait for the official paper. Then we come back here and pack up. Then we go down there for good, if we get the paper. Lord, when I think of putting this apartment back into its original shape. It's a wonderful apartment, top floor of a big old private house, and furnished in late 19th century Versailles and authentic 1902 . . . gilt, petit point and mirrors in the Salon Louis Quelquonque; real leather wall-paper in the dining room, with real sags in it.

And the vast master bedroom is all covered with green material, with a marble relief of a Knight of Malta over the fireplace, and the walls peppered with hanging bibelots, and big pieces of Boule* cabinetry, and moldings, and big French windows looking over a beautiful garden. Kitchen upstairs, big and sunny, with the steeples of Ste. Clothilde for a view, all newly painted green and white. And a guest room, newly fixed up, overlooking the garden. It is a wonderful place, old-fashioned as anything, but with most comfortable Frenchy old charm. And wonderfully located. I'd happily stay right here the rest of my life. We took down some 200 bibelots and chairs and sofas and put them in the attic maid's room. At first, we were very scientific, and Paul made wall sketches, and we numbered everything. But later . . . Mme. Perrier, who is 81, and was born in this house, and all her family live on the other floors, feels that this apartment, just as it was, was the symbol of *l'élégance française* in its true form, and perfectly beautiful. She is a darling, and we love each other. So, naturally, everything will be returned to its natural place. I wonder if that must mean 25-watt bulbs, too. But the thought of it!

And what to do with the cat; she can't stay with us in a hotel in Marseille. I guess she will have to stay with the Concierge until we find a cat house down there. We just want a house with a nice garden, a good wine cellar, a nice painting studio and a big kitchen.

---

* A famous family of cabinetmakers, including André-Charles Boule (or Boulle), a cabinetmaker to the king beginning in 1672.

I just hate to leave this darling Paris, though. Longer I stay, better I like it. Everyone is so nice in our neighborhood, too, and we have become such good friends. I must say, I do love French people, and had no idea they would be as they are. I thought the women would all be tiny little things, too chic for words; and the men pinching everybody and little, too. I was not prepared for this wonderful, earthy, human, sweet, gay, naturalness. Cozy, is the only word that expresses my feelings. Well, how lucky we have been to live here this long, and I shall never get over it. I know we shall like Marseille, too, but it will be quite different.

Nobody has written us anything about anything, cook-bookily. We are doing quite a bit of experiment on such things as French cakes using the electric egg-beater, which cuts everything down from 20 min. to 3 or 4, and is quite remarkable. And Simone and I have been on a quenelle research for some time, off and on. I have been sure they could somehow be made in the Waring mixer, and finally happened to run into a discourse by dear old Carême, which put us on the right track. Problem, what to do with the water you had to use in the mixer, to puree the fish. Squeezing is too difficult, and the towels usually split. So we have hit on the bread panade, and made some yesterday which were the best I've eaten almost anywhere . . . light and just the right consistency. Instead of pounding for 45 minutes and adding this and that, and pounding, and then rubbing through one of those sieves, and taking about 3 hours . . . this takes about 15 minutes . . . using egg beater too, after the puree part. I'm really excited about it, and don't see why no one has never done it before. As it is now, they are such a damn nuisance to make no one ever makes them except professional people . . . Now, thanks to A. Carême and *les 3 G's,* quenelles are at last to be the food of the masses. Hooray.

With love to you both, as always,
Julia

PS: Our address will be here, until further notice.

FEBRUARY 8, 1953

Dear Julish:
This will be bits and pieces. A hectic day, and I find I turn to my Child correspondence rather with the feeling of falling into a feather bed. In a word, I enjoy it. Especially when the morning mail (to the boss) contains such items as one beginning "O Gracious Sir," going on—"The weight of a problem no editor or agent will carry for me, my own genius"—ask-

ing B. to give him a day in February, and if he doesn't, "it will be your own fault . . . if you discover Balzac in a bureau-drawer." Item two—a long fulsome letter from an old flame, very old—recalling violets and roses, passionate poetry, always knew you had it in you—some of which B. has no memory of and all of which makes him want to spew. Item three—misspelled but very sweet letter from five boy scouts of Westbury, L.I., on an illustrated letterhead which says The Lewis and Clark Commemorative Expedition. These babes plan to cover the whole trail next summer (on foot?) and want B. to be their advisor. Awards are all very well, but we are going down for the third time. If I sound more than usually addled, put it down to the rewards of fame.

Mary washing and waxing more floors. Mark, since this is Saturday, rushing back and forth to the Harvard Chemical Laboratory, where he is an honored and adored guest. He is a genius in chemistry. I expect this is only temporary, as he was and still is sometimes also a genius in music. This is no appellation of ours. It comes, first reluctantly and then enthusiastically, from his teachers. The Harvard Chem Lab is driving me mad. Mark—just 13—has complete run of the place. Graduate students and young instructors have over the year given him what looks to my untutored eye like three hundred dollars worth of chemicals, glassware and equipment. I implore them to use reason, and they simply go on saying he's a genius and he won't get into any trouble. And he doesn't. Not even a hole eaten in the clothes. This very afternoon his father has to drive him over the lab and bring back all the equipment a graduate student is leaving when he goes into the army. I am proud, sure, but I am also apprehensive and I think these young men are crazy to treat a thirteen-year-old as if he were 25. And with Harvard aiding and abetting him, I am helpless. Living with Mark and his pa, I frequently feel as if I were tied to the tail of a comet. But I must admit I have never had a dull moment.

I'm cooking between all these doings. My special spaghetti sauce* which takes about five hours. 5 large onions sliced and softened in nearly a cup of olive oil, then two cans tomato paste, two cans tomato sauce, quart can solid pack tomatoes—garlic salt, bay leaf, caraway, basil, Italian parsley. Then eighteen Italian sausages fried gently and simmered in the sauce for couple hours. These are the fresh sausages made of pork butt and Marsala and garlic, fat and marbled pink and white—nothing like ordinary pork sausages and they don't cook apart, they're quite firm and rich

* Avis DeVoto had gotten this recipe from the poet John Ciardi.

and garlicky. The onions cook away to nothing and the sauce is intensely tomatoey. Then imported pasta cooked at the last minute, al dente. Fresh Parmesan, of course. This is the result of an incautious trip yesterday to my favorite Italian store way the hell and gone up in North Cambridge, where I sort of lose my head. They have such wonderful meat—I bought six beautiful little calves' hearts and a whole calves' liver pink and delicate as a baby's bottom. Never ate better. Well it smells very good and all I have to do is remember to stir it a bit from time to time. I wish to God I had more time to cook. I will be no good at all to you as a guinea pig from sheer lack of time, but my heart is with you and I drool over what you say about quenelles and such.

Shallots arrived safe and sound this morning—I never had such a cute present. They were slightly flattened but quite intact. And they are the same shallots I get here, only yours are bigger and healthier. I have to pay ninety-five cents a pound for them—but of course they keep very well and anybody should be able to budget a dollar a month or whatever—after all you don't eat them like apples. They have to be specially ordered. I should think anyone out of reach of a good big city market couldn't get them, unless there were a French or Italian colony around. Certainly the nearest thing we have in the way of an ordinary American onion, available most of the year and nearly everywhere, is the red Italian—provided it is not too big or too old, when they get pretty hot. The taste and texture is very much the same as the shallot. I've never seen the white shallots you mentioned. When you are settled, I'll send you the ordinary white and the ordinary yellow onion, and a couple scallions which are the little green spring onions. These last come from California or Texas and are available nearly all year. They haven't the crispness of the shallot, and when they are picked too big they can get quite hot. Tell me when to ship the produce.

Also want to ask about mono-sodium glutamate—sold under the name of Accent and others. As you probably know, this is standard in Chinese and Japanese cooking—I bought it before the war as Ajinomoto powder, put up in Japan. It has been well advertised and has become wildly popular, for use in stews, sauces, gravies, mixed dishes, seasoning vegetables and so on. I like it, and think it does a good job of seasoning, bringing out various flavors. Wonder very much what you think of it. And if you haven't any, I'd better send you some because the American housewife certainly is sold on it. In fact the spinach I buy, which comes thoroughly washed and bagged in cellophane, now has in each bag a small envelope of seasoning powder which I suspect is mainly mono-sodium, which you stick in with the spinach as it steams, and which does a very nice seasoning job with no other additions except of course the butter. I'll try to remember to stick in an envelope of the stuff for you, when I buy spinach next.

To business. Note from Dorothy de S. to whom I sent a couple of your recent letters (about leaving Ives Washburn etc.)—

> I have read with pleasure all the enclosures from Mrs. Child who grows nicer and nicer as her pen enlarges one's view. But!! Don't think I will ever think you are "butting in" because to do justice to Mrs. Child would be pretty near a full half-time job and I am so pleased that you are doing it out of love for the subject and friendship for her that I can't tell you!

(This because I suggested to D. that perhaps you might go on with your principal correspondence to me, I to keep her informed as progress was made, and that I might lend a hand with the editorial job when the whole ms. is done—I hope I done right—I know she has to work on a lot of other books as well as yours.) She goes on—

> Because the boys couldn't ever give me that much time for a single book. Sometimes I think my function is that of a mosquito—to flit around buzzing and nipping as actively as possible so that the hand gets raised and the slap accomplished automatically! Anyway—here is something that interests me. Every once in a while in the last decade Lovell or Hardwick (Mosely)* gets to dreaming about our only so-called "standard" cookbook. [She doesn't name it.] It was first done years before I came by two elderly domestic science teachers. It was revised once—about eight years ago—with some suggestions by me. It is a very narrow and dated book. Six months ago I was given time to make an extensive survey of the competition and recommendations for meeting it. The recommendations were so extensive—so minute—and so backed up by facts from the other books that they decided not to touch it. But, for this present purpose, what I found out is that the favorite book of the American woman today is *The Joy of Cooking*. It has conquered the field. And I am interested to see that both Mrs. Child's friends—quite different in their feeling about cooking—referred to it. If she doesn't have a copy she should—and I will have HM send one. Will you check on this?

* Hardwick Moseley, an editor and the retired director of Houghton Mifflin.

So I'm checking. She's dead right about *Joy of C.* It is a damn good book and I use it a lot. You can find practically anything in it. It is a hundred light years better than dear old Fanny Farmer* of which I now own five different revisions, all wildly different. I myself feel that Mrs. Rombauer† uses too many repellent shortcuts such as making sauces out of canned soups, but you will have to hand it to her—she is very sound on a lot of things, she covers the field, and the brand-new cook simply couldn't get along without her. So say if you have got a copy, which I am sure you have. Of course you will never sell as well as she does, but D. and I think you have a very good chance of being the *Joy of French Cooking,* and have a very pleasant sale for about a hundred years.

I hope by now she has written you. She phoned me and I said I thought she'd better find out from you how you wanted the contract drawn up, to you alone or to the three of you. Also—and this is solely for your ears—once you have got half the ms. done or thereabouts, I think you can get more advance if you want it. Up to something substantial. I wouldn't advise drawing on future royalties unless you have to, but if you do have to, if the money situation gets tight for any of you, and if the ms. looks sure, go ahead. HM is as sound as the Bank of England—sounder!—very keen about your book, and I'm sure willing to invest. As I told you, $12,000 was invested in *Course of Empire*—and thank God it's all been paid back! Now maybe the author will make a little money on it, I hope, I hope. And how we do agree with Paul's remark about the maps. The maps are absolutely fine—but B. thought he was getting a bigger book and the maps were planned for a bigger page, and he is very sore about this size. Where the hell was I when you ordered the book? I shoulda sent you an inscribed copy, only I was wool-gathering. Hope you enjoy it. It isn't easy reading, but very meaty.

Fascinated by what you tell me about Paul, and makes me all the more eager to set eyes on you both. When oh when? He is so right about Anne Barrett. And oh dear me, I am subtle and worldly too, not to mention sophisticated and glamorous. The wildest amusement I get out of life is the approach of the young, who frequently approach me with timidity, almost with awe—I am <u>Mrs. DeVoto</u>, wife of <u>the</u> DeVoto, and I am not grey-haired and fat and comfortable, or particularly shy. And they never know that my insides turn to jelly very easily, that I stay out of very smart

* Fannie Farmer, the author of the influential *Boston Cooking-School Cook Book.*
† Irma S. Rombauer, the author of *Joy of Cooking.*

shops because I am scared of those duchesses in charge, and that when I find myself in a strange situation (such as the time I was asked to a dinner of the Neiman Fellows* because Elmer Davis had come up to speak to them) I can't hold a cocktail because my hand starts shaking like mad. This must have been because I was the only woman there. I certainly am not afraid of newspapermen or of Elmer. How uneven one's personality is, even at approaching fifty.

Now that I know Paul is a photographer I have a definite request to make. (Don't for the love of God send me any more French cookbooks, because I keep dipping in and then I don't catch up with the *Herald Trib*, the *Times, Time, Newsweek, The Reporter*,† the *New Republic, Frontier, Harpers, Atlantic, Spectator, Economist* and Gawd knows what else, all of which I am supposed to digest for DeVoto—in addition to four mysteries a week. I will take Savarin‡ west with me next summer—only time I'll have to read it.) I want Paul to take a photograph of you at the kitchen stove. With or without decorated fish. Point is—we have two living rooms—one opens into the other—this is a Victorian house. You know the kind—high ceilings, etc. We sit in the smaller of the two. Do all our living there. The larger one has the grand piano, my desk, etc. Also a wall covered from top to bottom with photographs of near and dear ones. Or special occasions in our lives. (Perhaps my favorite is a photograph B. took [he used to be a photographer too] of the birthplace of Brigham Young, in Whitingham, Vermont. The stone marker says—Brigham Young was born on this spot in 1801. He was a man of much courage and superb equipment.) Elmer is on that wall, and Adlai, and Robert Frost, and various odd characters we see all the time, editors, publishers, the young in various stages, etc. So now I must have you, at your specialty. (I have one of the *Life* photographer [Wallace Kirkland] who accompanied us in 1946 on the L. and C. trail—in the bathtub, covered with soap.) No posed photographs, no studio stuff. I like to keep it fairly informal. And ordinary newspaper photograph size, so that you can be seen. If Paul could get himself into the picture, so much the better. I really want this very much. And you said yourself you can deny me nothing. I am willing to wait, but not long. You've got to move, after all. I shudder to think of it.

---

* Recipients of the Nieman Foundation's yearlong research fellowships to Harvard, awarded to journalists from around the world.

† A popular biweekly newsmagazine favored by intellectuals.

‡ Anthelme Brillat-Savarin, the French gastronome and author of the celebrated *Physiology of Taste* (1825).

If we had to move out of this house after twelve years, I think I would rather cut my throat.

B. shuddered slightly when I read him the paragraph about Helen Cummings Cook, whom I have no memory at all of ever meeting. He remembers her—apparently they clashed at Williamstown. This may very well be due to the fact that most of the time DeVoto is quite unaware of people—he has strong likes and dislikes, and a splendid absence of tact. If she approached him aggressively, or with large doses of the very best butter, he very likely strong-armed her. At any rate, he was told later that he had hurt her feelings. I dunno. I wasn't there. But I wouldn't be surprised if you got an earful. People either adore DeVoto, or resent him violently. I am used to it, but even I sometimes wonder how a man can be so blind and deaf to the overtones of a situation. On the other hand, if it's someone he likes and trusts, he suddenly develops intuitions and understandings quite beyond mine—and he will sacrifice time and energy and emotion for those people that knows no bounds whatever. He has been known to empty the bank account, bring pressure to bear on authorities, arrange an abortion for someone's erring daughter. I sometimes regard these activities as somewhat excessive, but I would sooner try to interfere with the action of the tides.

Which leads me. I would expire with pleasure at having the book dedicated to me (plus *La Belle France*). But I'm going to forget you said it, and I want you to forget it too. I mean this. For God's sake concentrate on writing the book, and see what happens to the relationship between you and me in the meantime. You may wind up hating my guts. Anything could happen. Because I look at B. and Helen Everitt. She was his agent for years. She was the one who forced him to write *Course of Empire*. She was the one who arranged financing with HM. After her husband died B. acted as surrogate father to her children, kept her on an even keel, handled all sorts of details. She was the one who listened endlessly. He was the one she turned to in all the family jams, which were plenty. So last spring they had a terrible fight (I still don't know what over) and B. said the book dedication would stand nevertheless and my God it was an embarrassing situation. She finally called up and said they weren't feuding any more, and she appeared at the Award and at the 21 party, but it's all very difficult and shattering. And now she's leaving HM (whose NY office she runs) and is going to *Ladies' Home Journal* which has been trying to get her for years and where she will be making a fortune. All this is confidential. So don't go putting a hex on your book, or on our very satisfactory friendship. See what I mean? Besides which, chum, leave us be realistic—there must be fifteen people who will mean more to the book in the end than I do. So thank you very much and please return to your senses at once.

Well, it's just too bad I didn't know you in July 1950 when I was around at the Hotel du Quai Voltaire, with only a couple of my Paris friends in town. I had a fine time, I certainly did—I made several fine pick-ups—a French professor from Tulane, three Johns-Hopkins medical students, a Princeton instructor in architecture, who took me every place I wanted to go including some very strange dives—but it would have been so nice to have someone my own age at hand. Both Dorothy and I have a strange fascination for the sort of young man who though far from being a pansy isn't very apt to get married—you know the type. They like to hang around older married women because they present no threat to their security or insecurity—and rapidly become tame cats if you let them. Quite useful, these boys, but certainly not the kind you want to become a solid part of your life. The older I get, the more I prefer the society of people my own age. I like every part of growing older except what happens to your feet.

Now I must wrench myself from this typewriter, having nicely wrecked the afternoon—skim my sauce, tend to that damn genius, and ask the boy scouts how they plan to travel three thousand miles. Did I tell you—Stevenson is stronger in defeat than Eisenhower is in de head?

Love to you both.
Avis

[P.S. in margin] What pleases me most about the Award is that McCarthy is <u>furious</u> about B. & Archie [MacLeish]. Also says he doesn't want to tangle with B. and Arthur S. [Schlesinger Jr.] any more because they are too smart about politics. Getting sense?

I might have known there would be a postscript. Have to tell you because nobody else seems to be excited about it but you will know how I feel. B. is in the chips all of a sudden—*Colliers* paid $5000 for long piece about the Louisiana Purchase, gave him a commission for another long piece about the west—he's doing a special piece for the *SEP,** and a handful of smaller ones—all very well paid praise be because we've been scraping the bottom of the barrel what with the long pull on *C of E.*

* *Saturday Evening Post.*

My share of this loot is a new stove. Not to replace perfectly good Glenwood, but to supplement. Bought an apartment-size Magic Chef and had it hitched up tandem to the old one. So I now have my heart's desire of many years—eight top burners, two separate full sized ovens, plus two broilers. I couldn't be more pleased and proud. I feel like the Ritz Hotel, or Rumpelmayer's.* Several people who were here the other night said I should have put in for a fur coat. Hah! It has maddened me for years that I couldn't do say a roast of beef <u>and</u> Yorkshire pudding without letting the roast cool off.

Hotel de Rome et St. Pierre, Marseille

FEBRUARY 12, 1953

Dear Avis:

Our first day in Marseille, and the first time we've seen the sun in months and months. It is a wonderful city, so gay and noisy, and everybody shouting in that wonderful accent. We came down in a very swish train, the Mistral, which does the trip from 1 PM to 9 PM; and we were traveling *tout à fait* first class, in a fancy Pullman car with a little table in between us, where they served us lunch and dinner. Nothing but rain and snow and fog until we got here, and the sky was clear and the stars were out. None of the consulate people can imagine why on earth we picked this tiny little hotel, instead of the big palaces (picked it up in the *Guide Michelin*). We just <u>hate</u> those 1st class hotels, where you might just as well be in NY or Rome or any number of impersonal fancy places. This has flowered wallpaper, appliquéd plumbing pipes, a bidet in the bedroom, and everything it should have, and we are very happy.

Paul is now out with the men, being introduced around. Evidently Marseille is a place where women are seen but not heard, and stick close to home . . . a conception which neither of us approves, but which, in this instance, will be very useful . . . I'll probably have much more time to work than I did in Paris. We were met last night by the fellow Paul is replacing, who seems very nice, but gives us both the impression of some inner insecurity or weakness. His wife had a serious operation, which I suppose means removing all her "organs," and has gone slightly off her rocker . . . which could very well explain him, poor guy. The consul is supposed to be very nice indeed, but a "maiden" (unmarried) of 53, and quite a stickler for protocol, and quite fussy and easily offended. This

* Rumplemayer's, a stylish ice-cream parlor in the St. Moritz Hotel in New York City.

doesn't bother Paul, thank heaven, who is able to get along well with all sorts of difficult types. (Though one type which absolutely enrages him is the dogmatic meat-ball.)

Your young Mark sounds fascinating. After reading about him, I happened to pick up *The New Yorker* with the story about "Teddy" the genius[†] . . . which did give me a shudder . . . but I don't imagine at all Mark is that type. He must be fun, but rather an awesome responsibility.

Glad the shallots arrived, and they are the same type as in the US. I suppose they are so expensive because they are so little used except by a luxury market. But I imagine as the demand gradually increases, they may become more available. I really don't think there is a substitute. Like the sound of your spaghetti sauce, and makes water at the mouth. Accent, yes. I have some, and it is even available here in France, but I forget the name. I don't use it at all as I sort of hate the idea. But I am sure it is useful in the USA where vegetables have probably lost some of their freshness from being shipped under refrigeration for days and days. One thing I certainly noticed here was the wonderful flavor of veg. and fruit. Strawberries, for instance, are dreamberries, but extremely fragile. Beans are so deliciously beany. They haven't yet gotten on to the system of growing a tough variety that will keep well in the markets. But I'm fascinated that they put up a US pkg. with spinach. Speaking of spinach, there is a French system which I find terribly good . . . which is first boiling it, then refreshing it, then draining it, then taking little handfuls and squeezing out all the water (which can be saved), then making a brown roux,* adding a good stock, stirring in the spinach, dotting generously with butter, and stewing covered over low heat for about 20 minutes, adding a bit more stock and butter and seasonings, as the spinach absorbs the liquid. Then, [you] can butter and cheese a baking dish, mix in a bit of cheese with the spinach, sprinkle cheese on top, and more butter, and bake in hot oven until cheese melted (5 to 10 minutes). *Brown roux very light, just to give a bit of liaison, ¾ TB flour, ½ TB butter, 1 cup stock.

Joy of Cooking, I adore it, and always have. It is a wonderful book. I am interested to hear your research showed it to be the all time favorite. I also have the new edition, and find it even better than the old. Somehow, old Mrs. Joy's personality shines though her recipes, too. I met her last summer, through Louisette. She is terribly nice, but pretty old, now, about 70 or so; and just a good simple Midwestern housewife. She said she'd been

† "Teddy," a short story by J. D. Salinger.

in some way weaseled out of something like royalties for 50,000 copies of her book, and was furious. (We all have copies of Mrs. Joy.) I did have a copy of Fanny Farmer, but was so wedded to *J of C* that I hardly used it. Have you run into Mrs. Beeton,* also? I managed to get a copy of a 1909 edition, which is charming . . . all about how to handle the servants, what the 2nd footman and the dairy maid should do, how to draw up legal contracts, etc. etc. I understand a new edition of *Household Management,* which is the whole works of Mrs. Beeton, is being done in England. (And the recipes in Mrs. Beeton, 1909, are real French cooking, plus English stuff.) I don't wonder you decided not to do a new cookbook when *J of C* has covered the field so well, in cooking.

Wonderful decision, that the principal correspondence will be with you! I was just thinking, sadly, as I wrote you my last note, that this wonderful flood would probably abate. I think you are quite right about the dedication, though. I really think it shouldn't be dedicated to anybody but *La Belle France,* or something rather international. And then to have a nice pre-preface, like in *Course of Empire*. However, I, personally, would know to whom it was dedicated. I cannot envisage <u>ever</u> hating your guts, as I don't think anyone ever could. I hate only a very few people, one being Mme. Brassard head of the Cordon Bleu who is a nasty, mean woman; McCarthy, whom I don't know; and Old Guard Republicans, whom I see as little as possible. But neither Paul nor I are inclined to get into fights with anyone, including ourselves, as it makes us quite sick. I know some people love it and thrive on it. I can see nothing but an increasingly friendly and fun relationship, and I look forward to it with great happiness.

Now, for a most serious note. I realize with awesome seriousness that the real work is about to begin. I am determined that this book is to be as perfect as we can possibly make it; and that every point in the basic explanations is to be absolutely mastered and masterfully explained. I think the Sauce chapter is, on the whole, a damn good job, and sets us a pretty high standard for the rest . . . which must be even better. I mentioned to Dorothy de S. that I could foresee no definite completion date, but that we were aiming for a year. As I may not have a kitchen for some time, I suggested that a good plan might be to rough out the whole thing . . .

---

* Isabella Mary Beeton, known as Mrs. Beeton, the author of *Mrs Beeton's Book of Household Management* and a famous nineteenth-century English cookery writer.

which is a matter of "think work," for arrangement, logical groupings of things, etc. etc. It would then be possible to get a good idea of how big the book would be, etc. As to the matter of publishing in parts or sections, that is up to you. The important thing is to get the stuff written. We are now proceeding with the main sections, and will do the desserts, pastries, etc, at the end. I have still a lot of personal experiments to do on yeast breads, etc (have just been doing croissants, and brioches, etc . . . and there is a lot of stuff concerned with temp for yeast rising, what to do if you are in a hurry, how it acts and why, etc. etc.). Then there is further research in the Waring mixer and electric egg beater. We have bought another egg beater for the Paris branch, and will also get another Waring mixer. As I said to D. de S. . . . this whole field is wide open, that of using the electric aids for a lot of fancy French stuff . . . and we'll be presenting something entirely new. No sacred cows for us. In this connection, we had dinner the other night with Curnonsky,* who is 80, and at the party was a dogmatic meatball who considers himself a gourmet but is just a big bag of wind. They were talking about Beurre Blanc, and how it was a mystery, and only a few people could do it, and how it could only be made with white shallots from Lorraine and over a wood fire. Phoo. But that is so damned typical, making a damned mystery out of perfectly simple things just to puff themselves up. I didn't say anything as, being a foreigner, I don't know anything anyway. This dogmatism in France is enraging (that is really about my only criticism, otherwise I adore them). For instance, I am always having fun asking various people about La Vraie Bouillabaisse, and get as many dogmatic *vraies recettes de la vraie Bouillabaisse* as made by the *vrais pecheurs*† as I ask questions. But it is always assumed that I don't know anything, even where a fish comes from. Well, says one woman, *"Nous, nous de la vraie Méditerranée, nous ne mettons jamais les tomates dans la B, nous jamais."*‡ Balls. *La Vraie recette* from *le vrai cuisinier provençale*, Reboul,§ has tomatoes. I don't know why I get off on this tangent . . . but I irk, sometimes. Usually, because I have had to study up on everything to inform myself, I know more than they do . . . which is so often the case with a foreigner. I suppose, cooking being a French preserve, they be-

* The most celebrated twentieth-century writer on gastronomy in France, known for his work collecting traditional provincial recipes.
† "Authentic recipes for real bouillabaisse as made by real fishermen."
‡ "Those of us from the real Mediterranean never put tomatoes in bouillabaisse—we never do."
§ "The authentic recipe from a real Provençal cook." Jean-Baptiste Reboul is the author of the late-nineteenth-century French cookbook *La Cuisinière Provençale*.

come dogmatic. I don't know. But dogmatism founded on ignorance or hot air is hard to take. I was thinking it might be useful to have the book reviewed by a well-known and really knowledgeable French gastronome . . . but I am rather sure it would be shrugged off as being OK but not really French, you know. However, I may be able to find that man. But, except for the practicing chefs, most of the writers aren't actually cooks, and far too many of them concerned with pure chi chi.

Where was I? Yes. Plans. Will now start out to work in earnest, and get those soups into their order. Will just continue with the sections as outlined. Each section will arrive in a neatly type-written state, all checked and proof-read . . . and should be considered as not yet final, but as nearly final as possible until we get into the final stages of kitchen checking, and possible new ideas. In these periodic transmittals, everything will be extremely formal and professional (and worldly). You will not be peppered or annoyed by little business communications from us. And unless you (and I include HM) have other suggestions, we shall proceed on this basis.

Simca, Julia, and Louisette dining with the celebrated gastronome Curonsky in January 1953.

We shall not, at this time, try to do such peripheral things as the fish and meat-cut list . . . although I have gotten a good start on these; nor the wine . . . as they can be done anytime. As to illustrations, I imagine it would be sensibler to consider them when the rest is done. This is just work on the meat of the book. Although Simone and Louisette will continue working up the different sections, I shall not be able to get in much solid work until after March 1st, when we move down again for several weeks. Which is too bad, as I long to settle down with my teeth in it . . . but it just can't be helped. Certainly hope we can find a house quickly!

Now I've just come back from lunch with Paul and his colleague, Paul looking somewhat glazed. The consul sounds pretty persnickety and difficult. Ah, this bureaucratic system, once they get in they stay in just if they don't stick their neck out. I guess he came in with the old crooked finger and striped pants Ivy-League era. They were beginning to weed out that type a bit after the war, but with more Ivy-Leaguers coming in with the new administration, maybe we'll be returning to that effete type. I certainly hope not.

Ha, Helen Cummings Cook. I can see how she and DeVoto might clash, though I saw and talked to her only intermittently I know the type. But how could I say anything, as she might have been one of your great friends, or something. She was very interesting on the subject of deep freezing, saying what one did at home under ideal conditions was far superior to boughten stuff. Talked about enveloping chickens and turkeys in liquid wax, and they are wonderful. And that they must be frozen fast, etc. I'd like to taste one, just to see. She sounds madly energetic but exudes what Paul refers to as the "Cambridge blight." (A rather dry and virginal "dedication," plus negation of the flesh and personal embellishment.)

How nice indeed that DeVoto has got those juicy contracts with *Colliers*. Please send us copies when they appear. And I'm terribly jealous of you having a good new stove arrangement. If we ever get into the money I am going to have a kitchen where everything is my height, and none of this pigmy stuff, and maybe 4 ovens, and 12 burners all in a line, and 3 broilers, and a charcoal grill, and a spit that turns.

Must put a stop to this letter! But there is always so much to say.

You may be relieved to hear that Baudelaire is not at all enthusiastic about Brillat-Savarin, calls him a kind of insipid old brioche whose sole use is to furnish windbags with stupid quotations. I find it pretty tough and dull going. Grimod de la Regnière* is infinitely more amusing, and quite a more sophisticated type. I think they are both being terribly over quoted. "Tell me what you eat and I'll tell you what you are," and . . . "*mais on est nait rôtisseur*."† Phoo. There's a new review out called *Gastronomie le Neuvième Art,* and they quote those two boys on every other page, until it becomes very silly.

Yes, we'll send you some photos, but only on the severe condition that you will send us several of the two of you and the two boys. We visualize you as about 5 and ½ feet (you said once that you weighed 112 when your weight dropped, so imagine you would normally be about 125). With dark hair (I don't know why) but blue dress is nice with dark hair. DeV. I think is probably looking a bit like Stevenson, with that kind of face, but heavier set and more vigorous.

I can't pin any size on him. Now I think of it, too, I am just sure I heard a lot about him in the early thirties from my cousin-in-law, Noble Cathcart, when Noble was publisher of the *Saturday Review of Literature* . . . I was at that time living under the Queensborough Bridge being a career woman. I had intended to be a great woman novelist, but for some reason *The New Yorker* didn't ask me to be on its staff, and I ended up in the advertisement dept. of W & J Sloane‡ (just to gain experience). I was then just out of college, and was rather madly in love with Noble, though curbed my passion because I was so fond of my cousin, Anne. I do wonder if you ever knew her, she was a darling. Though Noble, now, worries me; he has somehow gone to pieces . . . or had, when I saw him 2 yrs. ago; and had even 5 yrs. ago. I somehow suspect that he has never quite matured. Now, it will be interesting if we do have this connection, also . . . and am quite sure we have.

Will reluctantly draw to a close, though there are always so many things to say.

With much love, *toujours,*
Julia

Are you Upper Bohemians? We seem to fit into that—but I don't like to eat off card tables, and I hate to sit on the floor (too tall & not well-enough padded).

---

* Alexandre Grimod de La Reynière, a nineteenth-century epicurean and the author of the journal *Almanach des Gourmands.*

† "But one must be born a roasting expert."

‡ An upscale home furnishings store in New York City.

Marseille

FEBRUARY 18, 1953

Dear Avis:

Here is a combined report from Pasadena Guinea Pigs, interspersed with comments by me. I find it fascinating. One thing emerges to me, is that they are not aware of the classical tradition, or rules of the game, or whatever it should be called, of French Cooking. And I imagine they are very typical of most average Americans. I think that where they feel the ms. is pedantic, is where I have attempted to explain what is the right name, or the traditional way of doing things. I think this all is important, that the names and the old methods be correct . . . thus the book cannot be attacked as being "not really French cooking" . . . and, furthermore, that traditional French cooking should be preserved. Who knows, the US may end up as the final preserve, as life is so expensive over here it is all pretty hard to maintain. Anyway, I don't consider any of the bridge passages, or introductions (the "blah-blah," as Louisette calls it) are at all in their definitive form; and that they will not be until the very end. They serve as "something," while the rest is being worked into shape; but will be very carefully considered and worked-over at the end.

My, as you said once, there are so many slips and errors, and dangling references, to be avoided.

Anyway, the guinea pigs feel we are on the right track. I wish my sister-in-law, Freddie Child, would write something. She has a group of people whom I don't know at all, and we would get a more impersonal attack. She is a wonderful gal, Freddie, but I suspect she feels we have "gone over to the French," and somewhat disapproves. It is very hard to put a finger on her. As Paul says, she is sometimes the primitive mother; and if one in any way infers that what she is providing for her young and her man in the USA, is not the best in the world . . . and this is all very submerged and delicate, and I am sure she doesn't realize this primitive reaction . . . an emotional wall rises out of her. Has this wall risen about our Sauces? I don't know. She is an awfully good cook, and they eat good in Lumberville, Pa.

We've just been to Perpignan and back, and then to Nice and back, seeing mayors, newspaper people, college types, etc. Ran into a darling pair of Fulbright American students in Montpellier. He is writing his thesis on Church history, and they have rented a little two-room shack in the middle of a vineyard, no heat, no comfort, but a fine view. They have both plunged into the life, entertain other students from all over, and do everything on a shoestring. Very proud indeed to think of them in this area as unofficial diplomats, and they have lots of pluck. Montpellier is difficult, and the French there do not accept foreigners, which includes other French foreigners, too. The Western part of this area is extremely interesting, and quite untouched touristically. I don't, personally, care much for the Côte d'Azur, it has become so resorty, though I love the back country.

Have just had a Bouillabaisse for lunch at a good restaurant, Campa. It had potatoes in it, and no tomatoes; and the sauce or soup, had been thickened with, I presume, potato flour. Quite good. But I like, myself, a bit of tomato, and a stronger-flavored soup. So many ways of doing it! After we are settled down here it will be fun to discuss details with the restaurant people.

Walking back from the restaurant I happened to see Reboul's *Cuisinière Provençale* in a bookstore, and have sent it to you. I hope you won't mind, but it only cost a dollar, and is such a good book. You may have it already, and if you do, you can pass it on. In the recipe for bouillabaisse he discusses why it is boiled at a fast boil . . . which settles once and for all the question for some people saying it should be simmered. Which is . . . to mix the olive oil in with the soup, otherwise it floats on top. (Also, as was pointed out by another chef, to break up the tender fish, which also thickens the soup.)

Back to Paris this evening, after cocktails with the Consul General. He sounds awful (don't never say I ever made a remark about him). Maybe I'll have a more favorable impression than I expect. But he appears to be thoroughly disliked by everybody, a persnickety, timid, neurotic, ineffectual man, who is, furthermore, uninterested in eating. We've gotten to know the fellow Paul is replacing quite well this week. He is a very nice chap indeed, and really going through hell with his wife. She did the last two days with us, and she is a complaining woman type, always picking on him, and life, and France, and the weather, etc. Their 6-year-old boy was with us, too . . . adorable, gay, gregarious, good-humored, speaking French with a pure Marseille accent, or English, interchangeably. I suspect they are about to separate, and what will happen to this wonderful kid? But how could anybody stay married to a woman like that and not go crazy. To think of going home at night to a room full of recrimination and bitchiness. It sickens me to think of it. And he is a nice, patient, serious fellow. How do they ever get hitched up to these people in the first place?

Was glad to see in the *Herald Trib* this morning that old Agnes Meyer* had hopped on McCarthy and Jenner† about their proposed investigation of the educational system. I wonder what Ike and the clean brooms are going to do about cleaning up that stinking part of their stable.‡

Hoping all is well in Cambridge, Mass., and with love from us both,

Julia

Have just been re-reading your last, about shallots, etc. That is just too bad, and complicates things. How about us going into the business, the DeVoto/Child Shallot Packing Co. Inc., which could include Mirepoix, meat glaze, herb puree, shellfish butter, detective stories, photos and paintings. Amen. History—a chemical factory & cooking school. Just a small cozy establishment, built on the un-American principle of not trying to make money, just on not losing it.

What a shame we didn't know each other in 1950!

Love—

J.

FEBRUARY 23, 1953

Dear Avis:

This is really going to be short! (yes?) We are in the midst of beginning to pack, which is consisting of sorting. Yesterday we did photographs, of which there are hundreds and hundreds, today it is books, of which we have a like amount. I never want to throw anything away and Paul wants to throw everything away; so between us it works out quite reasonably. (In one of Paul's throw-away moods, we threw out our marriage license, some years ago, which was going a bit far, I think.) Paul says most women like to keep everything because it is their nesting instinct. Maybe he's right.

That was a wonderful letter of yours of 2/17/53, and I can't even begin to get into it, though I am longing to. We have both read it twice, and loved it. And the photographs arrived in the same mail. At last we know! I love the looks of DeVoto, there is a wonderful gaiety and intensity about that face. Now, how do I know whether you look as I pictured

* An influential journalist, philanthropist, and activist, whose husband, Eugene Meyer, owned the *Washington Post;* also the mother of Katharine Graham, the paper's future publisher.

† William Ezra Jenner, a Republican senator from Indiana and a McCarthy supporter.

‡ Eisenhower's 1952 presidential campaign had urged voters, particularly women, to help Ike "sweep out" the mess in Washington.

you, now that I've seen you. You are dark, anyway. That is a wonderfully worldly expression you have on in the group picture, really superb . . . It is the face I always try to wear when I am in New York, with no success. My compliments! Paul thinks both of you are wonderful. I feel that we do not have the definitive Avis here, and find the passport one a bit pass-porty, though probably very good in a certain mood of contemplation, or would it be resignation to the fact that a P. photo is always to be a P. photo. I can see that Mr. P. Child will have to give you a good going-over (with camera) when at last the great DeV/Ch meeting takes place. However, I hope in the meantime, you will send along other tidbits. I forget other people probably don't have as many photos as we do. I've sent you along 5 or 6, which give a pretty good coverage. As you will see, your pen pal does not photograph as being as worldly as she actually is. Paul comes off very well in that respect, though he is far less so than he would lead one to believe.(!)

Specifications: Paul, 5 ft. 11, weight 175, very muscley. He has done lots of wood chopping; etc, and is a 3rd-degree black-belt Judo man (which is a remarkable thing).

Julia, 6 ft. plus, weight 150 to 160. Bosom not as copious as she would wish, but has noticed that Boticelli bosoms are not big either. Legs OK, according to husband. Freckles.

Oh, I saw Helen Cummings Cook again the other day; and I find I quite like her. She is interesting, and interested in all sorts of things. I would classify her as a "girl's girl" type; and don't think she would appeal much to Paul. I would have to see her quite a bit more to know how I would really like her (I find my first impressions are no good, and should know enough not to make them); but I feel she is a good and worthwhile piece of goods . . . who would really work on something, such as the Public Lands business.* That article of DeV's we read with great interest; and are all for it. It is wonderful you are jumping into it. Is there anything we can do to help? Can we send anyone any money? Let us know.

Much, much love, and much, much more later,

J.

New address: care of Paul Child, Directeur Regional, Services Americains d'Information, 5 Place de Rome, Marseille (B du R)†

---

* From the 1940s to his death in 1955, Bernard DeVoto wrote several articles on public lands and conservation of the natural resources of the West. Many of the articles were published in *DeVoto's West: History, Conservation, and the Public Good* in 2002.

† Bouches-du-Rhône.

FEBRUARY 23, 1953

Dear Julia:
I am going to try to keep this really short as I have to write to some more Congressmen. But I probably won't succeed. I mean this merely as commentary on your last with its report from the Pasadena guinea pigs. I think you are right in saying that they are not really aware of the basic reasons for French Cooking—and I would not give it a second thought if I were you. Don't compromise—you know what you are doing, and you don't want to turn out a hybrid cookbook. This is a discipline, and it mustn't be watered down. Nobody who reads the book is going to turn overnight into a French cook who does no other kind of cooking. We are all going to do a lot of American cooking all our lives. But when we cook French, we want a clear uncluttered classic line and no compromises. Here it is, girls, take it or leave it. As a good neo-Malthusian,* I <u>know</u> damn well that this kind of cooking, this kind of eating, this kind of life is on its way out. But let's preserve what we can of it, for as long as we can, before we are all reduced to proteins grown in shallow sea-water.

Which disposes of the Pasadena girls' remarks about basic white sauces. And I would forget about the bride, too. I doubt if there are many brides who know what food is all about and many of them never do learn. God, when I think of the food we ate when I was a bride. I don't suppose there are many women in this country who are going to master your book and cook that way all the time. But there are sure as hell thousands and thousands who would love to master ten or a dozen dishes, to educate and enchant their families and to impress their guests. The good French restaurants in New York are flourishing, thank you. B. says most French restaurants in NY are run by Italians, which may be true—nevertheless, they cook French. And they are multiplying. In NY and every other sizable city. We even have two quite successful French restaurants in Cambridge—Chez Dreyfus and Henry Quatre. I do not say they are very good—but they have a few excellent dishes, and they are always jammed. I <u>know</u> interest in French cooking is increasing all the time over here.

Use Finesse as often as you like—once you have explained what it means once or twice.

<u>Don't</u> cut down your introduction. Phooey on those girls. I would

---

* T. R. Malthus, a British economist, in his famous *Essay on Population* (1798), held that population growth will eventually outstrip food production.

like a much longer introduction and I want to hear all about Mère Michel* and any other personality you heave in. When I make beurre blanc for dinner guests don't you suppose I'm going to tell them all about Mère Michel and what Curnonsky says? Damn right I am. All this sort of thing fascinates me and I couldn't disagree more radically with your friends' idea that it leaves Americans cold. You have a very nice touch for this sort of thing and your taste is so good that there is no danger you will overdo it.

I don't agree either that there is general ignorance of herbs here. There are a dozen different flourishing herb houses (dried herbs) and they are widely given for Christmas presents and sold in all kinds of stores. Trouble you're going to run into on herbs in US is that fresh ones are very, very hard to come by. It's only by remote chance that I can occasionally get fresh tarragon. Chervil, dill, etc.—never. Plenty of chives, but very seasonal. I think you've got to assume dried herbs always. Except parsley, of course.

Agree with all you comment on next few questions. I like "liverish" and if simple American types don't know what it means they ought to read more English novels.

Can't remember whether you defined "nap." It's a good word and I like it.

Dijon-type mustard is OK—lots of stores have it, and several other mild or sweet mustards such as Bahamian. Most housewives have access to a specialty store of some kind—even a delicatessen—which has a dozen kinds of mustard. Everybody doesn't use Frenchs' or Guldens' by a long chalk.

You've got a point about the drum-seive. Sieve! I can't spell that or nickle. Nickel. Anyhow, do you mean one of those pointed-cone things that rotates in a frame? That was another thing I meant to chase down at Bazaar Français only I never got there. The wiggly basket is indeed terrible.

Yes on second pressing of shellfish butter and I hope you will hit hard and frequently on getting last bit of goodness out of everything. Americans too hell bent on saving time and labor, and what is easier than to wash a bowl?

Whole point of putting Vinaigrette in sauce section is to suggest its

* The woman chef and owner of Chez la Mère Michel, a bistro in Paris, who showed Julia how to make an authentic beurre blanc.

use on something besides salads. I have of course used it frequently on asparagus and artichokes, but tried it on cauliflower the other night and the family wildly enthusiastic, though usually lukewarm about that vegetable. There are so many hot vegetables it is good with for a change.

And I certainly do not agree that the book as a whole seems somewhat elaborate and pedantic. Nuts to that. You have got exactly the right formula for lightening it and I think the tone is wonderful. This is a serious book on a serious subject and it is written just exactly right.

My goodness. If they want a recipe for maple mousse they know where to find it.

I think these gals are in general very helpful, however. And all your guinea pigs are lightening the load for Dorothy and me immeasurably, for which we are most grateful. They will pick up a lot of little slips and errors, and then D. and I can just say we agree or we don't agree. I am filing all the reports very carefully.

This may seem infra dig to you, but the more I think about it the surer I am that you ought to put in a paragraph or so to the effect that in general you get just as good as you pay for — in food as well as in clothes, furniture, medical care and whatever. Not many women in my generation have to be told this, not after having cooked and eaten through two wars. But the young things with unformed taste buds do have to be told. They have to be told that although you can make an eatable cake out of margarine, you make the best cake out of the best butter. That when you say oil, you mean the best olive oil. We learned all this the hard way during the war, with imported olive oil at sixteen dollars the gallon when and if you could get it. We know all about flavoring soybean oil with a little olive oil to make it bearable. But a lot of the young things have never made a salad with anything but Mazola, heaven help them. They buy bargains in butter when eight cents more a pound would improve their cooking eighty percent. And if they have to pinch pennies that way OK — we've all been there. But they have to know that if they try to turn out a first-rate dish, it's got to be made of first-rate materials. And it's cheaper in the long run to buy the very best — there's less waste, and fewer failures.

How in hell are we ever going to make bouillabaisse in this country? Haven't got half the right fish. Maybe in Florida where you can get red snapper and so on. But in good old Boston, great fishing port, we can get cod, halibut, salmon, mackerel, smelts, haddock, whiting — swordfish and shad in season — plus lobster, shrimp, clams, oysters. Butterfish occasionally. During bad winter weather, only three or four of the fish.

Eels I suppose could be ordered from the city. Or you could catch them in the Charles—people do. And I've always understood the list of fish for bouillabaisse was fairly rigid.

I am drooping on the vine. Been out late four nights in a row, eating too much, going out again tonight. I must try to snatch a nap. The telephone usually rings.

Much love,
Avis

FEBRUARY 27, 1953

Dear Julia:

Pictures just arrived. Hooray! I am enchanted with them, and think I will have the big one of you at Cordon Bleu and the extremely handsome one of Paul framed for my Hall of Fame. But I love the others too and it is hard to choose. We must have an assortment along the lines of the big one, only with Simone and Louisette, all in action, for publicity for the book. How elegant to have a really first-rate photographer to take on this assignment, with no fooling around with the pix boys who can be very trying. I am very, very pleased with your looks, so warm and vigorous and handsome. And perfectly positive I have met Paul long ago. He looks completely familiar though I think he had more hair when I met him. Am I crazy? Thank you very much and imagine me poring over these daily. I am rather astonished that you are such a big girl. Six feet, whoops. I adore height in women—my oldest, dearest friend on earth (since I was six) is six feet tall and had snow white hair when she was 16. She was always mobbed by photographers and painters when she went abroad. She is now well over seventy and (in Freudian jargon) my "good mother"—at any rate I have always been far closer to her than to my own mother. I am going down to see her—perhaps for the last time—in Kennett Square, Pa., next month and I can't wait. I suppose purely because of Allie Harkness* I have always looked on really tall women with helpless admiration. Don't think I'm not impressed, too, with Paul's judo title. I have read enough about it to have some faint idea of what that means. Well, all this just makes me more eager than ever for us all to get together but for the life of me I don't see when or how. I am perfectly positive I will never get DeVoto abroad—he never has been and never wants to go. Terribly provincial of him. But as he always remarks, "Well, I've never been to

* A friend and mentor of Avis, who considered her a "second mother."

El Paso, or Independence, or the lower Green River." And he would go there first. Maddening. I don't suppose we'll ever be rich enough again for me to go alone. But you aren't going to live abroad forever, are you?

I'm enclosing copy of Stevenson's Gridiron Speech—which is circulating in this country like feelthy postcards. Don't return—I've made many copies. Read it aloud—better that way. There are some magnificent cracks in it. Blessed man.

He made a speech in Los Angeles last night, which wasn't widely broadcast, dammit. Apparently he took a good stiff sock at the boys who want to give away the public lands with both hands. Opposition to anything of the sort seems to be growing rapidly, thanks in part to B's blast in *Harpers*.* Good God no, don't send money. This battle is purely and simply a political thing now, with the sides lining up. Horace Albright was here the other day—he used to be head of the Nat. Park system, is a lifelong Republican, and for many years has been advisor to the Rockefellers in their land benefactions. And he says that the Republicans (who seem to be learning more about the facts of life every day—none of which is going to help them carry out their campaign promises) are now appalled at the implications of the national resources plank in the platform. (Pat Hurley's† dirty work.) He is fairly sure that they are not going to be stampeded by the stockmen's pressure groups, in spite of the NAM‡ and the U.S. Chamber of Commerce lining up behind them, plus some of the big lumber companies. He also says the Forest Service has been cleared—nobody is going to be fired. This is terrific news. Though I keep hearing also that McCardle,§ the recently appointed Chief Forester, is on the axe list. Honest to God the Republicans are too stupid. They actually thought that they could get rid of thousands and thousands of experts and technical people in Agriculture and Interior and fill their jobs with I suppose thousands and thousands of minor executives from the big automobile companies. They now ruefully admit that there just aren't any successful young business men who want to give up their thirty grand a year for Washington. Dopes. No, we are feeling rather cheerful about the prospects for public lands. B. had reprints of his *Harpers* piece sent to everybody in Congress, and he's had a good response from Washington. Three

* "Billion-Dollar Jackpot," in the February 1953 issue.

† Patrick J. Hurley, a Republican operative, a former envoy to Chiang Kai-shek, and the owner of United Western Minerals Company of Santa Fe, New Mexico.

‡ National Association of Manufacturers, a large industrial trade association.

§ Richard E. McArdle, chief of the U.S. Forest Service from 1952 to 1962.

or four freshmen senators and congressmen are now on the public lands subcommittee and really hot on our side, and B. is keeping them briefed constantly. B. also has a widespread network in the west who keep him informed on every move. Some of the stuff is very funny. Such as a speech of one of the stock lobbyists in Wyoming the other day who described B. as a renegade Westerner who teaches history at Harvard and drinks pink tea with his little finger stuck out. Boy, I'd like to watch this guy's face first time he gets a load of DeVoto. What is pink tea, anyway? All these people attacking DeVoto think he teaches at Harvard, where he hasn't taught since 1936. Source-work very bad, like McCarthy calling him "Richard."

. . .

Thanks for copies of letters. I will take an hour off and arrange all your correspondence according to date—it's piled every which ways now. It will all come in handy when the ms. is completed. Terribly pleased that you are already pondering on several other cookbooks. You've got yourself a career, my girl.

. . .

I have now got Beurre Blanc licked to a frazzle and I am getting bilious. Also have put on five pounds which on a figure like mine ain't good. It looks all right, but I like to be able to wiggle freely in my clothes instead of bursting out the seams. Also I have made your top secret mayonnaise with great success in spite of the fact that both my electric beaters broke down and I had to shift to the whisk. It's delicate and lovely and I am pleased. But I do so hate to diet. Blast you.

And now I want to make béarnaise, and there is no chervil dried or fresh. What to do? No fresh tarragon either. This is maddening. Because beef has come way down in price and we are eating many steaks. Do you think béarnaise can be made satisfactorily with dried herbs? Anyhow I want to try. Can't think why no dried chervil at least—it used to be around.

The cattle-raisers are screaming blue murder over this drop in the market, and I can't think when I've been more pleased. They just got so glutted with cattle they had to sacrifice, and I hope some of them lose their shirts after what they've put the public through these last ten years. Just plain greedy, plus blithely overgrazing the public lands. B. thinks there will be further drops in price, due to panic selling. Meantime, steak twice a week, and a big roast for the week-end. One of these days a pork chop will be a real treat.

Well I hope your *Time* informant is right about Luce. God how I distrust that man. And his bitch of a wife. I do think that is the most incredibly bad appointment and I'll bet anything you like that within six months of Claire[*] landing in Rome there will be an international incident. I had the great pleasure, couple of years ago, of listening to her debate at a Harvard Law School Forum with Reinhold Neibur[†] or however you spell him. He's an intellect. She is not. She was dressed like a demure little girl—virginal white silk shirt, golden curls. He was very worldly, very polite, demolished every argument she put up with the greatest politesse. She was reduced to batting her eyes and switching her sex around, and she was all but hissed off the platform. I am fond of the story that is going around, of her last meeting with the Pope, to whom she gave a great deal of bracing advice, and who, so it is reported, finally remarked gently, "I must remind you, Mrs. Luce, that I am a Catholic."

When you get your new address I am going to send you the worst, positively the worst cookbook I have ever seen. It is called *Stars in Your Kitchen*, and has recipes from Eleanor Roosevelt, Eisenhower, Bob Hope, and the Lord knows who else. I particularly recommend Mr. David Niven who discusses fish cakes in a very sophisticated (*sic*) manner. I didn't buy this horror. A publisher sent it to me. Better not show it to your French friends, who think badly enough of us as it is. But if you ever need an emetic—

Now I must go to town and buy DeVoto a new mattress. He wants a foam rubber one, though I have always thought it felt like sleeping on a cup of custard. I hate shopping. Except for food. Bless you again for the photographs. I think you both look absolutely wonderful. I would love to have Paul photograph me, and I know just what would happen. He would go crazy, as all photographers do. Marion Schlesinger[‡] has been trying to paint me for a long time and has given up in despair. I'm all out of drawing. The parts don't add up to the whole. But I look perfectly magnificent in an enormous hat with a veil!

Best love,
Avis

---

* Clare Boothe Luce, the wife of Henry Luce. She served as a conservative Republican U.S. representative from 1943 to 1947 and was appointed ambassador to Italy in 1953. Julia's earlier letter referring to Henry Luce is missing.

† Reinhold Niebuhr, an American theologian and a professor at Harvard.

‡ Marian Schlesinger, the wife of historian Arthur M. Schlesinger Jr. They were divorced in 1970.

USIS, 5 Place de Rome, Marseille (B du Rh)

MARCH 4, 1953

Dear Avis:

*Enfin, à nous deux** (as they say in the fish market). We are here at last, and I do think we are going to love it. We are nicely fitted into our little hotel room, with Mrs. C's two trunks of cooking library well-arranged, two work tables and a bidet. (How did I ever get along without that in my life. Would you like me to send you a portable one? Only cost $10, and sits on metal legs. Wonderful.) Have just been out looking at apartments, one impossible, the other quite marvelous, with a balcony overlooking the Vieux Port and such a view of boats and hills and sea and sun as to knock your eye out. Am looking at another this afternoon, which sounds a bit bigger, but is a bit away from the center of town. We think we'll get an apartment as pied-à-terre while we are looking for our villa with wine cellar, garden, view, studio, kitchen and many guestrooms. Just can't get along without a kitchen, and, besides, it is too damned expensive to eat out all the time.

Met most of the people at the Consulate yesterday (the boss is away), and they are an awfully nice bunch, serious, hard-working and normal . . . and warm-hearted. It was they who have given us the leads for the apartments. Finally met the Consul General, last day we were here two weeks ago. He is a deeply neurotic man, a bachelor who had a mother now dead, but who cooked him up to a fare thee well, I guess. He is evidently all right if you handle him with gloves, and means well, but needs to be supported and reassured. I don't think Paul will have any trouble with him, as he knows how to handle that type.

My, this is a fun city . . . all this sun, and the shouting and the gaiety. The people are great fun and warm and nice. I just like French people anyway . . . and thank god I can talk French. It would be dreadful if one couldn't make jokes and pass the time of day. Paul speaks beautifully, with no accent at all. I am fluent, but have an accent all right . . . but I don't care how many mistakes I make as long as I can talk and talk and talk.

Now, will reply to your three letters, starting with Feb. 17, and will not even attempt to say this won't be an enormous letter . . . pages and pages.

INTELLECTUAL. Paul had no college either, only 2 years of Co-

* "Finally, two of us."

lumbia Extension Course in New York. But he is an intellectual, as I interpret the word . . . meaning he is interested in ideas, and is ready to dig for information and is always trying to train his mind (like with General Semantics), and has a thirst for knowledge. Me, I am not an intellectual, though had 4 years of college. But the people I admire most are the intellectuals; I am trying to train my mind, which is sometimes a fuzzy sieve. (Seive, no it is sieve. 'I' before 'E' except after 'C' with the following exceptions "Eight Weighty Neighbors Freighted on a Sleigh.") But, except for La Cuisine, I find I have to push myself to build up a thirst for how the atomic bomb works, or a study of Buddhism, etc. Probably that lack comes from my "Upper Middle Brow" origins . . . that useless and wasteful class of society. I never was thrown with intellectuals and academicians until I joined OSS and went to Ceylon, where we had a fascinating bunch of people including missionaries, geographers, anthropologists, psychiatrists, ornithologists, and the like . . . as well as some bankers, lawyers, and others. I found, suddenly, that these professional people were what I'd been missing all my life; and furthermore that they stood up and wore better and adapted to our odd life and were fascinating to be with. They were interested in where we were, what the country was like, wanted to know the people, etc . . . while so many of the others just wanted a bottle of coke and hot dog and mother, and didn't give a damn about exploring these fascinating foreign places in which we were living. So I got an awfully late start. I was even a Republican! You, however, have had years of it. (Will go more into Upper Middle Brow and classification of types when we get to page 5 and the Upper Bohemians.)

CORRESPONDANCE. When I think of what correspondance <u>you</u> might have, if you write as long a letter to everyone who sends you a knife or a bit of public land, you must be flooded with it. I don't have any really heart-to-heart correspondants, as most people don't answer right away, and I've sort of gotten out of the habit . . . but it is such fun. I write religiously every week to my Pop, but I can't say much as we are on such different beams. Can't mention politics, or philosophy, so it is pretty dull going. He is a darling fellow, a most generous father, a real "do gooder" in the community (how he would hate that term) as he is on the Community Chest, the Chamber of Commerce, the Republican Committee, the School and Hospital Board. He just has everything it takes to be a fine citizen and a responsible member of his community; except he is violently emotional over politics (so am I, but I am trying to be intellectual about them . . . but they roll around in my stomach rather than coming out in a quietly poignant yet devastatingly unanswerable thought-piece). He could really be a world-beater if he had had more intellectual training, which would have opened his mind and would have made him more tolerant and enquiring. He is an example of how not to be, and how one must continually struggle for understanding and experience and wisdom. And he absolutely dismisses Paul, as artist and "New Dealer," which infuriates me and saddens me . . . as it means there can be no real affection and sympathy between father and daughter. Too bad. But so many of my friends are in the same predicament.

ENGLAND. We are very fond of England, and go there every Christmas to visit our friends the Bicknell's* in Cambridge. He is a Don of Architecture and an architect, and mountaineer and a lovely fellow with a big mustache. She studied ballet with Sadler's Wells,† now has four children and teaches children's ballet in Cambridge. She's a good cook, and they both love France and French food. We always have a traditional pre-Christmas feast in the kitchen together, with Sole Bonne Femme, roast pheasant, soufflé Grand Marnier, and great wines (always a Château Yquem 1929 with the soufflé). All the Cambridge part I love, and all their friends, one of whom is the Cambridge printer, and another the librarian of Kings. And we've stayed in London, and wandered all over it, and eaten around quite a bit, and driven up to Newcastle, and visited friends on a farm in Hereford. It is such a poetic countryside in summer, with those great trees, cows, hedges, thatched cottages. I find I had to read some Wordsworth, even. But that public food is so awful! I remember, when we made our first trip, we were driving up north and stopped at a beautiful Tudor Inn, which was truly oldey woldey and charming. Dinner, and we had boiled chicken with the hair still on, partially covered with a real honest to goodness English white sauce. I had always heard of it, but thought it was just a lot of French chauvinism. But this was really it, flour and water with hardly any salt, not even made with the chicken bouillon. It was a wonderful experience and I enjoyed it immensely . . . sort of like seeing one's first red Indian. However, as we always say, one doesn't go to England *pour la gourmandise* . . . but I find it depressing after a while. And then, there is an atmosphere (not among the friends we like at all, but a general English feeling) that I get of wanting to scream, because people are so *retenu*.‡ I look at these faces, so held in by duty and "what is

---

* Peter and Mari Bicknell.

† The Royal Ballet School in London.

‡ "Reserved."

done" and more by "what is not done," and all that tea, and that bloody accent, and that chirping. People are certainly nice, and polite, and sweet, and honorable, and my god what courage, and they stop for pedestrians and say sir and all that. But I become wild after a week. I admire them immensely. But my heaven, how I love to come back to this sweet and natural France, this warmth, these wonderful smells, this graciousness and coziness and freedom of spirit. I think maybe you like formality better than I do . . . Boston life, perhaps, which is a bit English. I never knew Scotchmen kicked their pregnant wives in the stommick! Probably drink to excess because of all that conformity they have to go through, which is very unhealthy, I think.

CATHCART. After he left the *Saturday Review,* Noble went down to their place on Cape Cod, Chatham, and bought a local newspaper and printing establishment. Then he sold that, I think, and I just don't know what he is doing . . . if anything. Anne was my first and favorite cousin, and I practically grew up with her, though she was about 5 yrs. or so older. She was first married to a fellow called Farrington, who right away had a relapse of TB and died in about 2 yrs . . . She nursed him all the way. Then she was in New Mexico doing something, and met Noble, and they married. Had 2 girls, now about 22 and 18. She had had, I think, a nervous breakdown when she was about 18, but we were too young to know or be told what was the matter. But she always seemed perfectly normal and wonderful to me; a most generous and darling older sister. Paul and I visited them at Chatham in 1951 in the summer and she seemed perfectly fine. Noble was being peculiar and bitchy and nervous. Next thing we knew, when we were back here, we heard that she had shot herself. Nobody seems to know much about it, and why. She must have been one of those deep worriers; and I don't imagine being married to Noble, who can be charming, helped very much . . . as he is still an egocentric adolescent. He was completely undone, naturally. The girls seem fine, however, and they make a very affectionate trio. They are perhaps too young to realize his shortcomings as a man; but he makes a wonderful companion for them. I just hope, with his tremendous concern for their welfare, that he won't make them feel tied to him. I think they're all coming over here this summer. It's a pretty sad little story.

FRESHNESS OF VEG AND FRUIT. That difference in taste is certainly remarkable. We know some people who farm in Bucks County [Pennsylvania], and swear their stuff is so much better when grown with organic chemicals. (I certainly never saw more beautiful veg. than in

Kunming,* where they were all fertilized with you know what!) French seem to use mostly animal manure, and their farms are small compared to ours. Also, things are very seasonal . . . They last two or three weeks, like for strawberries, asparagus, etc, and then it's over. There are things grown in hothouses, but they aren't the same. Tomatoes are really good only in the summer. I imagine, as they modernize, they'll come to our methods. But possibly not, as they have only ¼ the number of people to feed, and the distances are so much shorter.

SPINACH AND VITAMINS. You must tell me what to do about vitamins. As you know, I don't like them at all, and refuse to bother with them in cooked things. I think one should get one's vitamins in salads, and raw fruits, and what is cooked should be absolutely delicious and to hell with the vitamins. I think we have a phobia about vitamins in the USA . . . and that the average person would have a horror of boiled and refreshed green vegetables . . . and would rather have spongy stuff with all the water floating about, just because they think it has vitamins in it. That is why, I am sure, so many men hate to eat veg . . . because they are so damned hygienically cooked. (Don't like hygiene either, and feel it is very bad for one to isolate oneself from all the germs, then one cannot go out of the *cordon sanitaire* because one has no immunity. Enough hygiene to be reasonably clean and free from typhus and dysentery and cholera, etc, but enough is enough. I shall refrain from putting my feelings on this subject into print.) Your avocado recipe sounds divine, and I'll try it. What is a mincer? A thing with a lot of little wheels for chopping parsley? Cornish pastries sound heavenly.

READING. Have you ever started in on Balzac? I happened to pick up *Le Lys dans la Vallée* about 3 yrs ago, was amazed, moved, enchanted, and said mmm to myself, how could he have known so much about human nature way back there in 1840? So I read a biography of him by [Stefan] Zweig, and then more Balzac, and find myself *Une Balzacienne Engloutie*.† Most satisfying stuff, one learns so much about France, about life, about Balzac. Some of it is a bit fantastic, some is quite long, but it is the stuff that you can chew and live on. I have still 85 books of him to read . . . which will keep me for the rest of my life. But I love it so much I feel I shouldn't indulge. Besides, I have so much to read on La Cuisine. Comes in a nice Bible paper edition in 10 volumes, and wonderful for traveling

* Julia ran the OSS registry in Kunming, China, in 1945.
† "A devourer of Balzac."

. . . and very good for your French, of course. Have also dipped into Gide, as I felt I should know him a bit he has had such an influence on contemporary literature, etc. I didn't realize he was such a complete homo, and with very young boys, mostly Arabs . . . and even when he was in his late 70's. A bit disturbing when he describes the beautiful young men, etc. But a wonderful stylist; and interesting. Proust I have all ready to read sometime. So far I have never gotten beyond the church steeple and the cup of tea in *Swann's Way.* Friend of ours who was French officer in a German prison for 4½ yrs. said he found it a perfect place to read Proust . . . that you had to have plenty of uninterrupted time, and he loved it. Read him 2 or 3 times, and Balzac. I should think he'd be fine for your trip. Sounds like a heavenly summer that you are about to have . . . wish we could be along. I haven't been camping since I was 14, when we went for 2 summers in the High Sierras, fishing and riding, and packing on mules, etc. It was wonderful, but I never had to put up the tents!

UPPER BOHEMIANS and all that. Paul doesn't like categories either. I love them, though I don't think, natch, you can really and exactly pidgeonhole people . . . But the two categories of Upper Middle Brow and Upper Bohemian are a pretty good catch-all, to me . . . and I find I can put people into a sociological group, with qualifications, quite satisfactorily. Someone is not quite upper MB; or Upper B with upper MB background; and so on. It's that Upper Middle Brow class that disturbs me very much indeed. Here is a class with all the money, all the leisure, all the educational facilities right at its front door. And what does it produce but, on the whole, a lot of Old Guard Republicans with blinders on, and women who rarely develop out of the child class and create just about nothing. I think your Boston UMB is quite a different fish, as in Boston there is a cultural and intellectual tradition. But certainly the middle and far-western types I know are distressing examples of conspicuous waste of good human material. They are usually the salt of the earth, good people in the community activities and awfully nice people . . . but they are almost to a man anti-Semitic, anti-foreign, anti-progress, anti–phi-beta-kappa, anti–contemporary art and music; and living in a little compartment walled about by their money, their friends who are all just about exactly the same as they are, etc. etc. They don't think, create or explore . . . and are marking time until they die, playing poker, bridge and tennis and sitting on Boards. (What do people do when they sit on Boards, I wonder?) And they are the ones who say, "Well, McCarthy, you know he's brought this 'thing' in front of the people; and when there's smoke

Paul Child introduced Julia to France, but Julia confided to Avis that he was "far less [worldly] than he would lead one to believe."

there's fire, you know." I know I'm lumping the worst together; but as a collective phenomenon it is pretty dreadful.

Now I'll skip to your last, and put any business bits in at the end.

Glad the pictures arrived. Paul says it is possible that you could have met, maybe in Boston in the 30's, at Edith Kennedy's, or somewhere. She must have been a wonderful person. He did know Bob Hillyer, who must have been mighty difficult. But he says he is sure he would have remembered you, but then he might not, of course. He did have more hair then. He also has a twin brother, Charlie, whom you might have met in New York . . . Charlie has no mustache. Might well have been in Boston, as you seem to know many of the same people. He is a great friend of May Sarton's, and I have met her about 3 times, twice over here. She's nice.

When you go to Kennet Square to see Allie Harkness, why don't you look up Charlie and Freddie Child, who live right near New Hope. They often go to Kennet Square to buy mushrooms. You'd like them I think.

He is a painter, tried making his living doing portraits after the war and found it hard to count on that alone, so does fabric designs also. She is a darling, red hair, good goods. They lived over here for about 5 years in the 20's, and then bought a big old stone house in Lumberville. Have two ravishing daughters, both in college; and a most delectable red haired boy, Jonathan, about 11, who is someone I am quite mad for . . . He is also a genius, and a babbler, and very gregarious and independent. They love visitors, always are having housefuls, live very informally indeed, love food and wine . . . and are just great fun. Charlie is nothing at all like Paul, in fact, they are absolutely opposites. Charlie is a big talker and energy man, and terribly funny, and often quite showy . . . and sometimes needs a kick in the pants . . . and a damned good artist and a thoroughly creative fellow. They've heard all about you, of course.

CHARLES AND FREDDIE CHILD, Lumberville, Pennsylvania. Telephone, via Doylestown, Pa, SUGAN 2611

It's wonderful that the public lands business is being taken up by some congressmen at last. That's a wonderful piece of work you are doing, and please keep us informed. Is DeVoto going on a speaking tour?

That's the best news about McCarthy! Gee whiz, I hope it is true, and I hope someone can pin it on him for sure. Although I'd rather have him beaten by a general rejection or by revulsion, as it would be healthier. Have been relieved to see that the newspapers are running strong editorials about him, and thought the Ed Barrett* statement in our *Herald Tribune* was excellent. Trouble is that the folks want their comic strip, rather than sober thought, darnit. I don't see why there cannot be a combination of good Republicans, Stevenson Dems and Independents to run down that demagogic element in our national life . . . as I suspect the Taft republicans are giving McCarthy his head so that maybe Ike will be discredited. It is horrible how people will use anything as a political monkey wrench and to hell with the country.

Loved the Stevenson speech, and thank you for sending it. We have also sent for a book on his speeches, they sound so good. Wonderful man. Thought the reporter's article on the Stevenson supporters was very interesting. That is a good magazine. The article on France by Teddy White† seems good. I haven't read it all yet, but think he knows what he is talking about.

---

* Edward W. Barrett, a former chief of the overseas news division of the Office of War Information and a *Newsweek* journalist.

† Theodore H. White, a journalist and historian.

The other night we were out with some high level types from Army and State Dept. (we don't go around much with that group, so usually know very little gossip). It seems there is a Douglas MacArthur III who is in the foreign service, related in name only to the General. He was here in Paris while Ike was at SHAPE,* doing liaison work at the time NATO was being set up. Ike saw a lot of him, and has great confidence in him. MacArthur III is married to a girl called something like WahWee, who is the daughter of Vice Pres. Barkley.† Seems that when Ike got into the White House he kept calling up MacArthur III (who was transferred to the Dept. in Washington last fall), asking him this and that, and what did we do about this problem in NATO, etc. And then it seems the White House boys weren't quite sure just how to handle the patronage business, so they get hold of WahWee and ask her to find out a little about patronage from her Pa. So she asks the old boy, "Daddy, when you were doing politics for the Democrats, however did you arrange for patronage, did it come to the Governor of the state, or the Senator?" Etc. So she'd get a nice little picture, relay it to MacArthur III, who'd relay it back to the White House. And then the Dems. who want something tell it to Barkley, who tells it to WahWee, who tells it to MacArthur III, who slips it into the White House. So now they've given MacArthur III some high up job, like special asst. to the Sec. of State, and he's sent post haste for his secretary from Paris, and all's pretty cozy. I don't know at all whether there is a word of truth in all this, but it is funny.

Too bad you think you can't get over here! But we should be home on leave next summer, 1954. Where will you be then? We shall have to go to California for 2 weeks, to Washington for one week, to Pittsfield for 2 days; leaving us about a month for a vacation and business. Certainly one of our principal and most pleasurable stops will be in Boston . . . but when would be the best time. (We want to spend 2 weeks in Maine, also . . . if Charlie and Freddie are there in the cabin.) I hope that by then the book will be in good enough shape to have serious discussions with HM . . . but when is the best time for them? In principle, August is the dead month for Paul here, so ideally we'd leave sometime in July, and return to France sometime in Sept. We get two months, not counting travel time. And we shall certainly need to get home for "re-Americanization." The last time, after 2 and ½ years away in France, I certainly welcomed that

* Supreme Headquarters Allied Powers Europe.

† Alben W. Barkley, the vice president under Truman. His daughter Laura was married to MacArthur.

good old shot in the arm. Somehow one gets a worsening impression of the USA living over here, with the McCarthyism, scandal, etc . . . which make good newspaper reading. I forget how nice it is, how wonderful the people are, how vigorous and hopeful and helpful, etc. And, my, it is so good to see one's friends and family. I came back a much better representative of my country. But it would be unthinkable to come to the USA if you were in Africa, for instance.

And Paul is ready to photograph everything in sight. When people say to him, about themselves, "But I take a terrible photograph." He says, "Well, I don't." He is awfully good. He particularly goes for, besides portraits, design and architecture. Being an artist, his photos have great poetry and drama, as he loves light (sunlight) and shadows, and contrasts. There is a bare possibility that some of his photos may appear in *Life,* as Hank Brennan, art director, has asked him to try his hand at some photo studies of French cities and countryside, etc. I don't know when Paul is going to have the time, as he works 6 days a week from 9 until 7 or later. But it will be wonderful if he can work that in. His work should be known, it is so unusual and so beautiful. And he is a dandy art critic, too. None of this folderol with words and paint dripping impresses him at all . . . and he feels the present trend (this old, old trend of 40 yrs) is pretty thin stuff . . . that of non-objective "feeling," with no form, shape, and usually no technical knowledge. I shouldn't go on with this, as it is not my field at all, and I might well misquote him. But I think he is in the "avant-guard." Myself, I am a quattrocento Italian and medievalist.

Can't go on any more, too much to plow through as it is. Will continue later, still have: Narcisse Chamberlain* & *Gourmet;* Waring Mixers, Salad Bowls, Bouillabaisse; Guinea Pig #6, so will make next mostly business, with, undoubtedly, various philosophical and historical overtones.

Veuillez croire, Chère Madame et Amie, à mes sentiments inifiniment tendres et chaleureux.† (I don't have my French "Perfect Secretary" book with me, which is too bad. Most useful, all kinds of letters for every occasion.)

Very much love,
Julia

* A French-born reporter and editor who collaborated on many books on French cooking with her parents, Samuel and Narcissa Chamberlain, including *Bouquet de France.*

† "Please believe, dear Madame and friend, my infinitely tender and warm feelings."

c/o USIS, 5 Place de Rome, Marseille (B du Rh), La Belle F.

MARCH 6, 1953

Dear Avis:

Continuation of that monster letter of a few days ago, and mostly on business.

GUINEA PIG #6, from my sister-in-law, Freddie enclosed, with attached comments from me. I think this is the last we shall get, and has given us a pretty good idea of reactions. Next time, we might try some young and inexperienced cooks, and also some of the same ones, if they are willing, to help on the proof-reading and vagaries, if you agree it is a good idea.

"I" before "E", etc. LEISURE, must be added to those on the sleigh.

DOROTHY CANFIELD FISHER. Simone is writing her.

MRS. FAIRBANKS,* mentioned in D. de S.'s letter of a while back. S & L [Simone and Louisette] have asked her to lunch next Tuesday.

NARCISSE CHAMBERLAIN, ran into her at Babes Moffats the Saturday before we left. Babes has just married a young Frenchman, named François Giraud, and has come forth with a little baby, Monique. He is very nice indeed, she met him at MIT; he's an engineer, and is the type France needs more of . . . He wants to modernize things . . . but it ain't easy. I was asking Narcisse how BOUQUET DE FRANCE was going, etc. She said, "Watch out if you have anything to do with *Gourmet*." It appears that McAusland† (according to Chamberlain family) is the type that wants to clutch everything to his bosom. Sam had great difficulty in getting a release on the copyrights for his French articles that were published in *Gourmet* . . . I don't know just what the trouble was. And now, McAusland wants *Bouquet de France* to be a very, very exclusive book, only for the upper classes, or something too recherché for words. He will let only very, very few bookstores have it, only something like *Gourmet*'s Bazaar (wherever that is, NY, I suppose). Sam wants to put it into every store and all around. But not McAusland . . . he wants it sold by private subscription by his magazine. So, natch, it can't have a great big sale. I'll bet old Sam never had a lawyer give him any advice . . . That's what can happen if you deal with such people as *Gourmet,* I guess. I also asked (in all innocence) however they got all those wonderful recipes. Lot of them from "reliable" books, others from chefs, and then "adapted" them. Did

* Wilma Cannon Fairbank, an expert on Chinese art and architecture.

† Earle R. MacAusland, the founder of *Gourmet*.

not try them all out, but if relied on chef or on book, put them in. That's about what I thought. They are not well-written. It's nice for us, but I'm so sorry, for art's sake . . . as it is a fascinator. I'm now reading it carefully, and find it fascinating. But page 12, Poulet de Bresse. Flame the chicken before sautéing it . . . sauté slowly . . . add whipped cream . . . don't sound right. P. 571, Poulet à la Niçoise . . . cut a 5 lb. fowl for fricassee . . . cook for an hour. How old the chick . . . 5 lb. pretty big, might be an old hen, an hour wouldn't be enough, probably . . . and it would be pretty tough eating anyway. I am being very mean about this book, but I think they are big competition for us, and I want ours to be way ahead of everything in accuracy and depth and perfection.

BEURRE BLANC, glad that is working well . . . My, it is good . . . and you can see now why our stomachs have been acting up! Another lovely sauce is that Chivry, with poached eggs.

HERBS. French herbs are seasonal, too. In winter one gets only parsley. For Béarnaise, use dried tarragon, then can strain the vinegar before making the sauce, and add fresh chopped parsley afterwards. Tarragon is a definite taste associated with Béarnaise, so should be in if you want to be classical. Tarragon vinegar might also help.

My sister, Dort, has never sent me no bottled herbs . . . Could you be a dear and send me some, and I will send you a check to cover the cost. When I was in Washington, one of the houses, like Spice Island was putting out an oil pack, or some process where the herbs were a bit damp. Used it for chervil and chives. What I would like:

Tarragon, chives, chervil, basil, fennel. And, perhaps, one sample of several brands. I don't think the kinds of stuff you get in the A&P are any good, only the specialty herb houses.

I always thought French used piles of herbs, but they don't. Main staples are parsley, thyme and bay. Tarragon, chervil and chives when they are in season. In Provence, they add fennel and basil.

Our mailing address for packages (which come through the Government "pouch," so we don't have to pay duty):

PAUL CHILD, PAO, Marseille, France
State Dept., Washington, 25, DC
Postage as to France.

PRONUNCIATION: Think we'll just forget about it. I love the enclosed clipping.

SALADE BOWL (Wooden) Was very much interested in your salade bowl treatment. What you mean, boiling oil? Please enlarge. I love the looks of them, but that old rancid taste of mine ruins it for me. French don't use wooden bowls at all . . . and I thought it was the home of the wooden salade bowl.

WORKING PHOTOS OF 3 G's. Paul planning to take these when we go up to Paris for our final house-moving.

WARING MIXERS. Simone and I spent all one morning at the Salon des Arts Ménagers, which is a tremendous exposition or fair of household appliances and wine tasting held in the Grand Palais every year. Such wonderful fun it is, and everybody deliriously happy because can always refresh self with a bit of wine tasting. We looked at every Waring-type mixer they were demonstrating, and there were dozens. All selling for $75.00! They have come out with some new types, with Rheostats on them so you can make them go faster and faster, or slower and slower, by turning a knob. Ours just has a button control, and only one fast speed. Also the cutting knives, instead of all pointing up, point both up and down. In these models, many of them, you could puree such things as raw fish and use no water; and you could puree (or chop) carrots and onions for mirepoix, no water. There was one model, called, in French, "The Packard," which seemed the best, and the fellow said it was an American patent. Ours is an "Osterizer." Unfortunately we had to leave for Marseille, but we made the acquaintance of a very nice chef-demonstrator who said quenelles worked fine in his Packard. Simone is taking him some fish, panade, eggs, cream, etc, and he is going to make the mix, and she will then take it home and see how it is. So we are not the only people who have thought of this . . . No one has done an article on it yet . . . but they soon will. I think we should do a bit on it right away, to get ahead of everybody. But where to place it? *Gourmet* would be logical . . . but I just don't like *Gourmet*. I also hear that they steal your stuff, and it comes out months later under another name. (Wonder if that is true?) What a racket.

And: I also must know what the new models of US Waring-type mixers are like. Do they have the speed control, and the up and down blades? Who's making them, etc. I need names and literature and pictures. I hate to bother you with this, but if you could give me names and addresses of manufacturers, I would write them PDQ.

BOUILLABAISSE. Your question got me fussing around again in my fish research, but I have strongly resisted it, as it will take a good 2 or 3 weeks at least to synthesize all the documentation I have . . . so I cannot, at this point, give you an exact list of American fish to use. Certainly a real McCoy Bouillabaisse Marseillaise has to be made here, because of their special fish which don't appear to exist anywhere else. I think there is a lot of bushwah talked about Bouillabaisse and what the "real" types do. A "real" fisherman would make a "real" bouillabaisse with whatever he had on hand. And then there is the "ideal" bouillabaisse . . . about which people take pleasure in differing. To me, the telling flavor, besides the fish is the "Provençale soup base," which is just about always the same . . . garlic, onions, tomatoes, olive oil, fennel, saffron, thyme and bay, and usually a bit of dried orange peal. Then there is the number of different kind of fish you use . . . They are non-oily (no mackerel), some are firm-fleshed, some are soft-fleshed, some are gelatinous. Bouillabaisse made for 8 to 10 people is usually better than made for 4 people because then you can use a larger variety and number of fish. I am also of the opinion that you should first make a good fish stock, with the bones, heads and trimmings of the fish (Bouillabaisse Riche) . . . Many recipes call for just water . . . but to me it ain't good enough. Some people also say that if the fish are caught and eaten immediately, they have enough taste so you don't have to make a stock . . . Maybe that's true, but who lives by a fishnet all the time. It is certainly a very easy soup to make, only long part being the cleaning and chopping up of the fish.

BOUILLABAISSE FISH (a poor start, only)

French:	American:
RASCASSE. Tough with iodine & gelatin.	There is a California sculpin of same family *scorpoena*.* None so far discovered by J. Child elsewhere.
St. Pierre or John Dory. Flat fish, tender.	Don't have it. Substitute Flounder.
Congre. Firm flesh, and gelatinous.	Conger or Sea Eel. *Webster* says this is an important food fish, but US fish bulletins don't mention it. Maybe we throw it away, and it's a good fish. Too bad.
Baudroie. (Lotte de Mer). Firm flesh.	Anglerfish (also called goosefish, bellowsfish, allmouth, monkfish and frogfish). Powerfully ugly, used to throw it away but now keep it, and catch it in New England.

* *Scorpaena.*

Merlan. Soft flesh.	Do not have it. English call Merlan "Whiting," but it ain't our whiting, which is of the Silver Hake. Anyway, it is of the cod family . . . small cod.
Loup (Bar). Soft flesh.	We do not have the exact variety. Substitute Black sea bass.
Vive. Firm flesh. And little rock fish.	English Ballan or Weaver. Don't know about this baby.
Crabs, lobsters, etc.	Suggestions for New England Bouilla (non-oily fish): Cod, Haddock, Pollack, hake, cusk, whiting, scup (?), rosefish (?), flounder, sea bass, halibut, Angler fish and sea-eel.

I suppose one would have to go down to the wholesale markets and pick around, if one wanted to be particular . . . It would be fun, anyway. Why did I start on this, anyway, as it won't be much good as a list. Recipe in Reboul, which you should receive pretty soon, is very good . . . so I won't go into the nasty mess any further. I am just sure, however, that you can make a good bouillabaisse anywhere, if you make it right, whatever kind of fish you use.

Stalin has just died over here in this part of the world. Pretty scarey business. I'd hate to be living there as a middling to upper functionary!

Much love, as always, and write soon again,
Julia

MARCH 11, 1953

Dear Julia:
Only one page, so help me God. I'm typing notes for B's new edition of Lewis and Clark and it is hell what with his handwriting and L and C's spelling. Also got rheumatics in my right elbow, advancing age no doubt. Also Jack [John] Fischer, new editor of *Harper's*, arriving in an hour to spend the night. Also a million other things but one of these days I will write a proper letter.

Enclosed list of manufacturers of Waring type mixers is from my most obliging hardware man—best he could do on short notice but ought to keep you busy. Next fall I'll have to buy one myself and you can tell me which. Hardware man says annual housewares show for dealers only is about due in Boston and if I want to go he will smuggle me in. This I must do as I can't wait to sneer at the streamlined toasters, chromium can openers with everything but a cutting edge, etc. And if I do I will investigate the mixers further, but better not depend on me as I may not get there. Next time I go to Boston and it will have to be very, very soon I will send dried herbs you asked for from S. S. Pierce. Never even heard of fennel dried and doubt if it is packed. We get it fresh sometimes. Family won't play but I love it raw, dipped in olive oil and well-toasted poppy seeds. House of Herbs very satisfactory and will send some. Spice Islands I doubt if it is sold around here—that's grammar for you. Twin Trees the only house that packs dried chives with any flavor and color. I got a big bottle at Vendome Food Shop in NY and will send you some of it in a smaller jar. Dried tarragon most satisfactory of all, I think. Also am using some very fancy white wine vinegar which has an enormous spray of tarragon floating around in it, and if I can get it out of the bottle I will try that in my béarnaise.

Think I will send to Maison Glass for their *glace de viande*, though good old Bovril puts up the same fight talk which always makes me uncomfortable because I would so much rather the English ate their beef.

The hell with vitamins, dear. I was only pulling your leg. I take mine out of a bottle all winter. And it isn't Cornish pastries—it's Cornish pasties with a short *a* and they are full of beef and onions and things. Cornish miners eat them practically every day down in the mines.

I have had a lovely letter from the Lumberville Childs and I am sure as hell going to see them when I go to Kennett Square which will be middle of April. I am so excited about this trip as I practically never go anywhere, being tied to household, child, and typewriter.

The only other thing that is important right now is that I must warn you to be careful about what you say about McCarthy. B. and I can say what we damn well please, and we do. But Paul has a job. And he could lose it. The situation is just as bad as that. I'm beginning to realize what it must feel like to live in a police state. Velde* we think can be punctured—he backs down when presented with a bold front. Or when someone confronts him with a misstatement of fact, as good old Agnes Meyer did, hooray. But McCarthy's power is growing hourly, and unless Eisenhower or Dulles† or Taft cuts him down sharply, God help the country. You two are particularly vulnerable because you are connected

---

* Harold Himmel Velde, Republican representative from Illinois and chairman of the House Un-American Activities Committee (HUAC).

† John Foster Dulles, the secretary of state.

with State, so for heaven's sake, watch your step. Let somebody else take care of that bastard.

With a new, unvarnished, unwaxed salad bowl—heat up some good olive oil to the boiling point and give the inside of the bowl a good rubbing with hot oil on a clean cloth. Wipe off the excess. Do it a couple of times a day for two or three days. Supposed to seal the pores of the wood. I dunno. It works, anyway. Mine smells only faintly of good clean olive oil, and it's been fifteen years or more.

Something should be done about the pores of my head. Here I go over on another page, whoops. Omelet pan plus gadgets arrived three days ago. I got it all cleaned off as per directions and I must say it is marvelous. First omelet practically perfect. Second omelet jerked right out of the pan onto the stove. What a mess. Third omelet the butter was too hot. I hope this trend reverses itself because I seem to be getting worse by the minute. But a very fine piece of equipment and I have issued firm orders that no male in my household is to touch it, as they seem to specialize in ruining iron pans. Thank you very, very much.

The wooden spatulas are elegant—never saw any before—and the perfect instrument for mixing up the sticky bits when making meat sauces. The lemon peeler also very useful, and why is it that only the French have sense enough to put a really sharp cutting edge on such a gadget? But the little instrument with pincurls on the edge has got me licked. For extracting oil from peels? And if so, how? And if not, what?

And before you get moved into your new apartment, perhaps you'll have time to advise me on the proper sautéing of a fish, which you once mentioned. I am just no good at it. Mary is fair, as with smelts, but I once had an abysmally dumb black girl from Barbados who could fry a fish so all you could think of was angel food and *The Green Pastures*.* She was entirely without time sense and if at ten in the morning I asked her to cook some sweet potatoes for dinner she would cook them then and there and present them ice cold at seven o'clock. She didn't last long, but I still miss the fish she gave us.

And now, firmly, not another word.

Bless you both,
Avis

* A Pulitzer Prize–winning play by Marc Connelly reenacting Old Testament stories with African-American characters in the rural South; one of the opening scenes depicts a fish fry in heaven.

USIS, 5 Place de Rome, Marseille (B du Rh)

MARCH 18, 1953

Dear Avis:

It was awfully nice, as always, to get yours of 11 March. It would, of course, be nice to get an Avis letter every day along with the morning paper; but then, one would not get anything else done in life, but perhaps soup for lunch, would one.

We are at last moved in to our little apartment, and I am at this moment sitting in the dining room (our work room looking right over the port). There is a small two-masted schooner filled with fish right under my nose, little fishing boats parked across the way, sea gulls flying, a lovely view out to sea, and a bright sun overall. What luck! This we are subletting from a Swedish consul who is home on leave for about 6 months. It is all furnished in that clean, pale, rather impersonal Swedish manner, but we've brightened things up with a bit of nice local pottery, and are about to remove most of the fancy plates and pictures on the walls and substitute some PC photos. Though the kitchen is presumably "Fully furnished," it is more fully furnished like a eunuch, so I've had to fill in here and there. Lucky I brought a small collection of knives with me (never travel without them, as you never know what may happen, and most other people's knives are no good). But, my, I miss my old Jeanne La Folle, *femme de Ménage*.* I really haven't done any housework in years. I can get it all done here in half an hour, but I'm darned if I'm going to polish any floors . . . wouldn't know how to, I don't think. I'm not going to try and get a 12-hour *femme* until we know about that mystery paper.

Yes, we shall be careful about McCarthy because of Paul's job. But it is just unbelievable to me that this kind of thing can be going on in the USA. I enclose a newspaper clipping from, presumably, one of his supporters. Maybe you've seen this . . . I can't believe it. And the John Carter Vincent† case is terribly upsetting. We know them only slightly, but like them both very much indeed. Neither of us is in any position at all to evaluate his work, or to know enough about him to know much of anything. But certainly, during our 10 months in China, we never ran into anybody who did not distrust and dislike the Chiang regime (possible exception of that fool, Pat Hurley). From many reports running around, most of the money and equipment we sent to the Chiang group

* "Maid."

† An American diplomat and the former director of the State Department's Bureau of Far Eastern Affairs.

was played with on the black market by the Sung group,[*] Tai Li,[†] etc; and equipment was sold. For instance, all the rubber tires our chaps brought laboriously over the Burma road or the Hump[‡] . . . The Head man in the Kunming area passed a law that all Chinese carts had to wear rubber tires, and so they all had US rubber tires . . .. and where did they come from? Vincent lost China for America. Pooh. Since when was China ours to lose? Chiang and Stoopnagles[§] did the trick, helped on by the China lobby and Henry Luce. And in my opinion (that great political sage, J. Child), as long as we support Chiang, we further cement Mao to the Russians . . . and we can't win anything with that rotten Chiang. Vincent, as I see it, reported events according to his interpretation. He is considered incompetent now, but I'm sure his opinions will be proved to be the right ones later on. What a precedent for future foreign service reporting. The whole service will end up being like the Russian guys, who report only what they think the head guys want to hear, but are as careful as they can be to be as non-committal as possible. When, if ever, are Ike and Dulles going to act against McCarthy? That's what I want to know. In any case, I am in implacable opposition and when the time comes to act, I am ready, irregardless of the consequences. I just wish I were more eloquent, organized, intellectual, etc.

So glad you got a nice letter from the Lumberville Childs, and hope you do all get together. We, of course, are very jealous that they shall know you before we shall!

And glad Omelette pan finally arrived. And I do know all those troubles one goes through in the beginning. I find, too, if I haven't made an omelette in quite a while, it don't go so good . . . just like playing tennis. That little gadget with the pincurls is to make little lemon or orange zests, just hold it with bent side against lemon and pull down, scraping off pealcurls. For orange sauces, etc, and doesn't get in any of the white part next to the skin. Too thin and small for martinis?

Sautéing fish. Two methods, one for firm-fleshed fish, one for fish that break apart. For firm-fleshed ones, either fillets or small whole fish, like smelts. Just before ready to sauté, place a big bunch of flour in a big plate or paper. Dip fish in milk and shake off excess, sprinkle on salt and

---

* Reference is to T. V. Soong, who was a close advisor to Chiang and whose family was allied with him. His sister was Madame Chiang Kai-shek.

† Chiang Kai-shek's secret police and intelligence chief.

‡ The eastern end of the Himalayas, over which Allied pilots airlifted supplies to China.

§ Colonel Stoopnagle, part of a radio comedy team in the 1930s. Julia means that the corruption and ineptitude of Chiang Kai-shek's government brought about the Chinese communist victory.

pepper, roll in flour and shake off excess. Heat combination of butter and oil in skillet, enough so that you have about ⅛ inch in pan. When butter has just about stopped foaming, add fish. Do not crowd the pan, each must have a little space between. When brown on one side (about 3 or 4 minutes), turn over and brown on other side. Fish is done when fork will pierce through easily . . . and should be only just done, or will dry out. (Reason you flour them at last minute, is flour will become sticky gooey if let stand on wet fish, and then is lumpy when cooked.)

For flimsy fish, which would come apart with only a flour dredging, you have to *Paner à l'anglaise* (bread crumb & egg). Beat up an egg with 1 tsp. oil, 1 TB milk, salt and pepper. Roll fish in flour and shake off all excess. Dip in egg mix, so it is well-covered. Roll in fine breadcrumbs, and pat it all over to make bread-crumbs even. (This can be done well-ahead of time, and no harm done.) When time to sauté, use same process as above, using possibly a bit more oil and butter.

I'll try the boiling oil method when we get our salad bowl from Paris. Wonder if it will work on old bowls. Hope so.

And thanks so much for list on Waring mixers people, whom I shall write. Simca had a terribly funny time at the Arts Ménagers show when she went with her little basket of fish. There were hundreds of people about, even at 9:30 in the morning. The chef mixer wasn't too happy to see her, with his 100 spectators, but he went ahead anyway. In spite of the fact he had said the fish could be done without water, he added some eggs with the fish and then some panade. In one minute the machine was completely blocked and wouldn't move. After an exchange of looks with Simca, he just put the mixer under the counter and motioned her to come back later. Which she did, three times, but there were always crowds of people about, so they had no time to do any more. But, it is quite clear they haven't gone into the question at all, really. So Simca rushed home and tried out the quenelles again, two ways. She finds she can puree the fish dry, but it is more trouble, as it has to be scraped, etc, first. Only problem now is to time and try out the two methods, and pick the easiest. Bread method is easiest, I think, as in the other one you have to make a flour panade, which takes more time. Now, for instance, take *Bouquet de France* page 52, "Mousseline de Brochet Dijonnaise. Pounding fish which has first been put through food chopper, then force it though a strainer (drum sieve), force some more stuff through strainer. Work for 30 minutes in a bowl placed over ice." Wow . . . who's going to do that. This preliminary would, literally, take about 1 and ½ hours . . . useless recipe for the USA as it now stands. No one has either mortar or sieve, for one thing . . . which is too bad. Drum sieve (French "Tamis") looks just like the top of a drum, with fine screen instead of drum skin. It is used to make purees, or to pound the butter out of lobster shells which have been pounded in a mortar. Ingredients are placed on top of the screen, and you pound through the screen with a heavy wooden pestle. Takes hours, is hell to clean, but is extremely useful, especially for shellfish butters.

In one letter you mentioned importance of good ingredients . . . couldn't agree with you more. My, the things I've cheapened by using fat instead of butter, or this instead of that, all for a few cents of economy.

Enclosed a letter from D[orothy] C[anfield] Fisher to Simone. 2 copies, as you may want to send one to HM, she is certainly enthusiastic!

Glace de Viande from Maison de Glace sounds very interesting indeed. I think I'll send for some too. I've never heard of it being made for 84 hours! Sounds as though it might be a bit like Bovril; but is much fancier than any French thing. Important thing is, how does it taste . . . and I hope it's good.

You sound awfully busy! Keeping up with detective stories also?

Love to you both, wish you were here,
Julia

Paul Sheeline has been awfully careful & legal about our contract. Hope HM are not annoyed. I've told S & L I think once it arrives, there shall be no more questions, just signatures.

MARCH 20, 1953

Dear Julia:

Seems to me I have five thousand things to write about and I don't know where to start. I can keep fairly well abreast of you when I answer your letters the same day they arrive, but these last two weeks I've been sidetracked, what with a couple of visiting Elks,* several manuscripts to work on (DeVoto's productive streak continues unabated, with excellent effect on the bank account), an even larger amount of his mail than usual to deal with, and a sudden urge to buy clothes. Spring, no doubt. I have therefore bought myself three new dresses, a suit, a hat, and various oddments, and now all I need is someone's eye to knock out. Half the fun of having new clothes is provided by trips to Harvard Square in my new finery.

* Members of the benevolent and fraternal order, to which Bernard DeVoto belonged.

Most of the time I dress solely for comfort and convenience and speed, and my pals in all the stores are so accustomed to seeing me every day in a battered gabardine coat with a scarf tied under my chin that they nearly drop dead when I appear in a proper hat, a fur jacket and high heels.

When I was in Boston I went into the herb situation, and have to report that House of Herbs has swept the local scene and there are no other brands to be had, though I have had rumor that Spice Islands is carried in Framingham. If I ever get out to Framingham, I will send you some. There is no chervil to be had, anywhere. I am sending you basil, thyme though you didn't ask for it, tarragon, the chives I got in New York, and H. of H. chopped chives which is new to me and which I haven't tried as I have fresh chives flourishing in the kitchen. Fennel is apparently only to be had at drug stores, where I am informed it is bought to be used in poultices, of all things. I will procure some, to include in the package, though I fear it is probably the same quality as drug-store saffron. All these herb houses seem to be specializing in flavored salts—dill, chives, onion, garlic, instead of the straight herbs. I do use them occasionally, in spaghetti sauce, garlic bread and so on, but I don't like them for ordinary cooking because they are so damn salty. Seems to be what the housewife is buying, however. I will get them off as soon as I've got the fennel, and if you pay me I will spit in your eye. After all the things you have showered me, in the most reckless Santa Claus fashion, I can only live with myself if you let me send you what you need for research.

I had no more than started to scratch the surface of Reboul than *Larousse Gastronomique* arrived and I must say I have never seen a more fascinating book. This is an overwhelming gift, and I want you to stop right there. The whole damn family dips into it all the time, including Mark who is a fabulous Latin scholar and finds to his joy that he can read quite a bit of French. B. read aloud the section on *La Cuisine Americaine* and we were both enraptured by its air of moderation. We are comforted to know that one can cook a buffalo as one cooks a steer. The damn book is like popcorn, you can't stop reading it. My vocabulary is growing by leaps and bounds. (My block against uttering a single phrase is as strong as ever and bothers me to death. I don't see how I am ever going to get over it. If I had time I would plunge into a course. But there is no time, and I'm getting older and more set in my ignorance all the time and I hate it.) And I have to rearrange my bookshelves to accommodate *Larousse, Bouquet* and the *Gourmet Cookbook*. Let's not have your book that size. I am more grateful than I can say for this dictionary or whatever, and don't for

God's sake send me any more reading matter because I will lose my job at the *Globe* and DeVoto will have to hire a secretary if you do.

Also the photographs arrived and I will see that Dorothy gets her set at once. Simone and Louisette are darlings, so very French, so smart in their good little black dresses. Whose apartment did this dinner take place in, and what oh what did you feed Curnonsky? I want every last detail. I gather you had a sherbet in orange cups, but the rest is a blank. And desserts don't interest me much. I'm probably the only visitor to Paris in history who never once tasted a pastry—I never had room. I'm a cheese and coffee girl.

This will be patchy and inconsecutive—Mark's long vacation has started and I have to supervise him at various spring chores, cleaning up the yard, sweeping the cellar, raking the driveway, etc. His span of attention to business is brief, naturally. The vacuum cleaner is crashing around me and the telephone keeps ringing. Honest to God, the life we lead. Last night I was tucked up in bed cozily reading a whodunit when the telephone rang and B. answered. He was there fifteen minutes laughing uncontrollably the whole time. Call from Cleveland, unknown fan named of all things Utter, quite wholeheartedly drunk on beer. Under this beneficent influence, Utter couldn't resist the impulse to call B. up and tell him how much he loved him and everything he had written, and Utter had read it all, too. The kitchen knife and all the rest. Utter is an advertising man, who has to work a lot with soap powder, and he has an allergy to soap, and it gets up his nose, and DeVoto has got to write something about soap and allergies and advertising, and he didn't really mind being called up, did he? Politics and psychoanalysis and a lot of other things entered into this fascinating conversation. You can't say we have a dull life.

I made a beautiful omelet for my lunch, with chives and parsley, but I still have to use a spatula to make it roll. We are going to have poached salmon with beurre blanc for dinner. Honest to God, Julia, you have brought a revolution into this household. I wholly expect the completed book to cause a real revolution.

Small world division. May Sarton, who comes weekly to confer with B. over this wretched anthology they got sucked into doing, got off on the subject of France when I told her how much I liked that poem, and said—Did you meet Paul Child when you were there? We'd never mentioned Child before. (I don't usually see her as she is huddling in the library with B.) Floodgates opened, naturally. And you never told me that

Paul taught at Shady Hill[*] long ago. And I may as well admit right now, Paul, that we are terribly anti–Shady Hill, at least as it was in the days of K. Taylor.[†] She really went nuts the last few years she was there, and Shady Hill became a cult, a religion, with violent pros and antis. Our son Gordon was there for four years—'38 to '42 I think—no by gosh it was only three years because we bought this house in '41. And we succumbed to the Shady Hill spell for a while. We finally came to our senses and snatched our child out—about a dozen other families removed boys that same year. I think the school did him a lot of damage, though it is perfectly true that he was a difficult child and is difficult to this day. May says Paul was a Greek God. She doesn't usually talk that way, though. She likes you enormously and says you are just right for him. She is terribly disappointed that you have left Paris—she was about to send you some people and who knows perhaps you have had a happy escape. Thank heaven she is making a little money—her new novel *A Shower of Summer Days* is going well and she has just got a $3000 fellowship from Bryn Mawr. I am very fond of May and admire her enormously—she is a really good person and takes on dirty jobs that no one else will tackle, and she works like a dog and no matter how you feel about her basic problem[‡] she has made an honorable and respected life for herself.

I can just see me with a bidet in my bathroom, though I grant you they are wonderful and I was strongly tempted by a Sheraton one in London. Little boys in the household however pose a problem when it comes to personal privacy.

Yes I thought Buchwald[§] was hilarious. He is reprinted two or three times a week in the *NY Herald Trib.* but they didn't print this one and I think they were crazy not to. Dammit, the *Trib.* has never recovered from the election. B. has said for years that it is the best paper in the country except for four months around election when it is so offensively Republican that it is awful. But this time it hasn't recovered. The *Times,* God bless it, was completely fair (though Republican) right through the campaign, and always covered both sides of the question, and since the election has been openly in opposition on many occasions, and regularly chides the President for sins of omission (he hasn't committed much of anything). But the *Trib.* is still pretty well ducking the McCarthy issue (or at least

* A progressive day school in Cambridge.
† Katharine Taylor, the first director of Shady Hill School.
‡ Sarton was a lesbian.
§ Art Buchwald, a columnist for the European edition of the *Herald Tribune.*

being a good deal feebler on the question than it has any right to be) and has kept mum on a good many other things that the *Times* has no hesitation in tackling. The *Times* has come out strongly against the administration on Tidelands,* for instance, and had a field day with [Charles] Wilson, and runs editorials weekly on the public lands. The *Trib.* is still weaseling, drat it. The paper that causes me to stand up and cheer is the *St. Louis Post-Dispatch,* which switched to Stevenson in mid-stream. Anne Barrett subscribed to it and provides us with sheaves of clippings every week. But peace, it's wonderful to watch even the most rabidly Republican papers, such as the *Boston Herald,* gradually inching their position around to the opposition, as B. firmly predicted they would, back in the blackest days when Stevenson was getting one [column] inch to Eisenhower's ten and the campaign was being fought with twisted headlines. Do you know that wonderful verse of Hilaire Belloc's—

It's quite impossible to bribe or twist
Thank God, the British journalist.
But seeing what the chap will do unbribed
There's no occasion to.

Anyhow B. says the role of the newspapers is always opposition to the administration, and so help me they are getting there. Slowly. Even the weeklys. Weeklies. Though *Time* continues to be absolutely stinkingly offhand about McCarthy.

I hate to think what this McCarthy thing is doing to our shreds of prestige in Europe, especially in England. I know, for that matter, from the *Economist* and the *Spectator.* This is a very <u>intolerant</u> country. Always has been. People just forget about the Father Coughlins† and Gerald E. K. Smiths‡ and Huey Longs,§ and the long line of rabble-rousers way back to the Revolution. If we ever needed Conant¶ at Harvard we need him now. I am not a mad admirer of James B. but I'll bet my last dollar that if he were here now he would have spoken up unmistakably, and it is perfectly true that colleges all over the country have always looked to Harvard for

---

* Under Eisenhower, rights to drill for petroleum and natural gas were transferred from federal to state control.
† Father Charles Coughlin, a Catholic priest who broadcast anti-Semitic radio programs during the late 1930s.
‡ Gerald L. K. Smith, a clergyman and political activist who promoted white supremacy, fascism, and anti-Semitism.
§ The governor of Louisiana, U.S. senator, and populist demagogue who was assassinated in 1935.
¶ James B. Conant, the recently departed president of Harvard, who was appointed high commissioner for West Germany by Eisenhower.

that sort of moral guidance. Now we are handicapped by a Harvard Corporation almost exclusively cold-roast Boston, and heaven knows whom they will choose. Us eggheads want Charlie Wyzanski,* the freshest and strongest legal talent of the day—but I can't see that Corporation choosing a Jew. Twenty years from now, maybe. But wouldn't it ring round the world if they did? Meanwhile the Harvard faculty is disorganized and mute, and Princeton, of all places, comes up with a strong academic freedom statement signed by 300 faculty members, and last night George Shuster, president of Hunter College (Catholic) called for the colleges to investigate McCarthy, to discover "on what meat this Caesar has fed." Well, maybe something is stirring. At any rate, Eisenhower has come out against Velde investigating the clergy, and that man seems on his way out—he never had anything but effrontery, and now has been proved to be stupid. Jenner I think also can be faced down. And I don't want McCarthy shot though he is likely to be. I want Eisenhower to slap him down hard, hard, hard. But I don't think he will. At any rate E. stood firm on the Bohlen† appointment, but he still doesn't know his own strength. How much longer must we wait for all this leadership we were promised?

I am in a state about all of this. I comb the newspapers. I listen to the commentators. And I get into fights all over the place. If a Republican knows his place and hates McCarthy and wishes to God Eisenhower would get more aggressive about these bastards, well and good and I will admit him to the brotherhood. If he says nasty things about Truman (who is rapidly becoming the Man I Love although I have been sore enough at him in my time) or still thinks taxes are coming down and we can get out of Korea‡ and we ought to fire all the Democrats in Washington and don't worry, McCarthy-ism will blow over or alternately Where There's Smoke There's Fire—well, dear, I am no lady and I argue loudly and lose my temper and it's disgraceful. B. is the greatest comfort. He is a profound political thinker—he knows everything—and when he's in the mood to expound he speaks with the tongues of angels and I am awed and reverent and uplifted and thank God I got him. I forget that he only wants to eat baked beans and never moves a finger around the house. How I wish you could hear him. Not that he does it often and I never know when to expect it, but he can cast such light into dark places and he is always right. About politics, I hasten to add in a wifely way.

---

* Charles E. Wyzanski Jr., a Harvard graduate and Massachusetts federal judge.
† Charles E. "Chip" Bohlen, a Soviet expert who was appointed ambassador to the Soviet Union.
‡ The Korean War began in 1951 and would not end until July 1953.

I just haven't got the vaguest idea where we will be summer of 1954, but can promise right now to break my neck to see you somewhere. B. talks as though this will be our very last trip west as a family forever — on account of no money though he is making lots right now. Trouble with our summers is Mark. It is out of the question to keep him here in town as there would not be a soul for him to play with. I rather long for a summer in town — the house is cool and comfortable, and I could go to Harvard summer school classes. I definitely do not want to go back to Annisquam again since we can't get a house on the Ames Estate — family owners pouring back in droves — and I loathed the village which is exclusively sailing and social. Mark hates camp with a passion. I sent him to camp, albeit a very, very small farm camp and very relaxed, the summer I went abroad, and he never forgave me and we had unforeseen difficulties. I know he is too much attached to me and I want more than anything else to have him kick mama in the teeth and go merrily off on his own, but you cannot wean a child forcibly. You've got to feel your way and take advantage of his own impulses when they lead outward. So we never can make summer plans. I probably won't be able to tell you where we will be until June '54. I'll have to find out about Houghton Mifflin though of course there is always someone there — Lovell [Thompson] lives in Ipswich week-ends and summers and entertains all his authors in Ipswich, Paul [Brooks] lives in Lincoln, neither of them go away in the summer but take their trips to England and California during the year, prospecting. Dorothy [de Santillana] is going abroad this summer but I don't know about '54 as I think Stanley's* job will be over by then and her main excuse is to see him. So we must just keep it in mind and make plans later. We will have two beds — two rooms too — here until the boy [Gordon] gets out of the Army but that won't be until November '54. So Cambridge is definite, if we are anywhere in the East. If we are West, we will meet there.

I think you are carrying this Upper Middle Brow etc. thing too far, my good girl. I told you I distrust categories and I am pleased that Paul does too. It is too pat and all-inclusive, and if I know anything at all it is that there are no final answers to anything, and the older I get the less I know and the less final the answers are. It's just possible, you know, that those upper middle class Republicans you distrust so may have some answers that are better than we eggheads have. Could be. You can only

* Stanley Hillyer, Dorothy's son by her first husband.

approximate—and if you can do that occasionally you're in luck. I condemn all kinds of Republican thinking, but really I know better, because I'm continually being surprised by the good things about these people. And I've been living in the pockets of the intellectuals for thirty years and some of them are the awfullest fools. If I could write I would write a book called the "Care and Feeding of Intellectuals." There aren't any more final answers about how people's minds work than there are about how their marriages work. I used to think, in my innocence, that there were some awfully well-adjusted people who got on together like a house afire most of the time. Well, there aren't. Or if there are I never met them. All you have to do is know them well enough to find out how things really are. In the family I bet on for years and years—four wonderful children, fine position in the community, great sympathy between husband and wife—it turns out (and this I might mistake but B. never makes mistakes like that) the husband has got definite strong homosexual tendencies though I doubt if he has faced them, and the wife is a mass of well concealed panic. They function damn well, but at what a cost. Thoreau was quite right, you know. It turns me into a mush about the human race, I love them all so much. I am quite sure I could never love McCarthy, but I suppose I could love Eisenhower, the poor boob.

I'm afraid I sound even more feather headed than usual. Anyway I'm awfully glad I'm going to Washington, and to see Allie and the Childs, and I may leave a day early and stay a little longer. I'm getting stale. I always do this time of year. I keep my nose to the grindstone and put in long hours and rustle up good meals and do all the chores and run errands and get along with people—and have a fine time doing it and enjoy life. Then I realize, bang, that I'm tired and I don't want to wait on my family for a while and I wish I could go away somewhere and have people wait on me hand and foot, and dress up and go to restaurants and the theater and act like a woman of the world. I feel as if I'd been swallowed up whole by all these powerful DeVotos and I'd like to be me for a while with somebody who never heard the name. I can't tell you what it did to my ego to get along so famously with the British who never heard of DeVoto in their lives.

God how I envy you going to Cambridge so often. Ask your friends if they know Harry Hinsley and his wife Hilary—he's a fellow in history at [St.] John's and took me around, and up the river in a punt, dear heaven how beautiful it was and what fun. If you go to Cambridge for Christmas you have undoubtedly gone to a service at King's Chapel with all the candles and the music tearing you apart and if I could do that just once I think I would die happy.

I find the English reticence extremely appealing, and I don't think it's because I like formality more than you do. Formality, in this country anyway, bores me speechless. But a formality backed up by centuries of behavior charms me. I adore all their rituals. And the reticence is only skin-deep. They are like Vermonters, give them a chance and they will talk the arm off you. I have a dozen English friends who started off chirping, as you say, and *retenu* and doing what was always done—and boy, do they take their hair down. Probably because they thought they'd never see me again, though a couple have come over and stayed with us and B. liked them too and he has always thought he didn't take to the British. You can have scenes with English people, but they are nice quiet scenes. Under all that surface calm they are simply seething with feeling and emotion and warmth, and it pops out all the time. And I adore the way they regard eccentricity. They are infinitely more tolerant than we are, infinitely more than the French are.

I know it is all a question of whether you can communicate or not, and how much you have soaked up about a country. I have read English novels compulsively all my life. All of Dickens when I was thirteen and sick with scarlet fever and nephritis for a year. Most of Thackeray a year later. Jane Austen always. Trollope for the last fifteen years, over and over again. And all the others including Angela Thirkell.* God bless that ineffable snob who knows so damn much about people. I was all ready to be bowled over and I don't think I hit the pavement the whole time I was there. I love the processions and trying to get a look at the Queen and reading the fourth leader† of the *Times* and having a charlady call you "ducks" and the taxi driver who tells you very delicately that in England we have a custom called "tipping" and the lady bus conductor whose feet hurt so she snaps at you. Sorry but this is the passion of my life. And I insist that under all that surface calm they are boiling and seething. That's the way I like it.

Well I wish I had time to read Balzac. Maybe next summer. They must have some in the Worlds' Classics which are a good size. Send me three or four of the best titles. Do <u>not</u> send the books. I am not going to read French all summer. I've read very little French literature—had a

---

* A prolific twentieth-century British novelist.
† Editorial.

swing at Colette a few years ago but it evaded me. But one summer when we had a house on the campus at Bennington [Vermont] and I had a really good maid and Gordon was at camp and Mark wasn't born I had time for Proust, and it was probably the great reading experience of my life. It is hell's own work getting through *Swann's Way* and *Within a Budding Grove* isn't much easier, but then you get caught up and it's like being on an express train and you can't stop. Please anyway read the long episode called "The Dutchess's Red Shoes" which I believe is in *The Guermantes.*[*] I suppose what really bowled me over, me being what I am, is that every major person in the book winds up, in the end, completely different from the way Proust presented him in the early volumes. Swann, Charlus, Mme. Verdurin. But you have to have two or three practically free months to read it all. And I'll probably never have that again. And oh Lord it is hard to get past that tisane and those madeleines.[†]

I think I will call it quits for today. Maybe finish tomorrow, if Chet Olsen doesn't show up. He's Forest Supervisor out in the Rocky Mountains, in Washington now, coming up to plot with B. and also go into a protracted huddle with us and the Rochlins[‡] over our trip next summer. A darling. A Mormon, completely unintellectual, very shrewd and sweet and capable and innocent and natively intelligent and uncomplicated and shy and the kind of honest good Democrat that is the backbone of the party. His food tastes are exactly like B's, worse luck.

I give up. No chance to finish this properly so I will finish it improperly and get it on its way. This has been punctuated by chores—cleaning of the cellar with Mark's help from which I emerged begrimed from head to foot and can only say that I am glad we do not live in a castle with a dungeon. Ditto of garage. Beginning made on yard. Moths beginning to fly which means mad rush to get woolens aired and put away, the job of all jobs that makes me wish I'd never got married or had sons or, indeed, been born. Escort B. to have a wisdom tooth pulled. Spend an afternoon making marmalade while California navels are at their most perfect. And watch every morning with a kind of horrid fascination that Bastard McCarthy interrogating writers on TV. After three days of this, I no longer can see the slightest reason why these chumps refuse to answer questions under the Fifth Amendment. On the other hand the spectacle this morning of Langston Hughes groveling and smirking and agreeing with Mc-

---

* *The Guermantes Way*, the third volume of Proust's famous *In Search of Lost Time.*

† Proust recalls the taste of madeleines (cookies) soaked in tea (a tisane) in *In Search of Lost Time.*

‡ Gregory Rochlin, a prominent Cambridge psychiatrist, and his wife, Helen, friends of the DeVotos.

Carthy and ending up by coming to Jesus was enough to make you puke.* All very, very confusing and upsetting. I haven't got time to watch all this but I do. Returning your Philadelphia clipping, which I can't understand why it wasn't the *Chicago Tribune*. These are the goddamnest days. Watching McCarthy, no longer willing to wait for him to dig his own grave, want somebody to shoot him quick. Glad to have your word on Vincent and agree with every word, though my knowledge about China abysmal. Chiang, Stoopnagles and Henry Luce—may I use that? So far on these hearings McC. hasn't dragged in Vincent, but he drags in Lattimore† on every possible occasion. I feel awful about the Lattimores—we've known them, not well, for twenty-five years. Positive as it's possible to be that Owen never followed the party line, but he was undeniably foolish on many occasions. Oh God I wish this madness would subside, as I know it will, but it is exhausting watching all this go on. I do not enjoy seeing the Senate floor turned into a bear-pit.

Must condense if I am ever to get this off. Got your herbs off this morning. Sample of fennel will kill you. Billings and Stover‡ says it is used in infants digestive upsets. Upsetting just to look at this stuff. If you will send me address of your new apartment, I will send air-mail samples of onions before weather gets warmer, if you still want them.

Have you seen *Cooking with a French Accent*§ by Gerald Maurois, son of André?¶ Jack Fischer just sent it to me. Fascinating very, very uneven book—much sense and much vagueness. Will send you my copy if you want it, when I finish it.

Sautéed some big smelts by your method, great success. Now I know what to do with the trout we catch next summer. I, also, never travel without knives, and a pepper-mill. No I would not think the boiling oil method would work on old salad bowls, unless you scrubbed them like hell with soap and steel-wool and aired them in the sun for days.

Yes I think I would duck dealing with *Gourmet,* if I were you, if Narcisse Chamberlain is right and she should know. If you want to do a piece about quenelles, or something special like that, how about *Vogue*? Don't know if you see it, but they run a lot of very specialized cookery arti-

* In his testimony before HUAC, the poet disavowed any Communist ties and embraced Christianity.

† Owen Lattimore, an American author, a scholar of Central Asia, and an adviser to Chiang Kai-shek, who was accused of being a Soviet spy by McCarthy and indicted for perjury in 1952; all charges were dismissed in 1954.

‡ An old-style apothecary in Cambridge.

§ The correct title is *Cooking with a French Touch.*

¶ André Maurois, the pseudonym of Émile Herzog, a twentieth-century French novelist.

cles—*Harpers' Bazaar* also but not as frequently as *Vogue*. Don't have any idea how they pay, but feel fairly certain that they would jump at an article, or series of articles, from you. Speaking of quenelles, bought a can of quenelles de brochette* at Jordans† the other day, while herb-hunting—packed at Bourg-en-Bresse. $1.20 for a ten ounce can. Haven't opened it yet but will report when I do. Jordans food shop very fancy, and bought a number of Mexican specialties for B. who adores anything red-hot. Only Mexican thing I have found that is completely satisfactory in a can is tortillas, which are wonderful.

Dying to try my hand at bouillabaisse but know damn well that nobody would eat it except Mary and me. My family in a very recessive mood about food at present, unhappily. After they've eaten in restaurants all summer it will be a different matter.

Sent copy of Dorothy C. Fisher's letter to Dorothy [de Santillana], who promptly wrote to her making all the proper noises of appreciation. We are going to get a nice plumy blurb from Mrs. Fisher to help along the publicity—too bad she still isn't on the board of the Book of the Month Club—not that I can see them choosing a cookbook, but as member of the board her pronouncements on books carried great weight. She's a very nice old gal. Dorothy de S. was greatly pleased to see the letter. Maybe we can get her on the jacket, or maybe she will review the book.

Convulsed at your description of Simca and the Mixer people—mean of me but I always enjoy seeing demonstrators at the mercy of their machines. I hope S. could keep a straight face. After you have done all the donkey-work on the mixers, pass the word down and I will get DeVoto to give me the proper kind for Christmas. But it's a big investment and I don't want to do any of the research myself.

Going to have three days all alone next week, as B. is going to NY and I am going to put Mark on the train to join him, first time he's ever been to NY alone. Big jaunt for the men, with no women around to impede their simple enjoyment. While they are gone I can wind up the chores and indulge in some simple gluttony. I yearn to try that elegant Italian dish which combines thin veal, thin ham and mozzarella cheese in a sort of glorified sandwich, fried in butter. Ate it at Sardi's‡ once and never forgot it. B. doesn't hold with cooked cheese, unfortunately. Also some frogs' legs. And a few other trifles. I seem to be stuck at 122 lbs in spite

---

* Quenelles de brochet, quenelles made with pike, a classic French dish.

† Jordan Marsh, a Boston department store.

‡ A famous New York City restaurant.

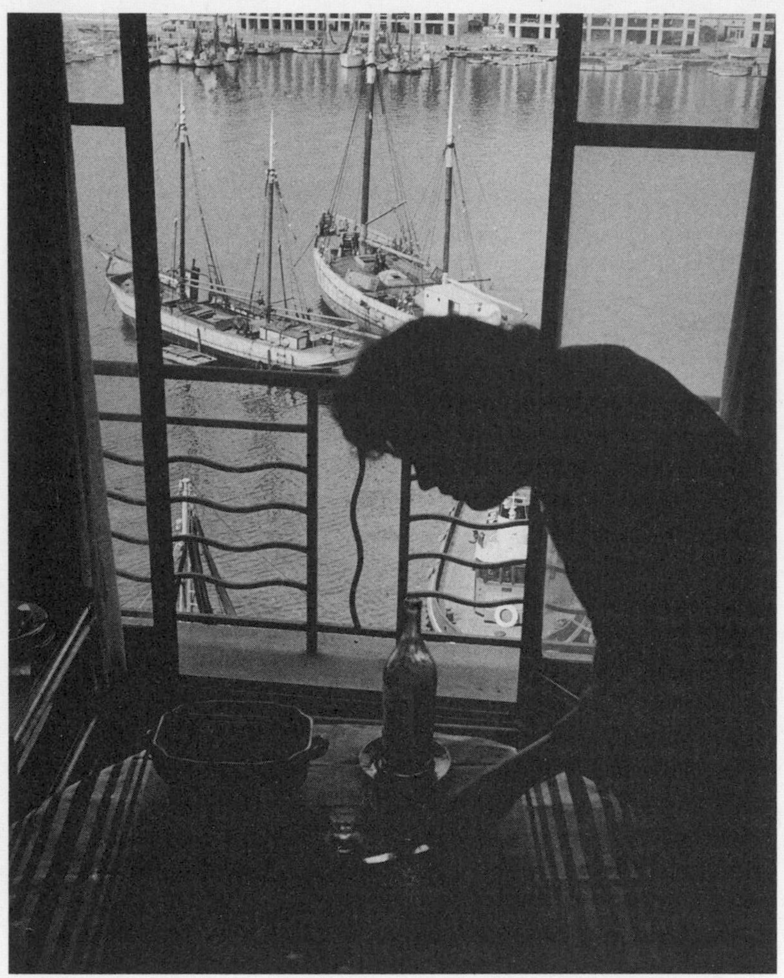

Julia preparing a meal in the Childs' apartment overlooking the Old Port in Marseille in late 1953.

of all this over-eating, but plan to whittle off a few pounds before I go off to Washington. I really can't wait to get away, I am so stale about my household and the spring jobs. Let alone lame all over from cleaning that ruddy cellar. Sure, I could have hired a man to do it, for fifteen dollars, and would have had to stand over him to see that he did it right.

Goodbye for now. I have to finish a detective story and write my column.

Avis

USIS, 5 Place de Rome, Marseille (B du Rh), La Belle F.

APRIL 8, 1953

Dear Avis:

This is to be only half a page, and on a torn piece of paper at that . . . but must make appreciative noises over yours of the 20th . . . which we have both read twice with great interest, as always. We are, suddenly, rushing up to Paris tonight, on 6 hours notice, as Paul's orders transferring him to Marseille have finally come through; so we must pack up our dear old apartment on the Roo de l'oooniversity; and it will be a true nightmare undoubtedly.

Paul talked to the head of the Information Service today on the phone, and the poor guy sounded pretty green and worn out. McCarthy's men have been on their tails all weekend. The paper yesterday said the investigators were "delighted with what they found," and then went on to say there was loads of duplication and waste of taxpayers money. Well, there is . . . and it is the [State] Department's own fault for letting it get that way . . . but it has been power-plays between ECA,* USIS, and an Empire† group in the Dept. and it is high time the proposed amalgamations of ECA and USIS took place. Why does the Dept. lay itself open, one wonders? The investigators have also said that all the books and information on the Far East followed the C.P.‡ line; and that it was no help in the fight against communism to have books on geology and gardening, and not one copy of the *American Legion Weekly* was to be found in any library. Well.

Have just come back from a nice Easter trip, but I like the French Riviera less and less. It is a cheap, tourist-ridden area . . . and if this were all I knew of France I would always go to England, where at least some things have dignity (and tradition!). The horrid transformation of the Côte d'Or has taken place mostly after the war.

I suppose you will be off any day now on your trip . . . You sound as though you need it bad. Have a wonderful time . . . and report, please, all you see and hear in Washington. Wish we could join you at Lumberville!

Heavens, I keep coming back to McCarthy. What would you do about it, if you were Ike and Dulles? Could you have a fireside chat with the people, telling them, in effect, that they had overwhelmingly elected

---

* Economic Cooperation Administration, which oversaw the Marshall Plan.

† The postwar State Department included many divisions that the McCarthy hearings tried to expose as a waste of taxpayers' money and as pro-Communist groups.

‡ The Communist Party.

you, and you were going to run things as you saw fit, and you were answerable for praise and blame alike. Would you openly repudiate McCarthy (after once having showed his hand)? What to do? What does DeVoto say about what is the solution? I would like to hear him, on that as well as other things. It is all too dreadful.

Much love, and an enormous letter coming in a week or so. Glad you received the *Larousse* and find it fascinating. This was Simca's and Louisette's idea, mainly . . . and is, in effect, a profound thank-you . . . as where would we be without you, indeed. I won't send anything more, except once in a while . . . I do love to send things, but they are usually of a most minuscule nature. Will look forward to the herbs, especially the fennel! Don't think, really I need an onion. Have no shallots here, this season, and have been using ½ amount of onion, hashed very fine or in garlic press, and it works fine . . . or I just leave it out.

Love again, from us both. (You really touched the old man to the quick with your remarks on Shady Hill!)

J.

USIS, 5 Place de Rome, Marseille, B du Rh

APRIL 24, 1953

Dear Avis:

At last I've gotten myself settled down physically and psychologically, and have established an ineluctable work schedule, and am feeling much better. For a while, what with all this moving and settling I was becoming frantic, schizophrenic, and even, perhaps, necrophilic, combined with anguish, frustration and ill-temper. But, finally getting into the book-work again, I find I work awfully slowly. There is so much that has been written, by people so much more professional than I, that I wonder what in the hell I am presuming to do, anyway. To think of these people who have been in the business since they were 10 years old. I keep having to bear in mind that what we are trying to do is to interpret the professional manner for the home cook, and that we are not trying to be an encyclopedia. My, there is so much to know; and I keep bumping into things, such as what are, really, the good cuts to use as stewing beef . . . and I don't personally know, and it is no good just taking hearsay out of a book. Well, I can see these self doubts and humble questionings will continue through to the end. I am physically incapable of bluffing, which is a terrible handicap, either in political or gastronomical conversation. Suddenly, wanting to make a strong point, I realize, "Do I really and absolutely know that fact?"

Just got in a letter from Charlie [Paul's brother] telling ALL ABOUT IT, and it sounds as though it must have been fun . . . certainly from their point of view. He says,

> Curious thing is that PJ [Paul and Julia] probably know a lot more about Avis, having never seen her, than we do who have seen her. After all, she was here only Friday evening and Saturday morning, which is not very much, and you gotta have some rest periods when nobody says anything, and you fellows have been writing these pen-pal letters now for months, asking and answering, comparing, analyzing and all that. We got along very easily and well, found we had an immense number of folks we knew in common, lived more or less in the same world of politics, music, art, food, and other basic matters. You just can't explore everything in that short a time. She is an excellent anecdotalist, gently witty in conversation, likes to talk, but will listen too. She is far-sighted, and puts on and off a pair of rather chic tortoise-shells. She has a deep, real, immense respect and interest in all that DeVoto does and is and thinks and writes, though she can be reasonably objective, too. All in all, a Great and Good Time was had by all and the bells of friendship are ringing o'er the welkin. She liked my work very much, to my surprise (you never know). An intelligent woman, hard outside, soft inside, real sincerity, real warmth, real integrity, real idealism, a person who yearns and believes and loves and is hurt. Us chickens hoping we also made favorable impression so to hold up PJ quality in her estimation, therefore v. much want to know DID WE DO GOOD JOB? DID she have a nice time. DID she get a good pic of sophisticaters, foodly-lovers and friendly faces? I may say, in this connection, that Our Wife was at her very best, which is very good indeed, making with the amusing little stories, giving with the utter charm, the inner glow, the friendly hand. The girls got along just dandy, and would state that if ever I saw a cornerstone laid, this was it. As one Wife-of-Artist to another, they indulged in those maddening little nods and becks and secret smiles of the Great Agreement.*

Certainly wish we'd been along.

---

* His description echoes a line from John Milton's poem *L'Allegro,* "Nods, and becks, and wreathed smiles," referring to the happy union of a god and goddess.

That nice little package of herbs arrived the other day, and thank you. The tarragon, thyme and basil are excellent, and just as good if not better than any fancy dried herbs you can get here. The chopped chives seem excellent too, and I'll compare their flavor with fresh chives, as soon as I run into some (season is still a bit early). The dried fennel doesn't have much flavor, but is perfectly useful. I think, actually, our US dried herbs are generally far superior to the French ones. One thing that helps them, too, is that they are properly corked . . . The French ones are cheaply corked and lose their flavor.

And the book arrived, too. It's lots of fun, I think, and rather wonderfully American . . . and not quite as bad as you had pictured. Someone sent me *Venus in the Kitchen* by Norman Douglas . . . recipes for aphrodisiacs of all sorts, very amusing.

Have finally your last long letter in front of me, which has been sitting unanswered for so long. I'm glad to hear May Sarton is doing well, and has a fellowship from Bryn Mawr. It is certainly not easy, I should think, to live happily in our culture with her special problem. She seems to have solved it and to be able to live with it without bitterness (at least on the surface); which is a pretty rare thing.

No, my good girl, you can never wheedle me to England for long. I was just thinking about England while I was getting lunch today . . . I just get bored there after a while. I do every once in a while get a thrill over the tradition that has gone on for centuries, and I adore our friends. But for me, it lacks a natural gaiety and spontaneity. Proper, good, admirable, courageous, dutiful, count-on-able, and all the sterling qualities. But it is not fun, for me anyway. Now, in France I am just happy, it's fun here . . . Walk out on the street, and it's fun. Look into a French face, and it's alive and moving. I just get depressed in England, winter, summer, spring or fall; and it is a purely physical reaction.

Balzac. *Le Lys dans la Vallée* is a most romantic and beautiful book, and in some ways not as Balzacien as most. But a wonderful starter. *Les Illusions Perdues* and *Splendeurs et Misères des Courtisanes* are a very long continued story, extremely Balzacien, and very interesting, and quite fantastic, and packed with contemporary history and customs. *Le Père Goriot* I haven't read yet, but is one of his most well-known. However, in reading Balzac, it's best to know something about his life, as you understand his mind and his work much more deeply. Stephan Zweig wrote a rather long biography which is quite good. After one B. novel, you'll probably want to know something about him. What a man he was, a real giant and genius, a great and tragic figure, a fool and a romantic and a penetrating observer, and an absorbing passionate character. Wish I had time to read more of him, but once in, it's too hard to break out. I wonder if you'd like him . . . People are usually fairly passionate one way or the other.

I don't seem able to get my usual letter-writing flow today, my mind sticking like glue . . . frustration over cook book, very likely. So I'll send this off, and continue in a day or two.

With much love, as always,

J.

USIS, 5 Place de Rome, Marseille (B du Rh)

JUNE 3, 1953

Dear Avis:

No word from you in WEEKS, and hope you are not sick or something, but imagine you are up to your nostrils in preparations to leave on the big vacation. Do you have to review X number of books before leaving, for your column? How awful, if so.

We ourselves are off for 2 weeks in Portugal June 12th; and this weekend make a business trip to Narbonne (near Perpignan) to meet the mayor and the head of the social/cultural club, who is a wine merchant. Should be fun. Then to Albi (that wonderful city with the great brick cathedral and the lovely museum with the Toulouse-Lautrecs), to lunch with a newspaper. Then on to Rodez to visit their library and give them some books. Should be fun and interesting, but a bit rushed.

Whom shall I write to, for fun, when you are away? No one is as interested in cooking & politics. I shall miss our correspondence very much indeed. All work and no politics makes a dull cook.

Speaking of politics, Paul and I had a bitter disappointment 2 weeks ago. A friend of a friend, with whom (the F of F) we have been having a joking post-card correspondence for several years, finally arrived in France on a long-awaited holiday. He is a writer. I think P and I fell into our own semantic trap, and romanticized him. Writers are fine, sensitive beings, aware of the world, the inner tensions, alert, inquiring, and thoroughly superior beings. Well. He doesn't even read the newspaper. He is not at all worried about McCarthy, saying the average man is more interested in his own work and his family than in what McCarthy does. He is not interested in eating. He is absolutely unaware of architecture. We took him on a little tour of Aix en Provence, and he didn't even look at a doorway.* We made the mistake of taking him to a tiny medieval hill-town, Gordes, where we spent the night. There is no running water in the town, and the WC was among the most pungent I have ever encountered. So every time he had to P he went in the car out into the bushes. So, it turns out he has been living in an entirely upper middle-brow world. Well. I have never been so slapped in the face by a wet fish . . . and I am sure, though we did our very best to be our most sympathetic, he must feel that we are a disappointment to him. As he is a widower, we were even planning to do a bit of matchmaking with a lovely friend of ours whose perennially adolescent husband has just run off and left her with 4 young children, he being caught in the trails of a siren 15 yrs. older than himself.

The work is going along at a fairly good clip, at last. I keep wondering about such things as GARBURE, page 513, *Larousse,* which is a thick soup made with Confit D'Oie (Preserved goose . . . put up in its own fat). Who the hell has any of that? I'd never heard of it before coming over here. Maybe there's some goose in the middle west, as I notice the Culinary Arts Institute's *Encyclopedic Cookbook* talks about goose quite a bit. I would think that kind of stuff would be more *Gourmet*'s dish of tea than ours. Of course, one uses preserved goose for the Cassoulet. Though easy to make, etc, I do wonder if it is not a bit too exotic. This Garbure can be made with ham, etc, but then is just the same as cabbage soup, or rather Mediterranean Vegetable Soup. Well, it may be worth a footnote. Some people might say, "How can a French cookbook not include a Garbure." Or would they. As we say, it ain't an encyclopedia, but a how to cook book.

Pressure cooker. I've been doing it quite a bit. It does work out perfectly well for meat stock, if veg. are put in only 10 min. before the end. But for most of the veg. soup, it just doesn't give the same flavor as slow cooking in an open pot. For Ordinary Meat stock it is wonderful, I must say. But, the horrible taste of overcooked leeks and potatoes! Ugh.

Am so glad we aren't doing book in sections. So many things we are finding to add to the Sauce chapter, little hints and corrections and additions.

During my fish research, I ran into the name of a book by Milo Miloradovich, *The Art of Fish Cookery.* She has done a tremendous amount of research, has all sorts of useful lists, very complete as to lean, fat, when,

---

* Aix has exceptionally beautiful doorways framing private courtyards.

how, where. Then, in Louis De Gouy's *The Gold Cookbook,* he has done even more research with a list more than 26 pages long. So there are at least 2 people who have done the work for us. So, we don't want to repeat on their ideas. I guess ours should just be:

Fish to poach in white wine such & such & such
Fish to poach whole
Fish to sauté meunière
Etc.

Then a list in the appendix of French fish and American Equiv. What do you think? These things, anyway, wouldn't be done until the end, so there is time to mull it over.

Was pleased to see in the last *Gourmet,* in their article on quenelles and fish mousse, that they hadn't used the Waring mixer. I think their recipes are getting much better, in the last months.

At 11:30 this morning, I had just come home from the markets, dishes unwashed, beds unmade, and Paul called up to say he was bringing six GI's home for lunch, members of the American Fencing team. So I made them a Soupe au Pistou, *Larousse* P. 871, with a few embellishments. I just thought it would be interesting if the most typical guys would like it. And they did, ate it all up and said it was a "very wonderful soup," and want me to give them the recipe. Interesting to see if and how these foreign methods appeal to the average American.

Did you ever send for the Maison de Glace's Meat Glaze? I wrote them and they said they thought it wouldn't travel. Too bad. Don't like the dried chives. Taste like hay with onion flavor. Tarragon awfully good indeed, and I'm using it a lot as it is hard to find around here.

What kind of sausages do you get at home? Like wonderful garlic sausages, that you eat hot, after boiling, and sausages you can put in a soup, and are all smokey, etc. Do you ever get salted ham, like ham hocks, etc. Like other kinds of salt pork than the salt pork that looks like bacon. Here it is called *Petit Salé*.

Do dried white beans, the baked bean kind, come in cans with no sauce, just sort of put up in a light brine? Do you ever get the white beans before they are dried? I don't think I've ever seen them.

Are our string beans all the big long fat kind? Do we ever have the little thin French bean type? In other words, do you always have to "French" them?

Do you ever get real European Sole? I ran into a fellow in Paris who is a NY wholesale fish dealer, Monsieur Prince, and he imports lots of European sole. I'll bet it is no more expensive in the States than here . . . Here it is a buck a pound, sometimes more.

Simca's garlic sauce for roast lamb (a deep secret!)

For about 1½ cups sauce:

Put a quartered peeled onion and roughly sliced carrot in roasting pan with lamb, to give juices a very dark color.

Peel one head of garlic (about 16 cloves). Place in pan of cold water, bring to boil, immediately strain and rinse. Repeat process. Put garlic, 3 TB rice and ¾ cup milk over low heat and simmer for about 30 or 40 minutes. Puree in Waring Mixer, or push through a fine sieve (the rickety kind). Add salt and pepper.

When roast is done, remove a bit of fat from the pan, leaving about 3 or 4 TB. Stir in the garlic mixture, and dig off all the coagulated brown juices. Squash up the onion and carrot. Thin out, if necessary, with a bit of strong meat stock. Do not bring to boil, or fat will separate. (Can add bit of tomato paste, if want to.)

Gee whizz, this is a good sauce. And you'd never know it was garlic, and it doesn't cling to the breath, either.

Now, I must to work. Garlic soup is wonderful too, just wonderful. You'd never know what it was. Thickened with egg yolks and olive oil. Yum.

Thank heaven Ike responded to Taft,* at last, and very well, I think. I was just beginning to think Taft was getting some sense, and wasn't so bad after all. But I guess it is high time these thoughts be expressed and discussed. Though I can see his point, and feel his feelings, and thought his speech very reasonable, from his point of view. Such a horrible report of a priest's speech in New York (*NYT*), supporting McCarthy. The way they say it is only the left wingers who are against him. I really read those things and scream from the stomach.

Much love from us both, and have a wonderful summer,

Julia

* In early 1953, Senator Taft had objected strongly to federal budget cuts proposed by the Eisenhower administration.

JUNE 11, 1953

Dear Julia:

I should be upstairs (whoops) packing, a chore I have not even begun to think about, and we're leaving Monday and this is Thursday. I warned you some time ago that it would be hell on wheels from then on, and it has been, which is why you haven't heard much. We have all been to the doctor. To the dentist. I have had my annual permanent. We have had a string of house guests—only one at a time because we only got one bed. Elmer Davis has been, delivered a PBK oration* that brought the audience to its feet cheering for five minutes (unheard of at Harvard). All about McCarthy and national cowardice and a few good stout pokes at "the only living ex-President of Columbia."† I gave a dinner party for him (hot night) at which were consumed fifteen lobsters (cold and out of the shell) with melted butter and lemon, peas done the French way with lettuce and shallots, thin sliced cucumbers and tomatoes vinaigrette with fresh tarragon, hot garlic bread, and ice cream with <u>both</u> hot fudge and hot butterscotch. (Everybody took chocolate except me. You never know.) I caught all the TV Coronation doings,‡ and I've sneaked into Boston to see the British coronation film which is superb. Mark has graduated cum laude from Fessenden and scored a great hit with the class prophecy. Mr. Sinclair Weeks§ was as patronizing as only a Republican can be, and quoted at length from Kipling's "If." We have been to Radcliffe Commencement where Sen. Symington¶ gave a nice speech but notably lacking in fire, and unfavorably compared with Elmer's. We then went to the Jordan's** (Pres. Radcliffe) to a lunch for 100. Lobster newburg and thank God plenty to drink. We have been on the fringes of a tornado which killed over forty people in Worcester—we were alerted by radio for two hours until the damn thing decided to depart the vicinity, ready to dash for the cellar which I privately think would be a death trap. Except for the tornado day, the weather has been superb—cool and sunny. Hope it continues as we cross the great plains. My son Gordon is I assume in Japan—he hasn't sent us an address yet. Elmer says I am unique among American mothers

---

* Speech to the Phi Beta Kappa society.

† The reference is to Eisenhower, who was the president of Columbia University from 1948 until he became U.S. president in 1953.

‡ The coronation of Britain's Queen Elizabeth II took place on June 2, 1953.

§ The secretary of commerce, who spoke at the Fessenden graduation.

¶ Stuart Symington, the U.S. senator from Missouri, who opposed McCarthy.

** Wilbur K. Jordan and his wife, Frances.

in that I am not at all happy about the proposed Korean truce. Looks like just another campaign promise to me, and a highly unsatisfactory one. My son, saved in Korea, may very well end up in Indo-China. National honor ought to mean more. Fat chance, these days.

Through all this and more, we've both been trying to get the correspondence dealt with. Leaving Monday before the mail arrives, and any that catches up with us will have to be content with a postcard.

Our new Mercury is perfectly beautiful and has something called Overdrive which I have not mastered. I am an absolutely superb driver—never had an accident however slight, never had a violation of any kind, never had no trouble. Completely relaxed and easy at the wheel in any kind of traffic. But what happens under the hood I do not know or care. We drove down to Annisquam last Saturday to try to get the last hundred miles on it before the 500 mile checkup and B. tried to teach me about Overdrive and we wound up not speaking. I'll get it, eventually, but I'll have to get it my own way which is by feel and intuition and blind faith.

Eons ago you had typhoid reactions and I never got around to saying how sympathetic I would have been had I seen you in the act. I know—it is a stinker. But now they tell us—at least I am so informed by Nancy Movius who is flying with two children first to the Dordogne where she meets her husband who is an archeologist, then to Rome, then to Karachi, then to Australia where her family is—that there is a new typhoid shot, very small, very minute, no reaction at all. And high time I say. I'd rather have typhoid than go through that again.

No I don't know Mrs. Fairbanks—don't believe Dorothy [de Santillana] ever told me about her. Dorothy is leaving next week—Giorgio left several weeks ago. Wonder when you will meet up with them? And I want a report right away when you do. Rather than give you assorted U.S. Forest Service addresses for unsettled dates, if you do write this summer send it here and my caretakers will forward. I will wire them once a week where to forward mail by air, and leaving them ten bucks worth of air-mail stamps. You probably won't get much from me, but oh how I will love to hear from you. Ask Dorothy about Douglas Dillon,* whom she knows. Ask Dorothy about politics. She's a fighting Democrat and one of the best. Your Pa's letter—! All I can say is, I know, I know. All those boys can say these days is—Believe in Ike. It's an act of religion.

---

* C. Douglas Dillon, the ambassador to France.

I can't go into politics in this letter or I'd never complete it and I haven't time. Gotta cut it short as is. But our little segment is coming along fine. We think the D'Ewart Bill* is dead. McKay† referred to it in public in Denver as "a lousy bill." Our side is marvelously well organized and right on its toes and so damn much smarter than those cowboys that it's funny. Public opinion in the West is outraged and very vocal. And B's July Easy Chair‡ will put the lid on the whole business. I can't wait.

This letter will get to Marseille while you are in Portugal—hope you have a super time and lots of interesting brand new food. Garbure I do not know. We are not much of a soup family. By the time we have got through our large daily allotment of meat or fish, no room for soup. Our grocery bills are enormous, since we live on high protein. But my, we are healthy. In summer particularly we live on meat and fish and raw vegetables and fruit. And we both look ten years younger than we are—fact, not vanity. I know some of it is due to heredity, as my parents both look much younger than they are. Still, meat and fish sure does help.

Delighted that you are getting on with the book, and when I get back I want to see all you have done. With you undecided as to whether to stay in France or return to the land of cowards, very, very important to get as much done as you can. Probably good thing I won't be bursting in with letters during the summer. Giving your food questions a once over lightly. Goose very hard to come by here. But understand it is to become more plentiful since some western growers have come up with new methods of breeding, growing and marketing. Hope so, since I love it. American cooks apt to dismiss Goose as being "greasy" which is nonsense if you cook them right. I have Miloradovich on Fish—lovely book. Useful to look things up in, but somehow I seldom use it. My obstinate family wants fish fried, usually—your sautéing methods absolutely invaluable to me and I use them constantly. And when we have sole done that way, they want a sauce I dreamed up—quite outrageous—mayonnaise, minced dill pickle, chives, parsley, clove garlic put through the press. This they like and want nothing else. But if I could ever change my family's tastes I know I would use Miloradovich a lot. Bought June Platt's and Sophie Kerr's new book, *The Best I Ever Ate*—some fine things in it. Finally broke down and wrote Platt blasting her on overcooking shrimps and

---

* A bill written by Wesley D'Ewart, a Republican representative from Montana, that would have given ranchers the right to graze livestock in national forests.

† Douglas McKay, the secretary of the interior.

‡ "Heading for the Last Roundup."

lobsters and refusal to let them cool in the broth. Got a kind of blah letter back saying she would try my method and it was so nice to find people interested in cooking. (I didn't really blast her—just sort of firm and plaintive at the same time.) No I never got round to sending for the Meat Glaze—too busy. Dried chives dreadful, I agree. The garlic sausages you talk about I think are the kind I put in my spaghetti—they're fat and firm and you can cook them forever and they don't fall apart. Italian sausages we call them—my Italian grocer makes them fresh twice a week. Wonderful. Chain store sausages getting worse all the time—oh for a slice of my childhood bologna which really was a sausage. Specialty stores apt to have quite good sausages of many varieties—and so have the really big supermarkets. Many kinds of salami, hard and soft, pepperoni, several kinds of liver sausage including knockwurst. But people mostly use the ordinary pork sausages for cooking—I don't care for them too much. Haven't eaten ham hock since we were young and impecunious when I used to cook them with dried limas. Couldn't face it now. We seldom eat ham—got out of the habit when I got anemia. Salt pork I always have on hand for baked beans and chowders—sometimes cook a thick slice of halibut with thin strips of salt pork on top, baked in the oven. Never saw white beans in brine. Chick peas (garbanzos) come that way and very good, better than you can cook them yourself. I hate the white beans we have around here—very small ones—and a friend always gets me the New York beans, much bigger and better to cook. Size of string beans depends on what stage they are picked at. Only one variety apparently. I try to get the small ones because I simply hate them Frenched—I like them cut or broken or with ends snipped off and cooked whole. Frenching seems to let all the flavor go. Only beans I know that are cooked before they are dried are good old New England shell beans, and I love them. We only get them for a few weeks in summer. The pod is mottled pink and white, the beans fairly large and also pink and white or red and white, very pretty. Delicious. But an acquired taste so I am told, and many people affect to despise them. You cook them twenty to thirty minutes, the mottling disappears and they turn a faint beige. Served with butter as a vegetable.

The sole we get mostly is just plain grey flounder—and very good too. Sometimes (at a price) we can get lemon sole. This not imported, but very choice. I probably could get imported sole if I tried hard enough—I know some of the fancier restaurants in town have it sometimes. But it's the hotels and restaurants that snap all that sort of thing up—as well as mutton—and the premium beef. But up until about ten years ago, when it burned down, there was a hotel in of all places Keene, New Hampshire—an ugly, white frame building and for the life of me I can't remember its name at the moment—that regularly served imported English sole and turbot. They had thirty different kinds of fish on their breakfast menu. Plus so many egg dishes that you wound up with the usual boiled or scrambled from sheer inability to choose. It was an absolutely fabulous place and people came from hundreds of miles around to spend a night there and eat and eat and eat. Donald Moffatt wrote it up in *The New Yorker.** The Cheshire House I think it was called. I shall never forget it. The dining room had horrible pine furniture and one of those pressed tin ceilings like the Greek's candy store in your youth. But the food!

Simca's garlic sauce for roast lamb will be my first joyful task when I get back to my two stoves in September. Thanks very much. It sounds beautiful. I must say I ate a wonderful piece of roast beef at a dinner party about three weeks ago. The gal had laid hold of a fine piece of premium beef, and tenderly injected it all over in every possible crevice and some she made herself with Lowry's garlic butter. This is put up by the people who make the famous Lowry's seasoned salt—and is a mixture of butter and garlic to be used on garlic bread. Worked out fine with the beef.

Well cookie, this is all very well but the afternoon is flitting away. B. is stuck in court all day apparently, waiting to be called as a character witness for our beloved family doctor who was accused by some madwoman of having engineered an abortion for her. He says he didn't and I believe him—he's too smart. I have got to lay out on my chaise lounge all the garments I am going to pack—denim dresses, nylon underwear, shoes, pills, bathing suit, riding pants, dresses for hotels, sewing kit, etc. Then decide how to get them into a bag and a half which is all DeVoto is allowing me. Having just seen a darling but somewhat addlepated friend off to Europe with sixteen dresses, three suits, a fur coat and even evening dresses, I think he's got a nerve. I've got to pack Mark too. And I can put it off no longer, even to write to you.

But oh horror, your pen-pal who turned up. Well dear, that'll learn you to be so trusting with strangers. However, I can cheerfully assure you that the DeVotos have none of this character's nasty habits. What do you do now—just forget to write? Let him down easy, or coldly chop it off? When he relieved himself from the car Paul should have picked him up by

* "Hotels Here and There."

the back of the neck and left him to hitch-hike. The idea! I am dying to know who this writer is. Come on, give. I always protect my sources. He sounds too ghastly for words and I should like to avoid him if possible.

Now work, sweetie. And write occasionally—I especially want to hear your impressions of the divine Dorothy. And the equally divine Giorgio. I love them both very much. I want to know if Paul falls for Dorothy. Because in spite of her size she has some whacking kind of sex appeal that mows the gents down. Worldly, warm, malicious, understanding, eager—and the loveliest voice. They love me, too, so you'll get a good report I think. Do write. As often as you can without breaking into the work. I'll make it up come fall. We both got hilarious over Paul's letter to B. about *The Hour.** A very good imitation of B's extravagant manner of writing—and this is an effect the book had on several correspondents—the perfect tribute. Wish to heaven B. had been less harried and in a mood to reply at greater length. But we loved Paul's letter.

Bless you both,
Avis

SEPTEMBER 30, 1953

Dear Julia:
All alone in the house, seems very queer. Mark at school. B. in New York to attend dinner for retiring Fred Allen.† No maid. Fairly busy with housework, but the last two specimens I interviewed were so obviously bad-tempered that I prefer pushing the mop around myself. Something will turn up.

I read the soup chapter immediately, telephoned Dorothy at the office and found she'd just finished reading it. We're both extremely pleased. Think length is just right—you have achieved thoroughness and maintained simplicity. Anyone with this much knowledge in his head can go on with the variations—the techniques are very good and clear. Don't see how anyone can go wrong.

There's only one change I want made, and Dorothy agrees with me. Anyway we want to put it up to you. If you've got strong reasons for not making this change, let's hear your arguments. I definitely want ingredients on the left, method on the right. Vague discomfort about reading recipes finally resolved itself into that—after all, normal reading direc-

---

* Bernard DeVoto's witty tribute to American whiskey and the dry martini cocktail, published in 1951.
† Frederick L. Allen, the retiring editor of *Harper's*.

tion is left to right—seems all wrong to read ingredients which one must assemble first, then jump back to the other side for how to do. What do you think? We think it would be more normal, more comfortable, less strain to reverse the order.

I jotted down a few minor notes as I went along—nothing of importance. On page 10, puree of fresh tomato, you left out the quantity of butter. Page 13—I don't quite like "madly extravagant"—"very" will do. Page 18—clams and mussels. Vast areas here where of course fresh ones are not available, but canned minced clams and canned whole mussels are—especially the clams which I think I mentioned to you and which are extremely good. You do after all in other recipes say when frozen vegetables can be substituted. Do you want to give a recipe using these canned ones? Shall I send you two or three varieties to try out? I would doubt that fresh mussels could be obtained anywhere except in New York and maybe west coast—and though they grow freely all over the place most people are scared to death to gather them—plenty of horror stories about pollution. Too bad, because they are so good. They haven't been too successful in glass or cans—but the clams certainly are.

D. and I think the quick country soups sound ghastly. All I can think of is the kind of gruel that was fed to me when I was sick as a child. Are these actually good? Everything else sounds so delicious and I simply can't imagine Americans making these things with browned flour and so on. We both think the garlic soups sound marvelous and we are going to try them. You've done a swell job of describing them—they sound perfectly possible for anyone and absolutely mouth-watering.

Note you spelled it safran for quite a while—then saffron which of course is the way it is always spelled here. Do you describe anywhere what you mean by a "fruity" olive oil? Wish you would go a bit into detail on this—Spanish, Italian, or what? Page 42—do you really boil rice twenty minutes? Should think it would be too long. Especially to put into soups where it would cook a bit more. But then I like rice al dente and find Dione Lucas's method very satisfactory—thirteen minutes and then over steam in a towel. What are "turned" carrots or turnips?

And so help me, that's all. I think you've done an absolutely bang-up job. My congratulations. Dorothy much impressed at your thoroughness and clarity and exactness and infinite care for detail. Me, too. We are both thrilled at the prospect of such a book and more than ever convinced that it will sell very well and for a long time.

I gave you a completely wrong steer on the index. B. does not make his own indexes, but until the last two he had to pay to have them done, a job running around $150. HM pays for them, on account of he can get away with murder down there and has this clause in his contract. Dorothy will advise you about this, but since you have had experience indexing, we certainly think you could do a better job than any one else, and save yourself some money. She is going to talk to Lovell [Thompson] about various methods—she doesn't know any more about it than I do. Sounds like a perfectly ghastly job.

Certainly I think the reader has to become familiar with the basic terms such as blanch, etc. If he isn't willing to make the effort at the beginning, what's he doing cooking anyway? And yes, we think the asterisk showing how far things can be done ahead is a very good one. In fact all your ideas are good. Wonderful woman.

Dorothy was also going to talk to Lovell about illustrations and diagrams and decorations. She thinks this should be decided on after the manuscript is complete, so that the editors can decide about price and so on. She will write after she consults the brass. We've got to keep the price of the book down so that it will sell to as wide a public as possible. Anyway there's plenty of time so don't get too involved before you know how much HM is willing to spend on illustrations.

Now complete change of topic. I hardly dare to believe this, but B. is talking about going to England next summer. I never thought I would live to see the day. I am simply staggered. But when he says he's going to do a thing, he usually does. Last year the Fulbright people wanted him to lecture at Oxford for six weeks, which he turned down. This year they wanted him to go to Cambridge for a year—this, natch, he considered so wildly impossible that he didn't even consider it. Lord only knows if the Fulbright's will ask him again, but they did seem very keen and of course give substantial financial help—travel expenses, living expenses etc. If I can get him to England, I can get him to France for at least a couple of weeks. And of course you said you would be coming over here—<u>late</u> next summer. So some place, maybe both places, we will get together. Meantime in the most delicate possible fashion I have let it be known where it will be passed on to the Fulbright committee that DeVoto is in a receptive mood—I would not let him know this meddling for anything in the world. He actually told his English Publisher Douglas Jerrold,* who was out at the Thompsons' in Ipswich last Sunday, that he

* A British friend of Avis.

might come, and Douglas is going right ahead to get us a flat at Whitehall Court. I have not, however, been able to convince B. that it would be a good idea to get ship reservations, which can of course be cancelled if the trip doesn't come off. Can you bear it? I would rather go to England than any place in the world. Should think Mark would come with us, and hope he behaves better than he did the last three weeks this summer when he was fed to the teeth with traveling and wild to get home to his chemistry pals. Anyway, a girl can dream, and I think a lot about it. If you come over here late summer, just how late do you mean? I don't know how long B. means to make this junket, but it's crazy to go for less than three months and I'd like to take off as soon as school is over for Mark, early June. But B. always wants to be back here about the first of September, he gets so itchy to get to work. If this comes off, he sure will have seen a chunk of the world this year.

No more now—promised to read a manuscript for Anne Barrett and must get on with it. Detective story written by Mrs. Howard Forbes, a sweet old gal of 80, who is briskly writing another one. Not, I fear, too publishable—sort of a cross between Daisy Ashford* and a Gothic novel. But you've got to hand it to her.

Blessings on thee,
Avis

OCTOBER 23, 1953

Dear Julie:
Herbs received lately—that basil really is something. Never smelled any so fresh, so fragrant. We're having lots of baked tomatoes these days, since they are slightly on the greenish side and the hot-house ones haven't come in yet, hooray, and the basil adds greatly. Orange peel tastes fine, don't know how I will use it. Fennel dandy too. Had some of the fresh last week—awfully good. Mark and I are at present alone, as B. is on his lecture trip and won't be back until next Tuesday. Find it hard to cook for two. Nobody likes hollandaise but me, nevertheless tried your top secret one again for the second time, for frozen asparagus which is very good and which I steam—decided it is not sharp enough for my taste though I stepped up the lemon juice. Next time I will do it reducing the seasonings. But certainly easy as pie. When the two of us are alone I find we eat

* Margaret "Daisy" Ashford, an early-twentieth-century English writer who wrote her novels when she was in her teens.

high but restricted—large porterhouse, or thick loin chops, or a flock of lobsters and shrimps, and practically nothing else. When B. is home he wants more of a dinner.

I have been in an absolutely blissful condition for several weeks. B. is well and productive and not worrying as he frequently does. Gordon writes cheerfully from Korea where he has struck up some fine new friendships with British and Indian troops. He is company clerk and sounds quite happy. Mark is going great guns in school. And I have at last solved my domestic problem. I have got the world's best cleaning woman, three days a week. She is a gold-star mother named Theresa Keating, a widow; she has white hair, false teeth, and a stomach ulcer. She has no sense of humor, but she is gentle and kind and the hardest worker I ever met in my life. Not that she sweats over it. She just works right on without pause and accomplishes miracles. She prefers not to be supervised, and God knows she knows how to clean things better than I ever could, and sees dirt I didn't know was there. She is upstairs now turning out the linen closet, a job I have closed my eyes to for many years. And she really uses her head. She does all the windows. The silver. Takes the stoves apart and puts them back together again. All this so expertly that I really am put to it to find something for her to do on the third day. Peace, it's wonderful. And she says she will come any time and stay with Mark if I want to go to New York, which I do next week when B. goes down to tinker that piece for *Colliers*. I am pinching myself. We really suit each other to perfection, and so far she hasn't turned up one single irritating habit. I think she must be the kind of mother we are all looking for subconsciously, and damned cross because we can't find her. She thinks it's wonderful I do all that book work and reading, and wants to spare me all household chores. Dearie me.

And inasmuch as I continue to be enchanted with my dishwasher and pig,* you can see that this big house has no problems for me as of now. Really those machines are wonderful, and I am prepared to give American industry a great big hug and kiss. I am usually out of the kitchen ten minutes after I've cleared the table, with a comfortable churning going on behind me as I march grandly into the living room and take up a book. With a guest from New Zealand last week, I had all the dinner dishes except the dessert plates in the machine while we sat over our chocolate mousses. The coffee cups couldn't go in, because they are lined with gold,

* Kitchen-sink disposal.

but a rinse and upside down on a dishtowel took care of them. All I had to do, when this newspaper gent finally left at one-thirty (by which time I could have strangled him) was to tidy up the counters. Yes, you certainly must plan to have both machine and pig when you return to this country.

I finally got to see the de Santillana's pink palazzo the other night, and I have never seen a more beautiful house anywhere. Not in its entirety, of course, but at least three rooms of it, and the others all very lovely. Dorothy is quite simply the most talented woman with houses I have ever known. Every house she has ever lived in has been rather spectacular, even when they were rather nasty houses, as when she lived across the street from us on Coolidge Hill.* Her Mount Vernon street house was simply lovely, and I was crushed when she sold it, but now I see why. The pink house is on a cove in Beverly, and I didn't see much of the outside of it because it was dark. However there was a fine moon, and Giorgio took me around to see the grounds. There is a long sweep of lawn leading down to the rocks and the sea, with fine trees—the house is Victorian with dormers in a gambrel roof, and is painted pink. There is quite a lot of land, and Dorothy sold some to Kenneth and Eleanor Murdock† who have built a very modern one story house rather too close to the pink palazzo, but Giorgio is going to plant many lilacs. She has one other piece of land, on a sand beach, that she wants to sell us, but I know damn well we'll never have the kind of money to build there for our retirement.

I am obsessed by that house. It belonged to Miss Fanny Mason, who was a musical elderly Bostonian—she died five years ago leaving seven million dollars to a young French musician named Dogereau‡ or something like that. Quite a scandal. Dogereau is now living in a small house nearby with his French peasant mother and seven (count 'em) sisters. Fanny also left him three other mansions, which he sold at once. He told Dorothy that he thought he had made all that money the hardest possible way.

I have never understood the de Santillanas' financing. Giorgio has only what he makes at Tech,§ which can't be much. Dorothy inherited from her parents, being an only child, but it couldn't have been a fortune. She did however inherit a great deal of absolutely superb early Amer-

* The DeVotos lived on Coolidge Hill Road in Cambridge from 1938 to 1941.

† Kenneth Murdock, a professor of English at Harvard.

‡ Paul Doguereau, a pianist.

§ Massachusetts Institute of Technology.

ican furniture—not the pine sort of thing but the copies of Chippendale and so on made by the best American furniture builders, and all this stuff has been beautifully cared for. Some of it came down from Dorothy Hancock,* who was an ancestress of D's. And now Giorgio has got some of his Italian stuff over—a pearwood credenza, an Empire sofa, and the like. Also, since his son is married to the daughter of the gent who at present controls the Venetian glass works, he has been able to have some magnificent chandeliers and sconces and lamp bases made. And as the rooms I am talking about have eighteen foot ceilings, they demand magnificence.

But it is the most clear, uncluttered, limpid, airy magnificence I ever saw. I could hardly tell you what is in the drawing room—very pale pink walls, long heavy white curtains from ceiling to floor, two very large Italian wallpaper panels in the Chirico style,† these in sepias, cream and so on, two long sofas slip-covered in oyster-colored chintz with a sort of pen and ink design and a bit of green and brown, and so on. Except that I never felt so elegant in my life and looking around at Anne Barrett and Lovell and Kay Thompson and Giorgio's publisher from the Chicago University Press (G. has just done a big book on Galileo)—they all looked extremely elegant too. Some rooms just do that.

And Dorothy does all this with only a part-time cleaning woman who sometimes will stay on and do the dishes. The cat ate part of the dinner while we were having our cocktails, so she was somewhat delayed while she whipped up something else, which was of course very good. She also produced the best crème brûlée I ever ate in my life. She's quite a cook. I didn't get much of a chance to talk to her—the guest of honor was Mrs. Hendrickson, the Methodist minister from Maine, who was having a few days of publicity in Boston (she was in *Life* a coupla weeks ago). But D. did say to me that she has been checking up on a lot of French cookbooks lately and she thinks you are going to beat them all hands down. She said yours was going to be a GREAT cookbook. She is terribly pleased.

Well why have I gone on nattering about that house I cannot think, except I can't get it out of my mind. Dorothy, unfortunately, is quite unable to help other people with their houses. The magic only works in her own, and takes plenty of time. She's been working on this house for five years and it isn't finished yet, so they damn well better decide to spend the rest of their lives there. I am so fond of them both. Remarkable people.

---

* The wife of John Hancock, the signer of the Declaration of Independence.
† Giorgio de Chirico, a twentieth-century surrealist, neoclassical Italian painter.

Dorothy is fatter than ever. I've never seen her eat anything, and she has the most delicate and beautiful legs and feet and hands. But she is undeniably massive.

Now sweetie I have a couple of great favors to ask of you. No hurry. You said casseroles and I take you at your word. I gave the New Zealander veal in cream with tarragon and it was lovely, but I had to be in the kitchen when I wanted to be drinking a cocktail in the living room. Please jump the gun and give me a couple of recipes for company that I will only have to take out of the oven and that will wait a bit. Chicken? Veal? Duck? I don't mind spending an hour or two or even more on them during the day, but no last minute fixing. It would be the greatest possible help.

Incidentally, I'd love to know how you prepare potatoes hashed in cream—you mentioned that at Lapérouse* they weren't too good. I make a potatoes in cream thing that is terribly good, but also terribly rich. Raw potatoes, diced fairly small, covered with heavy cream in a Pyrex dish and baked slowly until the potatoes are done. Then cheese (Parmesan or a very sharp Cheddar) grated over and run under the broiler. This is just too rich for some people, and too rich for anything but the simplest meal. The very best hashed in cream I ever ate were at the old Lafayette in New York—oh what bliss. With frogs legs and much garlic, and a green salad. Oh my. Breaks my heart that the Lafayette is no more.

The other thing I want is some more knives just like the ones you sent. I have nightmares occasionally during which I dream that one or both have disappeared, and what would I ever do? You sent me a paring knife and a fish filleting knife and I just reach for them automatically. I not only want two spares for myself, I want six or eight spares to give to my very nearest and dearest. In your second letter I think you said they could be sent one or two at a time without having to pay duty, though I must say I want them so much I would not mind the duty. And if you try to make this a present I will kill you. I am willing, eager, and determined to pay for them, and will send you a draft when you tell me how much. Don't be difficult about this, please. You have showered me with things and if you won't let me pay for knives I will be shattered. No hurry, but I do want them.

Now I must go and give my dear Theresa a bit of lunch, which is going to be lobster salad, since I had some left over from last night. She eats

* A Paris restaurant.

like a bird, with her ulcer, but she likes what I give her and I like her so much I would cheerfully feed her on turtle soup if she had a taste for it. And then I am going in to Bonwits* and try to buy three pairs of shoes. I brought a vastly expensive screaming orange-red dress yesterday and I think it has gone to my head. Red—at my age! But most becoming. Must remember to get plenty of sleep before I wear it, on account of it wouldn't go with bags beneath the eyes.

More later,
Avis

Marseille

OCTOBER 28, 1953

Dear Avis:
I was just thinking yesterday, as I was cooking up a big dinner, that it would be nice to get a letter from you, and then there it was when Paul came home. Awfully glad you have at last found the maid of your dreams, and she sounds perfect. I like particularly the aspect that she doesn't want to be supervised and has her own ideas of how to keep a good house. False teeth, white hair and an ulcer, and a sweet disposition. Sounds very good indeed, and I hope it continues. That means that you do all the cooking there is. My, how un-French. Funny how our kind of life is almost unheard of here. That dishwasher sounds wonderful, and I keep thinking about the convenience of it. Yesterday I washed all the dishes accumulated before dinner, then this morning washed them all from after dinner. If you only had to rinse and stack in the washer, how nice that would be. Do you have to rinse them before putting them in? The pig doesn't seem, here and now, quite as important, as we have a garbage shoot . . . but I can see, that with garbage cans, it would be a wonderful thing.

Love the sound of the de S. pink palace. The atmosphere of elegance you evoke appeals to me very much. I wonder who their Venetian glass fellow is. When we spent 2 weeks in Venice last spring before last, we met and became friends with the Venini family. We looked at all the glass places in Venice, and this Venini seemed so far and above everything else in design that we went out to the island of Murano to visit the factory. It is small, Venini himself does all the designing, making direct copies from the old classical stuff, or very good-looking modern things, and all

* Bonwit Teller, a department store.

in truly faultless taste. His daughter took us about, and we liked her so much we asked her to have lunch with us, and then that led to having dinner with the family, etc. etc. They have a beautiful apartment in an old palace on the Grand Canal, and seem to know everybody in Venice. We kept wishing we could stay there for several months and meet the life of Venice through them. And those lovely chandeliers, such grace and happiness in those flowing lines and colors. But you would certainly have to have an old-fashioned high-ceilinged house to put them in. We were quite ready to buy a small château and furnish it with Venini glass. Dorothy de S. sounds like quite a girl, and I [am] more than ever sorry we never got together this summer.

Always interesting what you say about the Republicans. (I find it rather hard to capitalize that word!) I find I have rather gotten to the saturation point in all of politics and cannot think clearly about them at all. I make the effort to be dispassionate, and never succeed. As to McCarthy, it does actually seem that he is blown into far more of a balloon over here than in the USA. Paul had a long talk with a young French professor just returned from a 4-month tour of the USA. The Prof. was all steamed up about McCarthy and kept asking questions everywhere there about him, expecting some violent reaction. Most people would say MacArthur? Oh, McCarthy, and shrug their shoulders. No interest. But he has been played up here as a neo-Nazi horror, in the local press. And we, of the foreign service, are still licking our wounds because of him . . . so we would naturally still be concerned, though much less so now.

We have had with us this week the new Public Affairs Officer for France, our boss, named Lee Brady. He spent some of his schooling over here and speaks French perfectly. Was a Cultural Affairs Officer in Paris in 1948–49, then spent a year and a half in Indo China as Public Affairs Officer. Recently he attended a course at the Army War College in Washington. He is intense, hard-working, intelligent, and a nice guy; and he knows his business. He reports that he likes [Theodore] Streibert, the head of the Info. Service, says he is vigorous, direct and honest and he thinks he's a fighter and a good man. So things may well be looking up at last. I certainly hope so. Certainly one of the main reasons for low morale in the Foreign Service has been the knowledge that, once attacked, no one in Washington would lift a finger in aid. Brady says the atmosphere in Washington is still yellow-bellied, however. But there is an awful lot at stake, meaning one's whole career in the government as well as outside it;

and it takes a strong and courageous person to stick a neck out. (I hope I shall be that strong. I have every intention to be. But then, we have nothing to lose. No children to support. No ambitions, etc.)

As for Kinsey,* I was baiting you, and then had been stimulated by various attacks on Kinsey plus all the fanfare. I looked up the Meninger† review in the USIS library, and agree with you, that it is extremely interesting and that he has brought up a very important point . . . that of the moral and ethical aspects. I shall really have no profound and final statements until after I have read the book. But I feel it is a most important study on the nature of man, and as man taken as an animal with his natural appetites and the way he satisfies them. The moral, ethical and spiritual sides of the question get into the abstract realm and discussions for the philosophers. But it would seem to be that, to make a solid philosophy, you need the kind of background Kinsey is providing. And if his studies do nothing else, they make a beginning for a realistic approach to sex. Like you, I am extremely curious about how people adjust to life and each other. Before marriage I was wildly interested in sex, but since joining up with my old goat, it has taken its proper position in my life. But, when people are queer, maladjusted, unhappy, bitchy, etc, I want to know why, so that I can understand them and get along with them if necessary.

Your book find, Thaddeus Snow,‡ sounds delightful, and I hope he is accepted. But what a history . . . I'm sure I read about that in the papers.

Will be delighted to get you some knives, 8 of each. Will also send you 1 very small parer, and one larger chopping knife. Will send you the bill for knives and postage. In return, I need some US flour. We have available at our commissary "Pillsbury's Best XXX flour," which I conclude is all-purpose flour. What I would like, is a list of what are the standard kinds of flour available. I think they are, cake flour, all-purpose flour, and bread flour. I think, also, that they are all made from hard wheat. The French is usually made from soft wheat, and that makes a difference. Then, I would like 4 five-pound sacks of BREAD FLOUR. We think that bread flour is going to be better for pie crusts and pastries than the all-purpose stuff.

---

* Alfred Kinsey, a biologist and the author of the Kinsey Reports, a two-volume study of human sexuality. In an earlier letter, Avis called Kinsey's *Sexual Behavior in the Human Male* (1948) "the dullest book I ever looked into."

† Karl Menninger, the psychiatrist and author who criticized Kinsey's findings as omitting the importance of love in sex.

‡ Thad Snow, a Missouri farmer, an advocate for sharecroppers, and author of an autobiography (*From Missouri*, 1954), which Avis brought to Houghton Mifflin.

Also want a tin of that new patent meat tenderizer I've read about. And 4 or 5 cans of your canned clams. (Couldn't find any in Paris.) So, if you could have these sent to me from Pierce's or anywhere, I would be very grateful indeed. They should be wrapped in 2 packages, 2 sacks of flour per each, and sent as follows:

> Paul Child, PAO, Marseille
> State Dept., Washington 25, DC
> Wrapped for overseas, and stamped as to France.

Then we can settle our accounts later.

Potatoes hashed in cream, etc. Potatoes certainly are very queer chemically, I find; I haven't put my thoughts in order on them as yet, but they are in my mind. And I have two learned treatises on potatoes from the Dept. of Ag., in which such things as, "The physiological changes taking place in potatoes during cooking includes hydrolysis of the protopectin and solution of the pectin with resulting increase in ease of separation of the cells." Hmm. It appears to be a complicated subject even for the experts. Yours hashed in thick cream sounds delicious, but mighty rich. I've been working on one raw, cooked in milk and butter. Rub garlic all over a flat baking dish about 1½ inches high. Heat the dish and heat the oven to about 375 to 400. Boil the milk. Then put in the potatoes and salt and pepper and add enough boiling milk to come ⅔ up the height of the potatoes. Dot with about 2 or 3 TB butter and set in the oven. Baste about every 3 minutes until potatoes have sunk down to milk level, then bake about ½ hour or until milk has been absorbed. Add more butter if necessary . . . they can take a lot. Have to use "boiling" potatoes that will hold their shape. They come out tender, buttery, garlicky, and have a nice brown thin crust on top. I have not tested this out for doing ahead of time. Works equally well using a nice meat stock instead of milk.

There is another version, Gratin Dauphinois, mixing in cheese between layers of potatoes and using same process. This I have not experimented on very much, I like the above method so much. It is said, also, that old potatoes, with their acid, can sometimes make the milk curdle. If so, set them in cold water and bring to boil, boil one minute, then drain. Then proceed with recipe. And the boiling of the milk first seems to help prevent it from curdling.

For creamed potatoes prepared ahead of time, I boil them first, then dice, then put them in a baking dish with cheese sauce and top with cheese, then they can be reheated any time and are very good. There is something about the potato being cooked, then covered with the sauce, that keeps it in good condition. But I have found that if they are kept hot, and particularly if they are kept hot and covered, they develop that awful taste.

Have found that mashed potatoes can be all prepared ahead of time very successfully. Boiled, then put through ricer, or "whipped" in the egg beater, then mixed with hot milk and cold butter, and cheese. Put in buttered baking dish, top with cheese and butter dots, and it can be re-heated any time. But you probably know all this anyway.

In fact, I have found with just about everything prepared ahead of time, that, once prepared, it should be let to cool, and not "kept warm," as keeping warm it loses or changes its flavor. Such as for chicken casseroles, cook the chicken so it is almost done, then set it aside. It will finish its cooking while it is re-heating. I am still experimenting with my meat thermometer and roasts done ahead of time. For instance, last night, I roasted my lamb until it registered 160F (we like it rosey, at about 165 to 170). Then I took it out of the oven about an hour before it was to be eaten. While first course was being eaten, the roast was in the oven at 325 about (my stove is far from accurate!), and thermometer had reached 165 or so by time I wanted to take it out. Then there is the possibility of roasting it so that it is just done by the time the first course is ready, and leaving it in the warming oven at not more than 165, if you have a warming oven. There is a French theory saying it is much better to get the roast done ahead of time and to let it sit for 15 minutes so the juices recede back into the meat. That if you take it from a hot oven and cut it immediately, all the juices spurt out. All of this I have to study much more. I usually always have a roast, as I love having a sauced first course; and these various systems have worked out perfectly well, and it never bothers me when we have dinner as nothing spoils. I do, however, go out into the kitchen during cocktails, once in a while, just to check up (mainly because I cannot depend on my lousy old French stove). Would love any ideas of yours on this.

Have been experimenting on Quenelles again, and have about gotten it down pat, using the electric blender and the egg beater. Had some quenelles in Lyon on the way up to Paris, just to see how they were, and I really and honestly did not think they were as good as mine . . . They were more floury. Do I dare say that? Made them last Tuesday, and they didn't hold their shape enough for rolling, so I poached them in little ramekins and then un-molded them. Delicious and light. Yesterday, did them again, making bread crumbs very dry and using regular French bread, and they did hold their shape, but were necessarily less light then the molded ones. Did you ever get an electric blender? After writing to all the people, the Oster company wrote such a nice letter, and said they would give us a complimentary price, so we have ordered one of theirs. It's the kind I have, and I like it very much. I find I am using it a great deal, and even put a lumpy Velouté sauce in it and it fixed it up fine.

I was wondering about American Sour Cream. Is it really heavy cream that has been soured, or is it sort of like yoghurt? I have heard people say that it curdles, making me think it is rather like yoghurt. I wonder if one wouldn't always be better off using heavy sweet cream, as it might probably be more like the French than the "sour" stuff. Or, if people didn't want to take the time to sour their own sweet cream, they might add a TB or so of buttermilk in with the cream. But I wonder if that would make enough difference. Maybe just a few drops of lemon juice; and a bit of butter, if cream not heavy enough.

Casseroles. I've just sent for CASSEROLE MAGIC, by Lousene Rousseau Brunner (Harpers $3) and it seems very good and sensible and is nicely written. Her chickens Madeira and Marengo are good. Too bad she missed a trick in not saying what to do about doing them ahead of time and heating up, though. As for chicken casseroles, all the ones starting "Poulets Sautés," page 890, *Larousse,* are good. Bourguignonne, page 891, is just like Coq au Vin. In any one of these you can do the whole thing, then set the dish aside, uncovered. Then, about 15 minutes before serving time, put covered dish in the oven at about 325, or so it will just be hot when you are ready to eat. It is important, however, that the chicken be just done, so that when fork tine is pushed into fatty part of leg or breast, the juice runs clear (or, you can let the juice run just lightly rosy, indicating it can stay a bit longer in the oven before serving time). If no juice runs out, it is overcooked. Bretonne, page 892, is awfully good, too; and *à la crème,* to which you can add mushrooms, same page, right hand column.

Other good casseroles, are Boeuf à la Bourguignonne, under Ragoût de Boeuf, page 202; and Blanquette de Veau, page 185. Your escallope de veau à la crème can also be prepared ahead, and then just heated up. Veal seems to stand this very well. Another good veal dish is Veau Orloff, which is roast veal carved. Then each slice spread with a rice soubise (rice and onion) and, if you like it, a mushroom duxelles. Then roast reformed, soubise spread on top and grated cheese. This is just reheated in

Living in Marseille and traveling in Province with Paul on his business trips gave Julia an opportunity to become acquainted with the produce of the South of France.

the oven then gratinéed on top . . . or can even gratinée the top first, so it is all done, and then let cool. Then reheat at about 350 for 15 minutes or until hot through. Can't remember whether I sent you Simca's soubise recipe, which is nifty. Will send you Soubise, a chicken sauté, and enclose a very old "pre-book" recipe on persillades, which is readable.

This damned stove I have. I could buy a fine French professional oven for $300, but that might be too professional and perfect for everyday housewifery cooking. But that is such a lot of dough, and I suspect it might not work in the USA, or that the American ones would be better. I've just been looking through HOUSE BEAUTIFUL, and all those fancy kitchens put in by millionaires, with burners on a work top, and ovens up in the walls, and brown wood paneling. They are beautiful, and some of them look quite kitcheny too. Then their articles on what all these marvelously efficient housewives do, such as "chain cooking" of things, doing a bit here and a garnish there and putting everything in the freezer. I wonder how it all tastes afterwards. It certainly sounds clever. I wonder if those people really do all those things. Wouldn't surprise me. I gather, from reading the mags, that the articles are pretty well vehicles for the advertising, and that the mag. is not free. It's a good little combine, certainly. Very much the effect from reading *Vogue,* that I am a frump. I suppose that is the purpose of all of it, to shame people out of their frumpery so they will go out and buy 48 pairs of red shoes, have a facial, pat themselves with deodorizers, buy a freezer, and put up new crispy window curtains with a draped valance.

During this letter I had to go up town and happened by the knife store, and got all your knives. I have taken you at your word and got you 8 of each, and you will be horrified at the bill, which is seven thousand two hundred and 30 francs, or some 21 or 22 dollars. Also got you the very small and the bigger tougher knife, as you ought to have them, whether or not you think so! But I suspect you are a megaloknifomat like me. Will send them in three packages, regular mail, and the bill will come later after the postage is determined. Well, actually, that's not too bad, 18 knives for $1.25 each, and look what you get.

Please send a copy [of] Mark's book when it is finished.

Love to you all,
Julia

NOVEMBER 5, 1953

Dear Julia:

Just a short one, as the Schlesingers are coming to dinner and I have a few things to do. In case you want to know, I'm giving them poached salmon with beurre blanc (they are mad about sauces), broiled tomatoes, steamed leeks, hot French bread, and for dessert some fine Comice pears and a Wisconsin Brie. Have found a Wisconsin cheese place that makes simply beautiful French cheeses. We had a Camembert Sunday that was as good as anything I ate in Paris. This is going to be a real comfort to me as I would rather have cheese for dessert than anything. Only I like the runny smelly kind best and they are hard to come by.

What do you serve for a "sauced first course"? Details please. With the way we all keep house nowadays, no maid or anything, the first course has practically disappeared in this country.

I went in to Boston yesterday and sent you some flour from SSP [S. S.

Pierce]. There is no such thing as bread flour in this country any more. At least, so says the manager of SSP, and so says the manager of the big supermarket down by the river where I occasionally spend three times as much as I planned. There is cake flour and there is all-purpose. SSP however has a flour of their own under the name of Swansdown—which has absolutely no connection with the cake flour of that name—which says on the package it is recommended for bread. So I sent you two five-pound sacks of that, and inasmuch as it will probably turn out to be identical with all-purpose flour made by anybody else, I thought two sacks would do. I also had them send three or four cans of minced clams put up on the West coast, which they swear are better than Snow's (which they don't handle) and one can of their own brand, which I find not as good as Snow's. They wouldn't send the meat tenderizer because it is in glass, but I will send that on my own, plus three cans of Snow's clams. Soon as I get a small box. Price for what SSP sent, hold on to your hat, $8.12. Overseas postage rates have simply sky-rocketed. The administration, dear. SSP is also sending their catalogue, which has a full list of all their flours. You only asked list of white flour—there are of course a lot of dark flours, some of them excellent. If you turn out to have more clams than you know what to do with, try mixing them with softened cream cheese, touched up a bit—makes a very good cocktail spread. Though you probably have no use for cocktail spreads. This has been very popular here for a couple of years and now seems to be on its way out. On the meat tenderizer, I haven't tried this one, though I bought some months ago. During the war, there was a very good one in liquid form that I used a great deal, because what meat we got was apt to be tough as hell. You could brush a round steak with the stuff, let sit five minutes or so, and it would broil very tender. But B. always said he could taste it. It's papaya juice, basically, the same enzyme that breaks down meat fibers in your stomach. It worked fine on tough old hens, too. But I got out of the habit of using it when the meat improved.

Couldn't care less about the price of all those knives. It will solve my Christmas problem, not that I ever do much about Christmas except for the young. But I have several dear friends, male and female, who would a helluva lot rather get two really superb knives than the most expensive black chiffon nightgown or Sulka tie.* Will be very happy to get the small parer and the big tough knife, though I have excellent specimens of the

* A silk tie made by A. Sulka and Company, a New York City haberdasher.

latter, two of them French and one, a medium-sized heavy chef's knife, made in New Hampshire. Gosh I do love cutlery.

I'll write at length pretty soon. Theresa here polishing everything in sight, three men doing windows and storm windows. Distracting.

Love,
Avis

New Address: 113 Bd. De la Corderi. Telephone Dragon 02-03
(For when you come to Marseille!)

NOVEMBER 23, 1953

Dear Avis:
No word from you personally in writing for some time, which is too bad, but I imagine you are busy, as who is and whom isn't. Yesterday, to my amazement and pleasure, all the flour and clams arrived. Seems to be that is very fast indeed. How sweet of you to wrap it up in those wonderful towels, one of which I am immediately using. Fine solid, soft, big things they are *Je suis touchee, Madame.*

What a horrible 5 days we have just been through, both snortling with head colds, and I got the curse on top of it. But at last the dust has settled, the cases are emptied, and most of the straw which keeps seeping out of corners has about subsided. If you could have seen Paul's face on Friday when he came here after the movers had left, and the living room was piled from floor to ceiling with 45 cases, 5 feet square, 8 trunks and miscellaneous things. He didn't even attempt to put up a stiff lip. How could any people have so much stuff. How do we have so many books when we had not even one book case in Paris, not even one. But we had 6 cases of books. I don't understand that at all. And, judging from kitchen equipment, one would think this was going to be a restaurant. Too awful. I don't think we are basically suited at all to this movable life. If only we had no "interests" things would be much simpler. Photographs, thousands of them, paintings, objects, Venetian glass, English pitchers, and endless cook bookery. Fortunately Paul is an unusually strong man, and is able to hoist cases, furniture, etc, without busting a gut. We got all unpacked on Saturday, then came the arranging (200 bottles of fine wine, too, I forgot about!). I have just removed the necessities, stuffed all the other things in our fortunately ample storage places and will sort it out bit by bit every once in a while, hoping to be able to be tough for getting rid of things. Arctic clothing for cold British houses . . . useless now . . . but we'll be needing it again, maybe next year, Etc. Etc. Sewing equipment, but I don't have time for that any more . . . I even took a few lessons in hat-making when we first came to Paris, and even wore three that I made. Never was any good at all at making clothes, though. Always looked as though I had made them. Awful, awful.

But I think this is going to be a very attractive place. It has wonderful possibilities, but quite a bit has to be done to give it the old chic and life and color. Kitchen has possibilities, though small. There is ample storage place, but no place to hang up anything within easy reach, and no good work top. But we are getting that fixed right away. Paul has a whole nice little room for his studio. I am fixed up in the bedroom with all my paraphernalia. Wonderful sunny terrace, where we lunched in our skins today.

It is nice to have our own plates and books and pictures, but it all takes so much time to fix. I am really getting discouraged about work, though. There are so many things pulling in so many directions. I have responsibilities as a consular wife, as a wife, as a house-fixer-upper and as a cook book. If I hire a cook I can't cook and be waitress. When I'm cook and waitress, I'm not cook-booking. I can cook-book if we don't entertain much. If we don't entertain, I'm not a wife but a cook book; and if we don't entertain I don't learn anything more for the cook book.

Then there is the continuing feeling of guilt from slowness vis à vis you, HM, S & L [Simca and Louisette], husband and country. But now that we are moved I am going to follow Simca's schedule of 5 solid hours of bookery a day no matter what happens (Sunday excepted). She is a terribly hard worker, and really quite a remarkable gal. My, it has all got to be so damned good and thorough and logical. I am just afraid all of you are going to get discouraged at the time it is going to take.

We are both very anxious to hear what you have to say about the Truman/White business.* I feel Truman has muffed it. I think he should have admitted his judgment might have been at fault, or that he was too new at the job, or that the evidence was not sufficient, or something. I wonder if this affair might serve to end the Truman dominance over the Democratic Party . . . which would certainly, I think, be a blessing. Anyway, it is an awful business, all the hate and chop-licking and face-saving and pure partisan politics that it has brought out. Maybe the Republicans find that by rousing the country to all these circus affairs the other things they are

* HUAC subpoenaed the former president Harry Truman, and Eisenhower's attorney general charged that Truman had knowingly betrayed the country by naming Harry Dexter White director of the International Monetary Fund in 1946 after he was warned that White was suspected of espionage.

doing may pass unnoticed, such as moving in on the public lands, TVA,* etc. etc. This is really the dark ages.

Very pleased indeed with our new Consul General. Have seen him and his wife but briefly, but liked very much their general atmosphere of ease, good humor, simplicity, yet adequate and comfortable worldliness. He is very swarthy, and may well be part Negro as he has served a good stretch in S. Africa and Liberia. Nothing has been said about this except from a silly old Embassy type we saw at Dijon, who said he had heard he was a "very good boy." (Negros cannot be men, to some people.) [Clifton R.] Wharton is all man, with confidence, aggressiveness and evident professional ability. What a relief it will be to have a normal, human and able fellow as our chief down here. I am sure I don't know where they dig up some of the drips they have as Consul Generals. Trouble is, it is so hard to get people out of the service once they are in. I was asking a British foreign service inspector how they dealt with the problem. He said that any time after five years they found a man was a misfit they told him so, and retired him with as much pension as he had earned. Very sensible. (I am always griping about something in the Foreign Service, it seems to me . . . First they fire them, then they won't fire them. I just want it to work right.)

We've just had a lovely dinner. A little partridge, *poêlé au fine champagne et au Madère*† and a Corton 1923. Yum. We do have a good wine cellar! This was a little shooken up from the move, but a noble thing. When we are alone we have nothing but a big main course and a bottle of wine. Rarely if ever any desserts, even for company. Cheese and fruit. And I rarely make any cocktail things unless I am feeling fancy and whip up some puff paste or something. I just think it is too much for one person to do. But I do love a first course, like scallops or fish in a sauce or a quiche, or hors d'oeuvres, or pâté, or something. So glad there is some good American Camembert type cheese. There is certainly no reason at all why we can't have all the cheeses anybody could think of, but I wish we would call them by local names. The Roquefort down here is simply delicious, very fat and soft, the Marseille preference. Camemberts and Bries have not yet come into their season, but should be right in a few weeks.

Can't seem to get this finished, but will dash it to an end. We've just

---

* Tennessee Valley Authority, which McCarthy attacked as a Communist threat and President Eisenhower disparaged as "creeping socialism."

† "Pan-seared in fine champagne and Madeira."

been over to the old place, polishing up the floors, furniture, shelf-washing, etc and it looks as neat as a pin. And things are looking very good here. All pictures are hung. Books arranged, but we'll have to get three book-cases as there is a large residue neatly arranged on the floor. Wonderful to have a man about the house with an eye for the right object in the right place. Paul has faultless taste and a wonderful color sense. It looks really attractive, and all it needs is 3 slip covers, 3 sets of curtains and some better lighting arrangements.

Had the new Consul General and wife for dinner last night, just the four of us for mutual brain-picking. A very nice comfortable pair, and very fond of each other, which I always like to see. 6 children, all grown, 2 married and 2 boys in Harvard. He comes from Boston, name of Cliff Wharton, about 54. They are neither of them the intellectual type. He is very outgoing and frank; she is terribly nice and comfortable and sensible and attractive in an easy motherly way. We're very lucky indeed, to say nothing of the USA . . . I did one of Simca's chicken dishes last night, Fondu de Poulet à la crème. Cooked it at 3 PM. Set it to warm up while we were eating our oysters and served it with some little stuffed mushroom caps and steamed rice. Very good indeed . . . I'll send you the recipe after we've worked it over a little bit more. It calls for 2 cups of heavy cream and ½ cup of cognac . . . which I think is probably a little too much of those expensive things for the USA. I used only 1 cup of cream, and ¼ cup cognac, but there wasn't quite enough sauce I thought. We had some lovely wines which were quite wasted, unfortunately. With the oysters a Pouilly Fumé 1941, a very unusual year with an astounding bouquet and maturity, yet bone dry. With the chicken, a Meursault 1947 which was a dream. With the cheese (a fat Roquefort), a Musigny Burgundy 1947, a little bit young, perhaps, but a glorious thing. I think, too, for those not wild about wines, 3 is too many . . . We could have left out the Musigny . . . but how did we know they were not bibbers. We were, actually, so pleased to get our wine out of storage, we went berserk. Had it only been the DeVotos, or the Cambridge Bicknells!

With the words of McCarthy's speech ringing in our ears, we went (by command) to the English church for the Thanksgiving service. They passed out copies of "My country 'tis of thee," and reading it over, it sounded so happy and hopeful . . . From every mountain side, Let Freedom ring . . . My heart with rapture thrills . . . Sweet freedom's song, etc. Them was the days. Woe indeed.

What news of your English summer?

Much love, and many thanks. I'll let you know about the clams after things have calmed down a bit. I know the Pioneer brand, and remember it as very good indeed.

I did want to ask you two things. What about cooking cheese. Do you get Parmesan and Swiss (or Wisconsin Swiss). Here Parmesan is about $1.25 a pound, and Swiss about 90¢. Another thing. In roasting big birds and veal and other things, they use here *Bardes de Lard,* which is fresh pork fat cut in long thin strips any size you want. Do we have that in the USA? I suppose, if not, one could use fat salt pork, and cut it in strips.

Love again,
Julia

FEBRUARY 12, 1954

Dear Julia:
Only a note. Trying to spare my eyes. Still struggling with them or should I say again, as I thought I'd got them cleared up. Not pink-eye. Either overuse, or a slight infection, or maybe allergy though I've never had an allergy in my life except maybe in Yosemite when heavy fall of pine-pollen coincided with a summer cold. Except I don't have summer colds I dunno. It's a bloody bore. Especially steaming them with hot towels which drives me mad. If it is allergy, probably face powder as I never use rouge or any kind of cream. Got to try going naked with no face powder but how I dread it. What a blight. Pity they don't teach seeing-eye dogs to read aloud. Just feeling low is all.

Please, please try to come in May or June and by all means come here long as you like, invitation heartily seconded by B. DeV. It would be so much better than August which is apt to be hot as Hades. DeVoto says sternly none of this paying guest stuff though. It would be wonderful. DeVoto of course works damn near all the time, but shuts himself up in the library. We three or we two can do the stores and markets, gosh I might even get into Faneuil Hall market after all these years. And cook like crazy. My kitchen if I do say so myself is lovely to work in, what with two stoves and the dishwasher and pig! You must do most of the cooking and I will help and watch, because I will be so self-conscious cooking in front of you. I will get over it. Everything makes it better in May or June, especially because all the Houghton Mifflin people will be available especially Dorothy. Dorothy and Giorgio will be at Beverly (now in Boston for couple months) and you must see that house and them. You can get all the business end done—make up and layout and all. Mark will be in school until about 10 June I suppose, but is no particular trouble when he is around. He is kind of a charmer. He is mad about French and gets very good marks in it. A real flair for languages, which his ma sure hasn't got.

You can have the guest room and Paul will have to make do with Gordon's room which adjoins it with a closet between. No double bed for you. Paul is warned that he will be surrounded by Petty Girls,* hundreds of sensational paper back novels, and a quite sizeable library of mixed classics and science fiction. Please work hard on this. You can tell your family that business keeps you in the east so long. It's the perfect out.

Knives have now all arrived, and I will enclose check for them minus rice and plastic bags and postage. Bless you. Eleanor Edmonds (wife of Walter the author)† to whom I gave a pair says she's never owned anything like them in her life and murmured something about ordering a lot for her next time I ordered. I didn't go into this however. My butcher reports his wife ecstatic. Also very much pleased with the two new spatulas which seem to me better than the others you sent. And the dear little truffle. I am going to make a truffled chicken. For something special, not just the unthinking DeVoto males.

Speaking of Edmonds—Eleanor's old Uncle Charlie died a couple months ago. Wat is an executor or trustee or whatever. Seems the old boy devoted a large part of a large fortune to buying liquor. All the whiskeys. All the gins. (Thirteen cases of Old Tom gin turned up.) And wine. Fabulous wines. More than you could shake a stick at. He didn't seem to drink any of them. He just bought them. Every room in his house was full, including the bathrooms. These were catalogued, and when Wat showed us the catalogue we nearly collapsed. There must be fifty legal sized pages. Well, the executors are taking some of these things, and much will be sold. Wat, being the wine drinker of the three, has come into some of the most gorgeous stuff you ever heard of. He came over to cocktails here Sunday and gave us three bottles of Mouton Rothschild '29. I'm sorry to say it won't be here when you arrive as we are drinking it. What Wat has got stashed away in the cellars of 12 Berkeley Place I would much like to know, but we are having a joint dinner party there next week with the Rochlins and some of the best bottles are to be brought out. Benny who is a rye man is absolutely speechless at some of the stuff Wat has got.

---

* Pinup girls painted for *Esquire* by the artist George Petty.

† Walter Dumaux "Wat" Edmonds, the best-selling author of historical novels, including *Drums Along the Mohawk*.

Wish I had an Uncle Charlie. Eleanora Sears* (Paul will identify) is bidding furiously for the Old Tom, which not being an admirer of sweet gin I would not have in the house.

Glad I didn't buy a rotisserie, though I have been tempted. I'll wait until the bugs are out. I want one that is infra-red to do steaks in three seconds or whatever.

Dione Lucas is an alcoholic, so I'm told. Goodness she makes a nifty appearance though.

ADA—oh gosh we'll have to talk about that when you get here. It's national, founded by Elmer Davis, Francis Biddle and others, and is a thoroughly admirable organization. Middle of the road, liberal, pro-labor, and now suspect simply because it is liberal and rather heavily intellectual. The run of the mill Democrat has come to think of it as a bunch of college professors and pinkos. B. was on the state board but never at all active. The local hassle is because a state senator named Furcolo,† originally endorsed by ADA and elected largely with its help, was invited to address an ADA meeting and used his opportunity to inform them that they were wrecking the Democratic Party in Massachusetts and should disband. Trouble is, he wants to run for the Senate or for Governor, he hasn't said which yet. Leaves us local ADAers in a hell of a fix. Trouble is also that there was something of truth in what he said. Mass. ADA has not tried hard enough to work with the party at large. More resentment at ADA in Mass. than anyplace else. If Furcolo ran for either office, don't know what I'd do.

No I haven't got an extra copy of DeVoto's knife piece. But I can easy make you one when my eyes recover. He would be delighted to let you reprint part of it or all of it, but better see how the book shapes up and what HM thinks.

Blanquette just as I thought, but what you didn't say was how about the sauce, which is where I was puzzled. Also how in hell do you bake or broil mushrooms so they get dark brown? Every time I eat them in a restaurant they are brown and juicy. Every time I cook them they stay very pale no matter what heat the oven is or how long I cook them. Can't understand it. I peel them, dip in olive oil or melted butter, butter the dish, put plenty of butter in the caps, or mince up the stems, stew in but-

---

* Emily Sears, the wife of Henry Cabot Lodge Jr., the U.S. senator from Massachusetts who was defeated by John F. Kennedy in 1952.

† Foster Furcolo, who would lose his bid for the Senate later in 1954, was a political enemy of the Kennedy family.

ter, stuff in the caps. Still pale. This is one layer deep and I don't cover the dish. What gives. Glad you sent recipe for the Tapenade. I did it in my beautiful wooden mortar last Sunday, before you sent the recipe, using my imagination. Guests liked it. I didn't. Think it was too salty and I don't really like anchovies. Think I should have desalted them. Will try it with tuna next time, and not so much anchovy—I used half a small tin. Will try your terrine when warm weather comes—go nearly frantic trying to get a meat course in hot weather, after running the gamut of jellied chicken, cold lobster, shrimp, salmon, etc. Just bought a new Escoffier. Great fun looking it over. My, isn't he inexact. Cooking times and whatnot. Had an Escoffier in French years ago but it seems to have vanished; didn't use it much anyway as never can be certain of my translations. All his quantities seem to be hotel size. Oh God I forgot to mention the valentine which is absolutely enchanting and I can't tell you how touched we both were with the poem. I am going to have it framed with glass on both sides when I can find someone to do it. My fondest blessings on you both and I only hope you had as much fun doing it as we had reading and seeing it. You <u>must</u> come in May or June. This meeting has been put off too long. Now I must steam my eyes.

Much love,
Avis

Rice, bags, postage came to $3.47.
Elmer Davis book on its way—it's superb.

Washington's Birthday
FEBRUARY 22, 1954

Dear Julia:
Gosh I hate holidays. No school, no marketing, no routine. All caught up in my work including changing all beds and washing all odds and ends of socks and so forth, no cooking to do on account of going to Rochlins tonight for broiling steaks in the fireplace. Eyes not too good though improving because I'm on cortisone drops now. Three boys in the basement developing pictures, Mark's latest passion as B. gave him a quite good camera for Christmas only don't ask me what kind. I only know the price which was $55. B. tackling his big natural resources piece for *Harpers*. So I will just gossip with you for a while.

I have done the blanquette and it turned out absolutely beautiful. The sauce was a poem, like gold velvet. Pleased with myself. Tasted divine and family rather lukewarm. Damn. To satisfy my passion for cooking French looks as if I will have to hire myself out a couple days a week. What to do. I can't throw a big dish down the john as you've been throwing eggs.*

Few small items showed up on the blanquette. First, couldn't get breast or neck of veal — my butcher says no demand and since he's a chain store he only gets legs, for roasts and cutlets, and chops. Hadn't occurred to me we would run into this but I think it will show up here and there all over the country. Harvard Square is rather well off, housewives don't have to buy the cheaper cuts. Think when you get here we must ask my butcher, Tom Mahoney, up to lunch and pump him on this purely economic angle, among others. He is a pet. So anyway, I bought about five pounds of the leg end, which included probably three pounds meat and quite a lot of bone, including the knuckle. I had him bone it roughly and break up the bones. Could have boned it myself perfectly well only my cleaver simply is not heavy enough, nor my aim good enough, to deal with such bones. Am contemplating investing in a small meat saw. But then though I have a variety of cutting boards, I haven't got a block, and don't want to ruin boards. Wouldn't, of course, with a saw, but would with a proper cleaver.

So, having been warned by you that roasting veal would be dry, I had it on my mind. I made a fine stock out of the bones, with some meat glaze, and at the end when onions and mushrooms were done, added part of their juices, leaving only enough to keep them moist. I did all this well ahead because we were going to a cocktail party we couldn't get out of, and thought I could do the whole thing ahead of time except for the sauce. The point I'm laboriously working up to is that I just left the meat pieces in the stock to cool, and they were perfectly tender and quite juicy. I have mentioned this to you before when dealing with salmon and shrimps, and I do think it is an important point. It can't just be my imagination that by leaving these things to cool in their various stocks they keep their juiciness. Gosh Wat Edmonds just called up and is getting us a case of Château Léoville Poyferré '29, I hope I've spelled that right, out of the estate. Whoops. So we are managing to latch on to some of this stuff.

Well so I got back from the cocktail party which was at Bill Langer's† across the street and though wonderful food was provided, the drink was

* Julia was working on the egg chapter of her book.

† William Langer, a professor of European history at Harvard, and his wife, Rowena.

sickening sweet Manhattans which neither of us could choke down so I was perfectly sober. And all my little pots waiting on the stove off the fire and covered were tender and juicy and the sauce put itself together beautifully. I used two egg yolks and a cup of heavy cream to two and a half of meat juice—creamier than yours but very, very good. So when you do the recipe for blanquette why don't you point out how much of it can be done ahead. This is so important for Americans who just can't stay in the kitchen for two hours before dinner, if anyone is coming. I've got a sizable quantity of the blanquette left over and am going to try a couple methods of reheating, for my lunch and for Mark's supper.

We had a nice long chin with John Carter Vincent at the Langers, and wife—both very talkative and nice. Do you know them? Far from being cast down, both are very chipper. They've bought a house here. We are going to have them over. But before then will be seeing them at a party John Fairbank* is giving for Owen Lattimore who is coming up to speak at one of the Harvard organizations, I forget which. There was the usual attempt at an uproar, to bully the organization into calling him off, but it was half-hearted and dear old Harvard stood firm. John Fairbank's wife is Wilma Cannon who is Marian Schlesinger's sister, and I like them both very much. We have known both the Lattimores for many years but not very well and haven't seen them much. I hear they are in terrible shape, both emotionally and financially. Consensus of opinion seems to be that Owen was rather a boob about this China business in many ways, which doesn't make much difference in the framework of the general problem.

The excitement is Sec. of War Stevens's† stand against McCarthy, and I can't wait until tomorrow when they meet, if they do meet. Never would have thought that Stevens would take such a firm stand. He is a cousin of that enormous Stevens family we have been so involved with for many years—owners of the Ames Estate outside Annisquam where we spent seven or eight summers. All Republicans, rich, interested in nothing much but hanging on to their money and their parsimoniousness has always been wildly entertaining. I don't know if I ever told you about Uncle Robert Stevens, not this one, who at his house in Florida had a Persian rug of such beauty and value that he kept it upside down on the

* A former adviser to Chiang Kai-shek and professor of Chinese studies at Harvard. He was accused by McCarthy of being a Soviet spy, which led to years of Senate hearings. The charges were never substantiated.

† Robert T. Stevens, the secretary of the army, clashed with McCarthy in hearings over accusations that Soviet spies had infiltrated an army laboratory.

floor so it wouldn't get worn out. Harriet Robey,[*] Ted Stevens'[†] cousin, and her husband [Alec] and family are theater mad but always sit in the highest balcony because you just don't spend money on good seats. Harriet and Alec have been building an extremely beautiful modern house down on the river at Sudbury, and at the housewarming B. was talking to her brother Ames,[‡] the most conservative of the lot, about Sec. Stevens and asking him if Stevens would stand firm against McCarthy, and Ames said he didn't have enough guts. It looks different now, doesn't it? Have you ever in your life, or in the extent of this long obscenity, heard anything more outrageous than McCarthy's handling of Gen. Zwicker?[§] He sounds now really insane.

. . .

Makes me so furious that Eisenhower could have stopped all of this, any time, by a good strong fighting statement.

Must stop now and make B. an absolutely calorie-less lunch. Two poached eggs on toasted gluten bread. Or a large salad with lemon juice. He still has two pounds he wants to shed, bringing him down to 152. Madness it seems to me.

All right I won't be silly about cooking with you. Trouble is I am worldly only at times, and at other times am seized with the most quaking shyness. Try for May or June. Sept-Oct can be ghastlier than August here, in fact last September was. We have just bought three air-conditioning machines but two of them go into the library and the other in B's bedroom, since he falls apart in the heat and we don't particularly. God they are expensive and the whole business with installation and extra electric lines is going to cost about a thousand bucks. I'm going to get this off now. Seem to be talked out for the moment.

Best love,
Avis

Cheers. The blanquette reheats with no difficulty whatever, in a glass double-boiler with only half an inch barely shimmering water in the bottom. I stood over it however, anticipating difficulty, and removed it when

---

[*] A prominent social worker and author.

[†] A Harvard Law School graduate before Eisenhower sent him on a mission to the Alaska Territory, where he eventually became one of the state's U.S. senators.

[‡] Ames Stevens, the director of Ames Textile Corporation in Lowell, Massachusetts.

[§] General Ralph Zwicker, a decorated World War II veteran berated by McCarthy and accused of protecting Communists within the military. Zwicker's subsequent refusal to return for questioning and the army's counter-assertion that McCarthy had tried to procure favors for an aide who was drafted ultimately led to the Senate's Army-McCarthy hearings in the spring of 1954.

just heated through. I think that like all stews it is even better the second day, and there was no sign of curdling with the sauce. Feeling a bit let down even, as I thought, in my ignorance, that it would misbehave.

Have just given Mark some of the reheated blanquette. Maybe he'd spoiled his appetite last night. He now announces he thinks it's a fine dish and would be perfectly willing to have it once a month. I'm so damn pleased it reheats well because I can have it for company now, without last minute worries.

FEBRUARY 27, 1954

Julia:

I very nearly sent you a cable. You'll never know how near. And I never sent one in my life not being an international type. Just to say how absolutely wonderful I think the egg chapter is. It came this morning. I started it this afternoon. Interrupted only to get dinner and clean up after. Just finished. I am absolutely overwhelmed by the amount of work you have all put into this. Swept off my feet. Knew before how good the book would be but never felt it quite this way before. Masterly. Calm, collected, completely basic, and as exciting as a novel to read. This can't just be my own reaction, because I'm nuts about food—I think many, many people will feel the same way. And should think the famous pros like Lucas would be green with envy, to see how the job should have been done. You must have sweat blood over all this, but it doesn't show in the reading. Can't tell you how much I've learned. Things cleared up in my thinking that were always fuzzy before. Wish to God I could get Dorothy on the phone immediately, but she's at Beverly and undoubtedly entertaining or being entertained, and probably not quite sober. Damn. Can't talk to her properly while she's involved with other people. MacLeishs and Murdocks down there this weekend. However, will call tomorrow, or Monday at the office, and seriously talk promotion and whatnot. I want to be coony* about the book when it comes out. Want big splash on household page of either *Herald Trib.* or *Times*. *Trib.* as you know is that dame Paddleford,† knows a lot, ghastly (for me) style, but the big noise in the newspaper food world. Dorothy or I must sound her out very care-

* Slang for "canny."

† Clementine Paddleford, called "America's best-known food editor" by *Time,* wrote books and newspaper articles chronicling American food.

fully. Don't know if she ever writes about cookbooks. Have a hunch if she knew about it ahead of time she wouldn't be beyond swiping credits and ideas. Know less about the dames at the *Times*—Nickerson* and Casa-Emellos† or whatever. But we must walk carefully and find out which is the best pitch for absolutely the right publicity for the book. However this is all far, far ahead. Think Paddleford could be most useful for you. If she's a right kind of gal, and I have quite an in at the *Trib.* with Irita Van Doran‡ and others. I do not want this book reviewed at Christmas time in a paragraph with twenty others.

Now I jotted notes as I went along so we may as well clear them all up. I'll have to talk to Dorothy about those deleted recipes. Question of bulk of book enters in. Me, I'd love to have them. She'll have to make the decision. I'm only a hysterical bystander. Page 5—cleaning eggs. Wire wool—what do you mean? First place we never have eggs that dirty. Second place there's steel wool only nobody uses it in kitchens any more—only for stripping paint and so on. Death on the hands. We have things like SOS and Brillo—all impregnated with soap which you don't want. Anyway dirtiest egg one could possibly run into would require only gentle rub with fingers under running water, or at most a soft brush. Time on soft boiling eggs. Size egg I use, quite large, takes five minutes cooking to suit us—that is with no extra liquid that hasn't been incorporated into the egg. I could be so wrong—maybe what we do for breakfast is *mollet*§ eggs. But if I boil an egg four and a half minutes, there's half a teaspoon or so uncooked gunk. We must discuss this when we can boil them in my kitchen. The egg poaching directions are absolutely wonderful. No one has ever explained it so clearly. Ditto on molding. Page 10 Oeufs Cendrillon. Knowing nothing about it—should think, if baked stuffed potato really got cool, doing ahead, couldn't be reheated plus egg without getting the egg done too hard. Can you clear this up just a bit? Should potato be partially reheated before putting the poached egg on top for finishing? I know I get quite frenzied about reheating things without drying out or toughening. Which is why I am so grateful to find that poached eggs can be reheated. Always wondered in restaurants how in

---

* Jane Nickerson, a *New York Times* food editor.

† Ruth Casa-Emellos, the director of the *New York Times* test kitchen who worked with Nickerson and later Craig Claiborne.

‡ Irita Van Doren, the editor of the *New York Herald Tribune* book review section.

§ "Soft-boiled."

hell they could produce twenty servings of Eggs Benedict at once. Now I know. On reheating such things as baked potato in this recipe, how about wrapping in foil part of the time? Have you used foil? I think it is quite wonderful for reheating as things neither sweat nor dry out. If you have not got foil, will be happy to send you quantities for trial. It is used more and more in USA—under the broiler, because it keeps the broiler from getting messed up and can either be washed off or thrown away. Covering pans for which you have no proper lid. Wrapping things to be gently reheated in oven. Such as a slice of roast beef. And so on. Maybe you should say something about foil. Two weights—the heavier which is used for things in the freezer is the best. Not expensive and can be used over and over again. Reflects heat but doesn't heat up itself, so you can use it to hold a fish under the broiler and remove the whole business without burning your fingers. You probably know all about it but you never happened to mention.

Page 12—my invincible ignorance. What does *à la tripe* mean? Here it is just going to mean real tripe, should think. Middle page 13. You mean potato ricer instead of potato masher, don't you? Incidentally I am very much in favor of your "Final Assembly" idea. An enormous help. Middle page 22 on additions to eggs on canapés. Good additions—this is Eggs Benedict and probably not called that in France at all. Enormously popular here, for lunch, at the better restaurants, where it is usually served in half a toasted English muffin. I do it a lot when I am not thinking about extra pounds. Wish you would specify thin piece of ham—sliced already boiled ham, available everywhere, is good just warmed in butter. We tend to think of a round of ham as rather thick, as for ham and eggs. Actually, what I use when I do it is a couple (per muffin half) of very thin (machine sliced) pieces of Canadian bacon, which as you remember is really very lightly smoked ham, with just a tiny frill of fat around it. It is not too salty and blends very well with the egg and hollandaise. And just the right size and if you don't have ham on hand, as we never do, very convenient to buy just the few little slices. Somewhere around here, sorrel—you know, I have never seen it or tasted it. Just isn't in any market I've ever been. Probably only in New York. Anyway, think very few Americans would know about it or where to find it if they did.

The shirred eggs all sound absolutely divine. Sorry to be so girlish. I want to try them all and that's the test of a good cookbook. I have got eight fine little French clay shirring dishes up on a top shelf—they have handles—and going to get them down and use them. Fell for them years ago at Bazaar Français and never used them more than twice. Then you go into your broiled egg dishes and there is a slight confusion in my mind about the two different utensils. Think here would be a good place for a drawing, don't you? And you get your dictionary because I'm too busy—but I think shir has one r, and shiver has one v. If I'm wrong, please inform. The word I cannot spell possibly is nickle nickel. Anyway about these egg cooking dishes, there is now a great and increasing variety of wonderful oven-proof dishes on the market and that's one of the things we've got to look into when you come over. All sorts of ramekins and cocottes and god knows what, all safe in the oven. All the way from Woolworth to the most exalted and expensive. Loads of wonderful enameled iron jobs. I can't resist them but am not going to buy any more until you get here. The Dru quart and a half saucepan I just invested eight bucks in is positively the last.

Scrambled eggs is just plain wonderful and I have learned more than I ever thought possible. I think you are dead right in going into these basic things so thoroughly. Indispensable. And before that baked eggs wonderful also. Next week going to do some of them. Your word unctuous page 32 drove me to *Roget*. I know perfect well what you mean, but you've used the word a few other times and it's one that lingers in the mind and if people go to look it up they are going to be a bit puzzled I think. But what to substitute? *Roget* gave me no help at all. And nothing has come to mind. I think most people think of unctuous as bearing on people, like Soapy Joe [McCarthy]. Page 36—cordon—we would use border. Then you talk about tomato cases and say root end. No root on a tomato. Stem end?

I can't get over how good this all is, how much work has gone into it, how clear and instantly communicable it all is. I ought to be a horse player—my hunches pay off. You and Thad. Gosh I'm pleased.

Your redoing of the omelet techniques improves it enormously—much clearer, not so cluttered. Page 42 what do you mean canned peppers? Pimientos? Canned peppers, actual peppers always in brine, and generally in Italian stores. Nobody apt to think of pimientos as peppers though of course they are. Spice Islands has dried red or green peppers and pretty good—freshen in water and use—I use them in stews without freshening. Actually, green peppers, fresh, nearly always on the market. Sometimes rather expensive, but you can find them. Fresh red peppers only in summer, and not generally found. I use a lot of peppers. We like them roasted, peeled, tossed in French dressing.

And that's about all. Oh Julia this is so good. You are going to scoop the field and last. Can't conceive of anybody doing all this work again for a long, long time. Don't know how you three divided all this up, but will you give Simone and Louisette my heartiest congratulations? I am terribly proud to know you all. And to think all this came into my life on the point of a knife!

You know, it's funny. By the time we develop real taste in food, and begin to learn how to prepare it, digestive disorders set in and weight piles up. When I think what I could have done in my youth, when I ate like a horse with no bad results at all, with the knowledge I'm getting now, I could cry. We had a joint dinner party at the Edmonds night before last—Rochlins, Sargent and Elizabeth Kennedy* and us, and we had quite a discussion of this sad fact. Not a fancy party—I did a prime roast of beef, Eleanor [Edmonds] who is famous for her Yorkshire pudding did that, Helen [Rochlin] provided canapés and salad, and Elizabeth made a terrific sort of torte, her own invention—about eleven layers with mocha cream and whatnot, the sort of thing I only eat when she makes it. And a good deal of wine. Wat [Edmonds] is drawing heavily on his inherited cellar. But none of us could tuck it away any more. Oh dear. You'll love Elizabeth. She was educated in Paris, and she was young and fat and hungry all the time at the convent or wherever, so she read cookbooks. And now she's one of those girls who simply can't force food down, but she does quite wonderful cooking for four children and her friends. New Years Eve they always come here and she brings a new and fabulous dessert she's spent two days making, and she eats one spoonful. Sarge is registrar at Harvard.

Now I'm going to knock off until tomorrow because I've got to plunge into the political situation, the blackest week I remember, and we've got a visitor coming in a minute, just returned from Mexico who has missed all the horror and has to be briefed.

Sunday afternoon. And I'm sorry I can't make this a good long session but people are coming for cocktails and there's a lot to do in one way or another. And before I forget it—your line page 2 about eggs absorbing odors. Little extravagant is the way it hit me. I'd say just cheese, leave out the tarpaper and skunks. Eggs invariably stored in refrigerator over here.

* Sargent's wife, Elisabeth, was one of Avis's closest friends.

Might mention garlic and onions, which sometimes wander into a refrigerator. Or salami.

I assume you've kept up with the situation of the last ten days* as well as you could, and have read *Times* and *Herald Trib.* with avidity. I've read about nine papers every day, including *Christian Science Monitor* which has been wonderful. Eyes not so hot, but what the hell. I can't send you clippings because B. is collecting the cream of the crop.

. . .

Much love,
Avis

Paris

MARCH 26, 1954

Dear Avis:
A report on knifery, and I have just spent such a nice hour in a wholesale cutlery place, and they are all steamed up about dealing with your Cambridge hardware now. Madame's grandson is out of town, but he is to write me all the details on his return next week. It appears that no export license is necessary, that orders can be sent by ordinary express shipment, FOB,† and that payment may be made by dollar check. They have several qualities of steel, some lighter & some heavier. The best are "Nogent." They will send a booklet of photos and price lists. They think it would be a good idea to send a shipment of samples, so your fellow could see what was available. They suggest that sharpening steels might be useful.

Now. I hope they really come through with the dope—but they seem very much interested. My, what a mouth watering place, with knives & scissors of all sizes all over the place.

Maison Paul Hory—121 Rue Vieille du Temple.

Paris is heavenly. Such fun to wander about in it again. I think, more than ever, that I shall never get over Paris, and never find anyplace more to my tastes. Every thing about it satisfies everything in me, and I shall certainly come back here to spend my declining years.

Paul has finally gotten some precisions on our future—which are a bit sad. The big wheels have decided they must make up some kind of a permanent personnel program (about time). And nobody is to stay in

* Broadcast live on television from April 22 to June 17, 1954, the Army-McCarthy hearings were seen by an estimated twenty million viewers.
† Free on board—that is, free delivery.

a post (country) over 4 years. We've been here 5½. So it is certain that we are to be moved. Probably get home leave April or May, and then be transferred. Germany & Middle East have been suggested. Paul is suggesting Italy & Spain. So our dear little apartment—for which we have spent some $500.00 (fools that we were) we'll have to leave. Too bad.

And all this means more delays on book work—which is enraging. I am more concerned about Simca than anything else. Had a splendid cooking day with them yesterday and our fine old Chef [Claude] Thillmont—who was at the Café de Paris for 25 years. We did the most delectable Duck à l'orange, that was so tender and succulent, and about the best thing I ever ate. Beautiful apple tart, also. Louisette, who is vice president of a Franco-American club has been shepherding around 25 delegates of the Federation of American Women's Clubs—and has pumped them full of our cook book (that far-away thing). I think she will be able to line them up for some publicity when the time comes—nice little market of some staggering figure of 64 or 40 million women. I forget which.

Lots on McCarthy in the French press. Enclosed an article from *Match*. The French *Life* type mag., same from *Figaro,* French equivalent to *Herald Trib*. I'm so afraid this congressional investigation is somehow going to come to nothing, and they'll all back down & no great exposure will result. All depends on the Admin. and the state of their guts, I guess.

Much love,
Julia

Marseille

APRIL 11, 1954

Dear Avis:

No time to write after all. I have such a hopeless feeling about the book with all this prospective house-moving, etc. Honestly, we are both just sick about this move. Paul is so beautifully set, with a studio room, and his office running well, and his contacts around the region just blossoming into real usefulness. And I have everything I need for my work, room for all my reference and research material. Kitchen all fixed up and workable. But, as we keep saying to ourselves, this is Paul's career, and if he wants to stay in it, we've just got to resign ourselves to abrupt changes. Trouble is, neither of us likes to move around at all. We dig in with our paints and pots, as though it were for a century. Looks quite probably as though we should be sent to Bonn, Germany. It has the one advantage of being close to France and Paris. But, having lived here for so long, I have a horror of Germans and Germany. It makes me retch to think of them, and I would never, willingly, set foot into Germany. I am probably exaggerating, as usual, but I can smell the concentration camps and human soap factories from here. However, it would be a big and challenging job for Paul, as he would be Exhibits officer. And it is a compliment that they are considering him. However, he doesn't want a big and challenging job of that sort; and doesn't want to specialize. He likes the generalized job of Public Affairs, or of Cultural Affairs. We are actually expecting our orders any minute, and think it will state Home Leave first, then transfer. When I think of all those mounds of boxes and books, and pans, and sheets and shoes, and packing straw. Oh dear. And what about my book. I think the only thing to do is to pack a big trunk with all my necessary research stuff, and just take it with us, wherever we go. That will be fun. WOE.

Last night I decided I would make a nice broiled *Loup de Mer* (sea bass, Mediterranean variety), with a BEURRE BLANC. Although I hadn't made a BB for a good year, having had quite enough of it a while back, I naturally didn't look at the recipe, and I found to my quite hurt surprise that it wouldn't BLANC. After the guests had departed, I looked up the recipe, and found that the trouble was, without question, TOO LITTLE VINEGAR. So I bought myself a buck's worth of butter this morning, and made several batches more. It was clearly a question of too little vinegar, I am sure. I also tried reheating, cooling, fixing turned sauce, etc. etc. And I felt the recipe ought to be re-written while I had it fresh in my mind. You have had far more experience, by now, with B.B. than I have. So how do you feel about this revision? Changed the introduction, as it sounded a bit hysterical, and also left out mention of Mère Michel . . . as that would involve asking her permission, and other attendant difficulties. You will probably have other suggestions.

How are you feeling?

Much love,
Julia

PS: Have you ever seen a little gadget that you put in the bottom of a saucepan that prevents milk from boiling over? (Definition of a wife: a gadget you screw on the head that does all the housework.) Milk gadget is a fluted plaque—3" diameter. And a wife gadget?

APRIL 30, 1954

Dear Julia:

I must get the knife check off even if I can't write a letter. Bills bother me out of all proportion—I do feel it would be much more sophisticated and worldly to let people whistle for their money, in the manner of an English duke with his tailor, but alas I cannot do it. I am already down to four spare pairs of knives. Goodness I wish Roy Dickson* would get a move on and get some ordered, but he is as harried as we all are, what with everybody planting gardens and freshening up their places. He promises me to get at it soon.

What is killing me and messing up my life is these goddamn hearings. You just try taking five or six hours a day out, looking at something that keeps your blood pressure over the boiling point. It has a horrible kind of fascination and I cannot stop. B. keeps wandering in and stays longer than he should, looking and listening. The neighbors drop in. It is a glorious day out and I should be putting in a couple of hours in my so-called garden, but I think and hope they are going to manage to put the finger on just who (hope to God, Cohn) ordered that photograph cropped and then blown up. And as Mr. Welsh who is rapidly becoming a national hero is opening up this afternoon in cross-examination of a shifty character named Sourwine,† I am now leaving you. If it gets dull again, which it seldom does, I will try to write more.

You see how fast I got back—this is Saturday evening. I am so completely bushed that I'm going to get this off and fall into bed with a good detective story, if I can keep awake that long. I really am so damn exhausted, tired to the very bones, that I am at the stage where small trifles make me ready to blow my top, and I can't seem to accomplish much. So finding my dear Helen Rochlin in very much the same state, because she dashes at things the way I do, we have made a tentative date for around the middle of May, after next week anyway, to abandon our clamoring families and go down to say Rockport for about three days and sit on the beach and not talk to anybody and not cook or telephone or read. We did

---

* The owner of Dickson Brothers, a Cambridge hardware store.

† Roy Cohn was the head counsel for McCarthy, while Joseph Welch represented the Army and eventually succeeded in discrediting McCarthy. Welch charged that Cohn had tried to obtain special privileges for David Schine, a McCarthy assistant, after Schine was drafted into the army. McCarthy introduced false evidence by showing photographs that were later found to have been cropped by Cohn. Jay Sourwine was the chief counsel for the Senate Subcommittee on Internal Security, one of the many committees formed to investigate Communist activities.

it once before in Vermont and just what the doctor ordered. The boys will just have to make do without us.

I have to go to New York next week with B. who is getting an honor, a fairly silly one but I thought I'd go just the same—that is I will if I have the strength and can get an evening dress re-hemmed and pressed, which right now is doubtful. Some guy named George Macy who is a rich promoter of books (Heritage Press) appointed a committee to name the ten contemporary writers most likely to become classics, and lo and behold there was DeVoto, along with Hemingway, Faulkner, Sandburg, Sam Morisson,* Rachel Carson, Frost, Van Wyck Brooks, Steinbeck—is that ten? So now I call him Little Sir Classic. This is just a publicity gag, we think. But the committee was good—Cantwell† of *Newsweek,* Irita Van Doren, Harry Hansen‡ and a couple others I can't remember. And it will be a fine food fine wine dinner at the Pierre.§ Maybe I can stay down another day and go to Bazaar Français.

B. is in Syracuse and I spent practically this whole damn day typing up notes for him and swearing like a trooper and making lots of mistakes. Notes are hell because so many numbers and I never learned them.

Yesterday afternoon was far from dull, and I am now mad in love with Mr. Welch and the country is absolutely rocking with his crack about the pixie, enclosed because I don't know how much of these proceedings you are seeing. B. always leaves the library when I scream out that Welch is on, and comes running, so he was on hand for the pixie-fairy¶ and I thought he would have a stroke, he laughed so hard. Welch is absolutely entrancing—his intonations are positively devastating. The cameras immediately swung to Cohn-McCarthy—McCarthy tittered in a rather embarrassed way, then did a double-take and began to look mad—Cohn looked stony. The hearing room rocked with mirth. And it's all anybody is talking about today.

It begins to look—as if McC. is panicky. He is swinging very wild. Me, I'm betting on Welch, plus a few others. He is unbelievably quick, very polite and courtly. The contrast between his cultivated, measured manner, and the slick film over the basic savagery and arrogance and com-

* Samuel Eliot Morison, a Pulitzer Prize–winning American military historian.

† Robert Cantwell, the literary editor of *Newsweek*.

‡ Longtime book reviewer who wrote for *Harper's* and the *Chicago Tribune*.

§ A fashionable Fifth Avenue hotel.

¶ In using the terms "pixie" and "fairy," Welch was implying a sexual relationship between Cohn and Schine.

plete lack of morals of the other side is very, very clear over TV. I wish to God you could be here and see all this going on, but I promise you wouldn't do much thinking about cooking, or even powdering your nose. I keep starting off to the bathroom, and hear a few words as I leave the room and come right back unable to miss a moment of it.

A friend of my army child's, one Phil Rulon was in yesterday — home on a month's leave before going to Japan — he's been at the language school in Monterey soaking up Chinese. He says that until this army ruckus blew up there were plenty of guys in the army who liked Joe — now they hate his guts, back [Secretary of the Army Robert T.] Stevens to the hilt and are sore that he appeased so long. Stevens has been quite a good witness, makes a good impression. Not wildly intelligent of course, but dignified and fair and honest. Nevertheless, the impression of appeasement lingers.

American asparagus is green — some fat, some thin, purple tops. I don't peel — I remove scales and scrub a bit.

I have a sizeable booklet about foil for you but am perfectly baffled about how to send it — one of the small things I reported that floor me these days. Easy enough to do it but somehow I can't get at it. Says you can wrap up hot roasts, hot chicken and so on, and put directly in the icebox. Cook with it like parchment, if you've ever used that. I use it mainly for broiling. Will send if you are not coming soon, otherwise save for you. I dunno about coming to Maine — Mark may have to go to summer school. Seems to be flunking math, all his own fault because he skims over his work so carelessly. Marvelous marks in anything he likes. Serve him right to have to make up.

Mark Childs* knows a very great deal about the conservation problem and has written marvelously about it. How in hell did she ever read my reviews?† They are quite awful I assure you and I would not dream of sending you them. After four or five years seems I repeat myself most monotonously.

I don't set the beurre blanc right over the pilot light. The pilot is under the enamel top of the stove, which makes it just right for holding the sauce. Just comfortably warm to the hand.

---

* Marquis "Mark" Childs, a Pulitzer Prize–winning political writer and the author of a book on the Mississippi River.

† The reference here is unclear; it is probably a typing error for "he."

I shall now collapse, having not even touched on many things of interest to us all.

Love to you both,
Avis

Marseille

MAY [N.D.], 1954

Dear Avis:
Wonderful that you can put us up in July, and that will work out wonderfully well with our schedule, in fact everything is dove-tailing in marvelously. We shall do our peregrinations about New Haven, Hartford, Pittsfield, etc, and then arrive in Cambridge between the 12th and 15th (letting you know more precisely later). Then, filled with vigor and hope, we can take the train from Cambridge to the badlands of the West. My, we are so anxious to meet you! But I guess we can contain ourselves for a few weeks. What is your telephone number? Thank heaven Dorothy will be back. It would be too awful to miss her again. And we don't care how hot it is, either.

I have sent you off 12 each knives, as I thought you had really better have more than less. Hope that is OK, but I suspect it is.

12 at frs. 230	2760
12 at frs. 515	6180
	8940
Less 10%	895 (because you are such a customer for them)
	8045
Postage added	8435 frs. $24.10

If you think of anything else, let me know. We leave here the 11th, then will be in Paris (Hotel Pont Royal, 7 Rue Montalembert, Paris 7) until the 17th.

It sounds to us, my girl, as though you were having a *Coup de Foie Français*[*] . . . all the symptoms. That is a little malady we know only too well now, though we had never ever had a touch of liver trouble before coming over here. I remember in one of your recent letters you said you made a whole raft of Hollandaise for asparagus, and then ate it all yourself. We said, Gee whizz, she must have an enviable stomach to be able to

* "A liver attack."

do that. I just happened to be going over my old pocket diary for 1949, and amid such fascinating items as "Minette Pussy eats raw rabbit's eyes"; "Julia starts in at Cordon Bleu"; 4 days later, "Dortie (my sister then living with us) has stomach trouble"; next day, "Paul has stomach trouble." Two days later "Julia is bilious." It took us a good two years to learn to live with our stomachs, and now we do fairly well. But any time that biliousness happens, we immediately cut out all alcohol, which appears to enrage the liver even more, and stick to plain food *à l'Anglaise,* and take some of the French medicines for the liver, and I think the US KAO PEPTIN is quite good for that. It goes away in a day or two. Almost all the French people we know have a touch of liver quite frequently. As some one said, the French want to change their organs to fit the food, not vice versa.

But so glad, anyway, that you are feeling generally better. It is hell to be tired. Maybe the McCarthy hearings make your digestion poor, also. How about his ordering all the gov. employees to tell him anything he wants to know! It is incredible. Certainly hate to be missing all of this.

We are about to start in two weeks of hell. This coming one, I go over everything in preparation for the packers; next week the packers. All these horrid decisions about what to put where, what we will need when, etc. etc. Paul has nicely resigned himself to allowing me a trunk for cookbookery, which we will take wherever we go, and will weigh a ton. Winter clothes for October in Allemagne? Do we need evening clothes? Such a nuisance, but the last time on the boat (we are sent first class by the government) we did not have evening clothes, and I must say I felt rather frumpy, not being dressed up ever. Is the price of unfrumpiness worth the trouble. Etc. etc. I hate decisions about things.

Have managed to get in a few furtive licks lately on the poor old chicken chapter, and enclose a draft "master recipe" for another type of chicken, in case you may have time to go over it before we meet. It, again, is "total recall," and will be followed by brief outlines. I am extremely partial to this method of doing chickens, and it seems to me it is not very well known in the USA.

Do you happen to have seen a new book put out by Little Brown, TIME TO ENTERTAIN, by Charlotte Turgeon. It is a very good little book, and she has beaten us to the gun on "preparing ahead." That, of course, is inevitable, and surprising that some one hasn't done it long ago. Oddly enough, she was a classmate of mine at Smith, though I only vaguely remember her. She sounds like a smart and efficient girl; from the book-wrapper it appears she spent a year in France, and worked at the Cordon Bleu. But the book is more on American cooking, with French touches.

Heavenly weather here now, after a rainy May. Now it is clear and blue, hot in the sun and cool in the shade. We sunbathe on our terrace during lunchtime. Wonderful little apartment, I must say, and how we hate to leave it. But we feel so lucky to have lived here in Marseille at all, as it is very special, and like nothing else in France.

DeVoto's new article in *Harper's** for Aug. sounds awfully good, and we look forward to reading it. Hope it will do some good!

Love to you both. It is hard to believe that the great encounter is about to take place.

Julia

MAY 17, 1954

Julia:

I haven't got time for a letter. I keep thinking I'll get a cable any minute saying you are on your way. It seems goofy to settle down to a long cozy chat about cooking and whatnot. Cooking! What's that? Barely have time to heat something through. House is now being painted and perfectly beautiful, we will look gorgeous (outside at least) when you arrive. Spring is killing me—set out tomato plants day before yesterday and can hardly move as a result, what with fitting them out with little collars against slugs. I'm too old for such foolishness. Fifty next Saturday. My God.

Maybe little more time now as hearings were recessed this afternoon for a week. We suppose the boys will now cook up a deal. God. I'm feeling too Goddish for words. All this stinks to heaven.

Yes, if you have time, bring eight more pairs knives, or send. Absolute flock of weddings is upon me. They melt away, accompanied by grateful tears from recipients. If you're on your way, forget it. Went to NY last Tues-Wed—to that dinner which was incroyable, speeches going on until nearly one in the morning, stupefying beyond words and Sandburg my neighbor even more stupefying, champagne flowing in rivers, photographers all over the place, God. But next morning spent hours at Bazaar Français and spent a lot of money too though things haven't come yet. They have <u>not</u> got the fish filleting knife, never heard of it. Imagine.

* "Conservation: Down and on the Way Out."

One smaller, many larger. But a lovely shop, very French with everything all which ways and nothing dusted and treasures here and there you have to dig out yourself and only two clerks. Had a good time.

But when are you coming? I won't be in such a heap when you arrive. I promise. Just everything happening at once as it always does in Spring. Doc says no infection in my eyes, nothing showed up in cultures. Funny because they still bother. Anyway no keratitis sicca. I'm just getting old, honey.

But cable me or something when you get the word. I am so anxious to see you both. We will hash over absolutely everything then. Hurry, hurry.

Best love,
Avis

PART TWO — 1954–1955

A Somewhat Foreign Land

After Three Years of Correspondence, the Childs and DeVotos Meet in Cambridge Before the Childs Move to Germany

We miss you terribly. It doesn't seem at all possible that less than 2 weeks ago you were all of you but words on paper. It did not then seem that love on paper would not blossom into love in the flesh, and it certainly did with an all-embracing bang. That did it for good. We are but your creatures. Your generosity in putting us up, wining us, introducing us, talking, sitting, ice-coffee-ing, and just letting us share your life for a while was heaven. And we felt so at home with you. You were exactly, I think, though now I can't be sure, our "Dear Avis," perfectly familiar.

JULIA CHILD to AVIS DEVOTO, July [27], 1954

At home in Plittersdorf in late 1954, Julia gradually adjusted to life in a U.S. government compound while mastering the German language.

PREVIOUS PAGE: The Childs' traditional Valentine sounded a plaintive note.

In the late spring of 1954, Julia and Paul took their home leave in the United States, visiting family and friends. In July, eager to meet Avis at last and confer with the editorial team at Houghton Mifflin, they drove their rented station wagon filled with Julia's pots, pans, knives, and manuscripts to the DeVotos' home in Cambridge, arriving in the middle of a Sunday cocktail party and offering to be "cook-butler or paying guests." As Avis recalled it in a short unpublished remembrance of Julia, Bernard was trying to finish a book and wanted no part of Avis's plan to have guests for a week or more. "Benny wasn't very happy about strangers. But he went up to Julia and he said, 'What will you have to drink?' and she looked <u>down</u> at him, because she's a very tall woman, and said, 'Well, I think I'll have one of those martinis I've been reading about [in *Harper's*].' Julia drank two or three without turning a hair, although she must have felt them. Bernard admired that enormously, so his attitude toward the Childs' visit softened a great deal." During the visit, Julia and Avis cooked and marketed together, and in August the Childs returned for another short visit with the DeVotos and Dorothy de Santillana at Houghton.

When Paul was posted to Germany that fall, Julia's dependence on the commissary in the GI-based village of Plittersdorf outside Bonn led to an even greater exchange of letters with Avis about America's developing supermarket scene and Julia's uneasy feelings about living in what she considered to be little more than an American community on foreign soil. Julia soon decided to use her time there to adapt American ingredients—cans of fruits and vegetables and a complete line of frozen foods—to French recipes. Cooking for the first time on an electric rather than a gas stove, she indefatigably tested the recipes that Simca forwarded to her and tried to make progress on the book. Julia also took classes in German and gained fluency in the language while dealing with shopkeepers and service providers. Paul worked long hours arranging exhibits and trying to master German, but both activities proved to be difficult. His dislike of Germany and his continued frustration led to a series of minor health problems.

As always, Julia was his mainstay in their humdrum existence in this American enclave, where their neighbors washed their cars on Saturday mornings and followed sports events enthusiastically. She knew how to bridge the gap by inviting neighbors and associates to their modern apartment for experimental "casserole sautés" and by roasting a turkey for Paul's office Christmas party. She also used the lure of the countryside and the expectation of plain good food to plan a series of weekends in Holland and Belgium, with the prospect of a trip to Paris in the early spring of 1955. More than once, Paul wrote to his brother, Charles, in Pennsylvania that Julia was a "natural." She could converse with everyone from the ambassador to the local sausage maker with ease and good humor, and very few people could plan a trip with more enthusiasm than she.

They drove to Frankfurt for a modern art exhibit that Paul had arranged and that was attended by the director of the New York Museum of Modern Art and Nelson Rockefeller, whose mother, Abby Rockefeller, had helped found the museum. They spent three weeks in Berlin while Paul organized a clothing and textile show at the Berlin Fair, an annual international event. When Paul was occupied, Julia enjoyed the feeling of being in a large city again. She found Chinese restaurants in Berlin and shopped for well-made woolen clothes. She told Simca that it was best to order only German specialties—sausages, sauerkraut, smoked pork, and beer, which was always good. German wines, too, she said, were wonderful. She ended every letter with the conviction that there was simply no place like France to shop and enjoy food and wine.

As the months passed, Julia gained respect for the energy she saw everywhere in Germany. On January 16, 1955, she wrote to Avis:

> Now I am beginning to see why such importance has been put on Germany. It is incredible to realize to what an extent they were battered and flattened, yet one does not have the impression, now, of a war-torn country. They are building like mad, so things don't seem to have been devastated, somehow. One's impression is of immense vitality, vigor, bustling activity, prosperity. What a difference to my dear old France, where, "Poor me; France is through 3 wars in a generation and the flower of our youth killed." . . . Same happened here, of course, whoever's fault it was; but I certainly don't sense any "poor me" here. The trains are beautiful, the automobiles are sleek, trucks are solid and fine, "things" well-made, everything seems to work, buildings are solid, and there seems to be a great surge of creativity in the decorative arts, such as textile design, ceramics, and furniture (compared to the paucity of creative activity in France).

When Paul was summoned to Washington in April 1955, Julia retreated to Paris to spend three weeks cooking with Simca and Louisette. She initially thought that Paul's being called home was to recognize his work and give him a well-deserved promotion. It turned out to be the result of an investigation by the U.S. Information Agency's* Office of Security, which had assembled a dossier containing the names of his liberal friends, his possible connections with Communists, the books he read, and the suspicion that he was a "treasonous homosexual." His phone calls to Julia recounted the Kafkaesque questioning and his efforts to clear his name. Although he was successful, the episode left the Childs dispirited by the poisonous climate that the FBI director J. Edgar Hoover and Senator McCarthy had created. Before leaving the United States, Paul met with Ed-

---

* The U.S. Information Agency (USIA) was created in August 1953 to encompass earlier information agencies, including the U.S. Information Service (USIS), established during World War II. In many foreign countries, however, the agency was still known as the USIS. The names seem to be used interchangeably in Julia's and Avis's letters.

ward Steichen in New York to arrange for his photographic show The Family of Man to be viewed in Berlin, turning a depressing trip into a positive accomplishment.

Julia's three weeks of cooking with Simca and Louisette in Paris also proved profitable. They analyzed the difference between French and U.S. chickens and concentrated on various casserole-type dishes, such as coq au vin and *poulet farci au gros sel* (stuffed chicken roasted in salt), as well as roasted and sautéed chicken. Julia's concern that Simca was doing the lion's share of the teaching at L'École des Trois Gourmandes was also confirmed. Because Louisette had a difficult husband, two children, and a household to manage, she could teach only three hours a week and spend less than six hours on testing and research for the book. Julia and Simca were working at least forty hours a week on it. The "coauthor" designation in the Houghton Mifflin contract posed problems for Julia, who did not feel that Louisette was contributing her fair share.

The center of Julia's attention was "cookery-bookery," and she increasingly relied on Avis to investigate ingredients available in the United States and to proofread her emerging chapters. Although the chapter on sauces had been written much earlier, she concentrated on revisions and bringing it up-to-date, and her letters to Avis reflect her detailed work on the poultry chapter. The manuscript for just those two chapters was well over five hundred pages . . . and counting.

Pittsfield, MA

JULY [N.D.], 1954

Dear Avis:

It was nice to get your letter when I arrived, and made me feel somehow more at home in this still somewhat foreign land. But we are gradually getting acclimated. It is amazing to me how much has changed in only 3 years. Everything seems bigger, shinier, faster, and I barely recognize anything I used to know. We've been hopping about considerably since

Washington: Bridgehampton LI, New Haven, Avon, Annandale on the Hudson where I left P, and now Pittsfield. Have had no memorable food so far, except a foot long hot dog decorated with sauerkraut, and very good it was. Nor have I set finger to knife except once, so I hope you need lots of cooking done as I am getting the old itch.

There is no need going on, as we shall see you so soon. My, I hope Mark can persuade Paul to show him a few [judo] tricks . . . He will never show me any as he says it ain't the tricks but the whole technique. But if Mark will flirt with me in French, I shall be flattered. I need practice in both. I'm glad May Sarton is around, as Paul is very fond of her. I thought she was planning to be in Europe.

Thank heaven you don't live in Pittsfield. It is loathsomely dull and we have too many old aunts here. It has always depressed me. New England at its worst, I have always thought. But I do love seeing my brother, and their four children, two of whom I have never seen before. They're cute and lively and affectionate.

I'm so dumb I can't think of who lives in Missoula.* Is it HST [Harry S. Truman]? But I won't say anything anyway, being in enemy territory. (No one is for McCarthy so far, thank heaven.)

We have decided to give up smoking on July 26th, so I am puffing away like mad until then. Smoke too much anyway, so it is probably just as well.

So much love until Sunday. We shall make a point of arriving around 5, so as not to burst in at the middle of the sacred hour.

It is not possible that we could hate each other on sight!

Julia

Just over the Continental Divide

JULY [N.D.], 1954

Dear Avis:

We miss you terribly. It doesn't seem at all possible that less than 2 weeks ago you were all of you but words on paper. It did not then seem that love on paper would not blossom into love in the flesh, and it certainly did with an all-embracing bang. That did it for good. We are but your creatures. Your generosity in putting us up, wining us, introducing us, talking, sitting, ice-coffee-ing, and just letting us share your life for a while

* Montana, where Bernard DeVoto was traveling for research.

was heaven. And we felt so at home with you. You were exactly, I think, though now I can't be sure, our "Dear Avis," perfectly familiar. We expected Mark to be something special, but had not envisaged the real thing in all its fascinating diversity, personality and gaiety. He's a remarkable boy, besides being a most lovable one. DeVoto himself is, to me, even more impressive in the flesh. Besides all that learning and all that devotion to his art and his country, there's that lovely and very human man. Well, it was all a most deeply pleasurable experience for us old expatriates and we rumble over the Western plains better fortified for our encounter with the enemy.* (I am determined to be more intellectual. I must not become so enraged that I spew gibberish. THINK, etc.)

We are riding on a pretty bright yellow train, the City of San Francisco. And there is practically no one on it. I guess they all fly nowadays. But we are having a nice time reading and looking out the window, and wondering how in the heck Lewis & Clark or anyone else ever got out here the way they did. After a day of giving up cigarettes I have decided what the hell, and am back on the old routine. I found that my smokeless yesterday didn't bother me at all. It just seemed, merely, silly to deny myself that pleasure. I think it is really the adult substitute for thumbsucking.

Have you run into the July 20th *Reporter,* with the article on Styles Bridges?† I wonder what effect that will have in New Hampshire. We also picked up a *Chicago Tribune,* to see how their new editorial page was working. It included one article from the *Post Dispatch,* under their heading "From the other side." One of their own editorials in a very heavily "ironic" style, suggested that a fine substitute for Cohn, one who would be kind to the witnesses, etc, etc, would be Alger Hiss.‡ Ha-ha. But so clumsily written that I bet a lot of their readers won't know it is all in jest.

May we please spend Thursday night, August 12th with you? We'll probably be coming in on the NYC train of 9:20 AM and would taxi to

* A tongue-in-cheek reference to her Republican family.

† With just a slender Republican majority in the Senate, Styles Bridges, a conservative Republican from New Hampshire who was the president pro tempore, led a group that warned Wyoming senator Lester Hunt, a Democrat, that unless he withdrew from the upcoming election, his son would be exposed as gay. Hunt withdrew in June 1954 and committed suicide shortly thereafter.

‡ A former State Department official who appeared before HUAC in 1948 and was accused of being a Soviet spy. Hiss denied the accusation but was charged with perjury and espionage. He was eventually convicted of perjury and imprisoned until 1954. The case remains controversial among scholars.

Berkeley Street. Our address, in case that would not work out with you (August 2 to 9):

Care of John McWilliams
1230 Hillside Road, Pasadena, 2, Calif.

Do get a good rest, and have fun.

A big kiss to Mark, and all love from us both,
Julia

JULY 31, 1954

My dears,
All I can say is, it's mutual. How lucky we all are. You might so easily have found things about the DeVotos that were quite unendurable, and, I suppose, so might we as regards you. Instead, all this fine love at first sight. God is good. I never had anybody in my house who was so completely effortless and easy, and whom I was so eager to see again. I will now never dare to reply in [a] friendly manner to any sort of fan letter, because this couldn't happen again. The best, of course, is that B. took to you so enthusiastically. He is a man of violent dislikes, he does not suffer fools gladly, when he is bored it sticks out all over him in great big knobs—frankly, I was worried. He never goes into ecstasies over anybody, except perhaps toothsome young virgins—getting to be a nasty old man—but there is no mistaking his whole-hearted acceptance of the Childs and more than willingness to see them again. If you lived here, every Sunday he would say, why don't you ask the Childs, we never see them. As he does every Sunday of his favorite people . . . Ah well, what could be sweeter?

Mark and I miss you terribly also. With everybody departing at once, we fell into a terrible depression, and the house seemed empty and echoing, and we were morose and without appetite. We miss the old man. We think it is fine for him to dash around the country, and we like the chance it gives us to sleep late and catch up with chores, and to see people he does not care for. But the fact remains that 8 Berkeley is damned dull without him. I haven't heard from him yet, but shouldn't be surprised if he phoned tonight. He should be getting into Missoula, and meeting you know who [Adlai Stevenson].* Wish I was there, in spite of its being

* DeVoto pressed the Democratic candidates to meet in the West to reconsider land and water use, and Stevenson agreed.

West. Wish I knew what happened at Grand Coulee.* No newspaper or radio comment as yet on the Resources piece.† Pity it coincides with the hoorah about Flanders-McCarthy.‡ Hope it doesn't get lost in the shuffle.

Enclosures, speech of our dear Louie Lyons§ (Curator of the Neiman Fellows). I send because I believe it will cast a great deal of light on the newspaper situation in this country. Want B. to see it too, but talked to Louie and he is sending me a couple more copies. I also offered to proofread the *Neiman Reports,*¶ they needing them badly, and was taken up. If there is time. They are usually left so late there is no time. But sloppy proofreading offends me.

Wondering how you are coping with family, and expect a blow by blow account when you return on August 12. Needless to say, we await you with open arms. If just for one night, will probably not ask anybody in, but selfishly keep you to myself so I can hear all the gossip. Anne Barrett and her Elizabeth over for dinner last night, and she reported only rave notices on Childs from everybody at HM. Of course.

A ghastly day, meteorologically speaking. Over ninety, and humidity maximum. I damn near keeled over at the Square** when I went down to buy the minimum food for the week-end, but recovered enough on my return to make some oeufs en gelée, with Mr. S. S. Pierce's consume†† how do you spell that. Bought several varieties to try out. We needn't be tied to Campbells. Did not do any great thinking about this composition, but they seem to be turning out fine. Too horrid to stir out again, as have closed the house up tight, turned on all cooling appliances, and am writing to sweet music from the record player. Could be worse. Next weekend Mark to Martha's Vineyard, and me to Annisquam. Find I could not endure this house alone.

Certainly, have seen the *Reporter* on Bridges, and understand that it has rent the state [New Hampshire] end to end—normal sale of *Reporter* in Manchester 24, but they sold about three thousand, and newspapers

* The largest hydroelectric dam in the country, on the Columbia River in Washington State. The event Avis is referring to is unclear but might be related to the plan to set aside 32,000 acres as the Columbia National Wildlife Refuge, or perhaps to her husband's visit to the dam.

† Bernard DeVoto wrote more than forty articles on land use and conflicts over natural resources, many appearing in his "Easy Chair" column.

‡ Ralph Flanders, a Republican senator from Vermont, introduced a motion in the summer of 1954 to censure McCarthy, which later passed.

§ Louis M. Lyons, a journalist who broadcast the nightly news on WGBH-FM.

¶ A journalism review published quarterly by the Nieman Foundation.

** Harvard Square.

†† Consommé.

reprinting chunks of it. I still think lots of Republicans are bitter about what is going on, but not so bitter that they won't vote Republican again. Anxious to hear what you gather about the California situation.

Mark starts tutoring in Algebra Monday, and I have hired a young man [Peter Koestenbaum] we all fell hard for. Twenty-five, married, wife expecting—German Jewish refugee who grew up in South America. Getting Ph.D. in philosophy, has taught German, French, Latin—knows Spanish and Greek, top flight mathematician and plays piano, violin, clarinet. All this sounds a bit oppressive, but he is also delightfully full of bounce and energy, healthily enthusiastic about people and sports and so on. I think he can handle that awfully verbal child without any difficulty, and may very well convince him that hard work can be deeply satisfactory, which is what Mark most needs to learn.

Here I am chattering on as if I didn't expect to see you for another seven years. Perfectly silly. Hurry back. We are dying to see you again.

Fondest love to you both, from both of us,

Avis

SUNDAY, AUGUST 20, 1954

Dear Julia:

Tuesday we are going up to Williamstown to stay for a couple days at the Inn—just the two of us—leaving Mark with his organ teacher, and may they not leave my kitchen in the same mess as last time. B. just wants to set and look for a spell, being dog-weary. The western trip was wildly successful, but too much in too short a time. He never learns. Me, I like to travel very, very lethargically only moving when absolutely necessary. We are now catching up with the mail, and I will write you in intervals of copying letters.

B. very low in his mind about politics. Thinks the administration has shown definite signs this last month of learning the ropes. Terrible blows to us on losing [Lester] Hunt, Moody,* and Johnson† (Colorado). Thinks public too apathetic about conservation, and the power issue, to amount to much in the way of votes. Why on earth can't some Republicans die in office? Got the same, other night, from Schlesinger Jr. Both talking about Eisenhower running in '56, and probably winning. Very depressing.

---

* Blair Moody, a former Democratic senator from Michigan, who died while campaigning for the Senate in 1954.

† The Democratic senator Edwin C. Johnson, who frequently aligned himself with Republican interests.

Did your elegant roast chicken yesterday, being in a cooking mood for the first time in weeks. Did it from memory of watching you in Maine, and halfway through remembered that you'd sent me a recipe. By that time I'd browned the chicken top of stove, so went ahead anyway without basting. Did this in the morning, almost completing the roasting, and let cool. Hadn't expected company, but some turned up—Beaumont Newhall, curator of Eastman museum,* and wife who is an art historian—they both do a lot of work with Ansel Adams.† So asked them to dinner in full confidence of having a good meal, and completed the roasting of the chicken by basting with cream. Terribly good, and I had tripled the chicken liver–cream cheese stuffing and it made a great hit. Wonderful party recipe. Thanks so very much.

. . .

I have put Friday the tenth down in my book. Only thing that could possibly prevent our meeting would be wholly unexpected return of G. K. DeVoto from Korea, in which case we would drop everything. But that seems unlikely. I will be seeing all the HM people at the Brooks' daughter's wedding on the first—wouldn't it be nice for Dorothy to have us out to Beverly, assuming she's back in the pink palazzo by then, as she ought to be? Won't be entirely happy until you see it. You are going to miss Giorgio in any event—word is he is staying in Italy for several more months.

Mabel Souvaine of *Woman's Day*‡ was here yesterday, and after her conference with B., I talked to her about you and your book. Told her about your south of France soups, and so on. She thought it sounded wonderful, and wants you to go to see her when you are in New York. They have an enormous circulation, way up in the millions—also plenty of money. You may want to do a few articles for them—anyway, it's great fun seeing their kitchens. Mabel said something about having you do a demonstration in their kitchen. After all, it's a service magazine, a "how-to" magazine. I think it would be good promotion for the book. The most surprising people write for it, and read it. You'll have to look at a few copies when you get here. B. thinks Mabel is the best editor in the women's magazine field, by a long shot. If you think well of this idea,

* George Eastman House, the world's oldest photography museum, located in Rochester, New York.

† Ansel Adams, the photographer famous for his pictures of the West, and a close friend of Bernard DeVoto.

‡ Bernard DeVoto wrote many articles for *Woman's Day*, both under his own name and under a pseudonym, Cady Hewes.

Even while vacationing at the shore, Avis devoted time to editing Julia's first draft chapters on poultry and sauces.

you ought to make an appointment with her a few days in advance. I don't know how long you plan to stay in New York. Or whether, indeed, this particular magazine appeals to you at all. Would be interested to hear the reaction of the Charleys.* Mabel has a pair of your knives—wildly enthusiastic—and goes abroad every year so she knows about French cooking.

B. brought a nice story back, about the first earth-man to get to Mars.

* Paul Child's brother, Charles, and his wife.

The Martians greeted him in [a] friendly spirit and took him around and showed him all the sights, which didn't impress him especially until they arrived at an enormous factory, acres and acres of it all full of dazzling machinery. This really fetched him and he had to admit we didn't have anything so impressive on earth. But, he said, he didn't quite understand what they were making here. They said they were making babies. This really floored him, and he said it certainly wasn't the way they made babies on Earth. So then how did they make babies on Earth? "I'll show you," he said, catching sight of a luscious blonde working at a machine, and he proceeded to do so. "Well what do you know?" said the Martians. "That's the way we make automobiles."

Thank you very much, the weather is lovely here, and we are all busy as bird-dogs. Mark spent all yesterday afternoon inside the organ at the Harvard Club, and has reviewed, successfully, a whole year's algebra in about three weeks. He is growing visibly, his voice is changing rapidly and he enjoys playing tricks with it, he can't button any of his shirts and I will have to spend an ungodly amount of money getting him outfitted for school.

Fondest love to you all, and hurry back. Let me know if you want any arrangements made, publishing or otherwise.

We aim to please,
Avis

. . .

Bernard*

AUGUST 23, 1954

Dear Avis:

Nice to hear from you, as ever, and I hope you have had a good time in Williamsburg [Williamstown] (a place I would never think of going as it is too near Pittsfield, but must be nice if one has no emotional fetishes in the neighborhood).

I return herewith the clipping slandering my hero. Ugh. We all retched over it, and boiled. This clipping, and the *Time* article are awfully good examples of "how to lie with words," so everything sounds logical and powerful to people who don't know the facts.† My, partisan writing

* Bernard, Maine, near Acadia National Park on Mount Desert Island, where Charles and Paul Child built a summer cabin and vacationed.

† Avis's earlier letter implies that DeVoto was attacked for his position advocating government control of public lands to conserve it against powerful timber interests.

is so dangerous a weapon. Why wouldn't a discussion of these techniques make a good Easy Chair . . . Seems to me B. has done it before, but I think it needs doing again and again and again.

That's wonderful that Friday the tenth will be OK with you. We shall spend Thursday night in Cambridge with people whom we knew well in OSS/Ceylon during the war. One is Cora Du Bois, an anthropologist, who has just been made head of the Anthropology Dept. at Radcliffe (not sure of her title or the details, but it is something like that). She is a wonderful old girl, with a remarkable mind. She was in the State Dept., but I imagine was advised to get out of it on account of you know who. And she spent a whole year on a little island, some 15 yrs ago, in the Netherlands Indies giving the natives Rohrshach tests, etc, and published her findings in a 12 lb. book called *The People of Alor.* Gregory [Rochlin] probably knows about her.

Friday morning, I thought I would see if we could have an appointment with Dorothy at HM, and see Lovell Thompson to talk over book layout and illustration . . . to get some final plan of attack which we can follow out for the rest of the book. Will ask Dorothy if we may take her out to lunch, and would you could you please join us? I shall write her a little noteski suggesting same.

I wonder if that Jean Carter* is the same as Ernestine Carter whose husband used to be English agent for Harpers pub, or Atlantic,† and who now lives in Washington as she is aide to someone in the British Embassy. If that's the one, and might well be, she wrote a charming little cookbook, and does a food column for an English newspaper or mag . . . and we did meet them in Cambridge at the Bicknells.

Wish you'd been here for the big boat arrival last week. 8 of them, dead tired from sailing in fog-bound Nova Scotia, and having almost cracked up in a violent storm the day before, and one of the 8, wife of an extremely nice chap, turns out to be a real alcoholic, brandy before breakfast and all that. She was feeling up all the men, biting the women, lurching about drunkenly in close quarters, and being hopelessly unbearable. She was still in a drunken state when they arrived over here, and we all had to be nice to her because everyone is so fond of her husband. But why has she turned into an alcoholic, all the Drs. Child were trying to

* Obscure reference from Avis's earlier letter.
† Julia is referring to the publishing houses Harper & Brothers and Atlantic Monthly Press.

figure out. She's gotten so bad you can't get through to her at all, seems to like her state of perpetual miasma, is happy in it, etc. etc. 3 children. God. Rochlins should have been here[*] . . . But I wonder, as she is about 42, if it may not be too late and her brain has rotted.

Haven't done your bread recipe yet, as was waiting on the kneading. There has been a great deal of cooking going on. Last night had the W. Lippmanns[†] (Lamb, scalloped potatoes, fresh wax beans . . . simple but tasty). Tonight a rather dismal but formal 2nd cousin and wife and Boeuf Bourguignonne . . . and a boatload of 4 people just drove by saying they would call us up from SW Harbor,[‡] so maybe they'll be coming too. (FHB)[§] I like the Lippmanns.

Paul has suddenly shaved off his mustache, this morning. 3 people have not noticed yet that he did it, and it is now mid-afternoon. I think I like it very much off.

Would love to meet Mabel Souvaine. Charlie & Freddie think she is wonderful. We'll be in NY for 3 days, Wed. to Friday, week after we see you. I like the sound of *Woman's Day,* anyway, and anything with a mass market is certainly what we want to hit. But I don't think we should take time out from the book to write any articles "at this time," do you?

We loved your story about the Martians; and Jonnie[¶] is already practicing up on telling it. Can't believe Mark's voice is changing already, and it's only about 10 days since we saw him. Please give him our organic love (that doesn't sound too good, maybe not).

This is a choppy letter, as I've been up and down, fixing sauce, temperatures, etc. The boatload is coming, only 4 extra. Paul Sheeline, our nephew, and probably the Donald Moffats.

Much love to dear Benny, and thank heaven he is not flattened out by his enemies, the beasts and bitches. And a generous two-quart measure [of love] to yourself from us both.

Julia

Everybody sends love.

---

* A reference to Rochlin's being a psychiatrist.

† Walter Lippmann, the Pulitzer Prize–winning political columnist for the *New York Herald Tribune,* and his second wife, Helen.

‡ A town on Mount Desert Island.

§ An acronym for "Family Hold Back," meaning that the family should hold back on eating too much to ensure that the guests get enough.

¶ Jonathan Child, Charlie and Freddie's youngest child.

The Gralyn Hotel, 1745 N St., NW
Washington 6, DC

SEPTEMBER 21, 1954

Dear Avis:

End of an epoque, and we are settled in, very comfortably indeed. This is such a good little hotel. We are in a housekeeping apt. with bedroom and living room, kitchenette which includes icebox, stove and sink-with-pig, all for 10 bucks a day. Couldn't be better suited to our needs. While it is good to be on our own for a change, we are sad to have it all over. We both agree that our very happiest times were with you. There was, of course, the sheer excitement of meeting you at last. Then, we just loved being with you, and felt so at home, and treasured every moment of it. I have never known anyone so selfless and so generous, and so creatively kind as you are. And I only hope that some day we may have the good fortune to lay ourselves out for a DeVoto, because we love you all.

New York was fun. The Paramount is not a hotel I would recommend, though it is quite OK as a place to sleep. It is just a great big barn, adequately run, but with so little confidence in its clientele that they won't cash a check for you. But it is sort of fun being in that honkey-tonk district. We had a very nice half-hour with Carl Brandt,* who is a sweetie. He made appointments for Paul with *Esquire* and *Holiday,* and told Paul to keep in touch with him. He just loves you and Bennie. Paul saw [Louis F.] Mercier of *Holiday,* who liked his photos very much; and a chap on *Esquire* who also liked them. They said that if Paul thought up a photo story to send it on in. Very useful contacts. I think it is a question of Paul having the time to do a project . . . and he never had time in Marseille.

We spent two hours at *Woman's Day* with Mrs. [Glenna] McGinnis the Food Editor, and Miss Anthony, the Executive Editor, both very nice. We talked about knives, methods, etc, toured the kitchens, etc. They said to send in anything when we were ready. Anything for them would have to be good and simple and elementary. We saw Mabel Souvaine only for a moment, as she was wildly busy. Very attractive, we thought. Paul said she looks the very picture of a successful magazine editor. She said she was so anxious for an inspirational piece from B, she just must have it. She doesn't like our knives! Says they stain so badly they must be poisonous. However, in the interests of better cutting techniques, we went to the Bazaar Français and ordered two knives to be sent to the cooking dept.

---

* Bernard DeVoto's literary agent.

Bazaar Français seems to have been re-arranged. It looked very neat and professional. They have everything anyone could want, certainly. So there is no excuse for anyone saying that such and such a piece of equipment is not available, because it is, if they take the trouble to send for it. Most of their French knives have Dione Lucas's imprint stamped on them. What a girl!

RESTAURANT NOTES.

Manny Wolf's Chop House. Good tough New Yorky hard-guy atmosphere, but steaks have no flavor. Cross it off.

San Marino, 53rd & 3rd. Sort of second drawer. OK, but we don't come to NY for that.

Pierre, 53rd East. Loved this and thought food was very good indeed, and wine very good, and atmosphere just right.

Robert, 55 West. Very nice atmosphere, but over-breaded smelts and sour Chablis. Cross it off regretfully.

Sempione, 49 & 2nd. Good food, carefully cooked. Simple bistro type and not too expensive.

La Lune d'Argent, 52 & 2nd. Looks nice, but there is a smell of bad fat which comes, perhaps, from Pizza place next door. Overcooked chicken en cocotte. Too bad.

Brussels, E. 54. Wonderful, we thought, and entirely and authentically French atmosphere and food. Good menu, fine wine list. The real McCoy. ($10 Per person.)

Paul spent all yesterday poking about the "Agency,"* and finding that at least half the people he saw in July are now gone for good. But they did know who he was and why, and have him down for Bonn. It seems that the Info Program for Germany is centered in the "Amerika Houses"† of which there are some 30 or 40 in Western Germany. They have tentatively placed him in the slot of "Program and Planning Director," which sounds like hot stuff, but they say the situation is so fluid he won't know his job until he gets there. There are some 164 slots to fill, it appears. They do have a language course here, and we are going to start in on it for two hours a day next week. They will be a great help. We shall be sailing probably between Oct. 10th and 15th.

This about sums us up, except that we miss you three very much in-

---

* U.S. Information Agency (USIA).

† Amerika Hauses, a series of small U.S. cultural and information centers established after World War II.

deed. Innumerable kisses to Mark and much, much love and so many thanks to the both of you. Wish there were a chance of a DeVoto coming to Washington! How can we bear not to see you again for so long?

Julia

PS: I do think, bookwise, that our business manager/promoter has done wonderfully. I feel we have good personal relations with HM; have reached a good understanding with Lovell giving us the ideas we needed for general style and illustration. Now, get the bloody thing done, and no more nonsense. And I like that HM Co. I don't think I could bear it if we had to work with some slick operators. They, HM, represent the good, solid, ethical way of doing things that I am fighting for (land of free, home of brave, etc.). Thank you, dear Avis.

The Gralyn, 1745 N St., N.W.
Washington 6, DC

OCTOBER 1, 1954

Dear Avis:

I'm not having no time to write to my little old pen pal or anybody else lately. So busy over our Washington house, paperwork for Germany, and German. And it is real Washington hell-hole weather, which doesn't bother me nearly as much as it does Paul, who really suffers with it. But we do have a fan here, which helps very much.

We've now had four days of German, and can say, "The beer's no good; where is the airport; do you want some matches, too"; and can count up to twelve . . . all in reasonably good pronunciation, and at normal conversational speed. This is a marvelous language system, I must say, and I wish I'd learned French this way. We attended two very well-given and erudite lectures explaining the linguistic theory of this teaching-method, saying that any average child has a mastery of the structure of language by 5 and ½, and only then does he learn to read and write, and he doesn't get into grammar until far later. (Something is wrong with this filthy stinking little piece of junk called a typewriter . . . a Hermes. About on a par with the 1954 Chevrolet.) Anyway, no grammar or verbs until they come naturally, and by the time they come along, you presumably know them anyway. And I find I like German. It looked sickening to me at first . . . but I think it is the living system that makes it attractive. And our German teacher is charming, a middle-aged woman with so much niceness, and such a warm, loving smile, I want almost to cry when

I look at her. She reminds us so much of our dearest French friend, Hélène [Baltrusaitis], who is one of the loveliest natures we know, that it is startling. There comes another psychological pattern that I shall [have] to deal with. How can Germans, who are, as I know, monstrous people, be lovely people? Or are they not monsters. If not, who made people into lampshades and soap? What would WE have become had we been Germans. These questions I will have to know and will have to find out while there. I keep wondering if here the McCarthy's took over, if much the same thing couldn't happen as did happen in Germany, as the McC's are on a par with the Nazi butchers. Where would we be, etc. Well, I have to find that out.

We are sailing on the *United States* from NY, October 15th. We shall be in New York, at the Vanderbilt (reservation confirmed), as of Monday, October 11, after spending the weekend with Charlie and Freddie. (Leave here the 10th.) Why don't you come to NY for a day or two with us? There must be some things you ought to do, and how we'd both love to be there with you. We are planning to do nothing in particular, except paddle around. The only drawback we can think of, that it would be too miserable to say goodbye to you, but that can be borne, if it was borne at Cambridge.

How nice of you to have a copy of Stegner's book* for us, I was intending to buy it, as that is Paul's meat-type of book. I guess you'd better hold it until we get a mailing address.

I wonder how that buffalo meat is, doesn't appeal to me very much, it is more the idea that I like. We had a lot of water-buffalo in India and Ceylon, and it was like very tough stuff. Maybe it needed marinating in red wine.

We'll ask Freddie about bulgur methods† when in Lumberville, will send her recipe, or Rachel's. No need to, now!

I haven't had much experience with little clay pots, or casseroles. But I have a friend, Rosie, who is a pottery maker, and she steeps them for a day in very strong tea. She says the tea sediment fills up the cracks. She does it for both glazed and unglazed things.

Would like to see both laminated block and Jap knife, both sounding very good. I should think the block, there where it is, would be just right. You really needed that. (Always good to have justification of some sort

---

* *Beyond the Hundredth Meridian* (1954).

† Bulgur, a coarse wheat, was beginning to be available in markets.

for kitchen expense. For a long while, I had the excuse of our cooking school.)

Very anxious to read B's new Easy Chair,[*] and the big *Holiday* one.[†] And the Alsop[‡] article. We've heard rumors of that here, but it hasn't yet hit the stands.

You asked how we felt now about the USA. I think, actually, that we are at last about adjusted. Our friend, Cora Du Bois, who has been doing work with exchange students and returning US types, was studying just that question, as it was a continuing problem, and everyone had it. One makes such an effort to integrate into a foreign country, that one becomes a bit foreign, and one usually has become well-adjusted to that new environment; so a new move is a wrench, a new series of re-adjustments, even if it is "home." Etc. Well, as anywhere, there are things one likes very much, and things one doesn't like. But what I do like about the USA is that one can find what one likes; and that big item, that the average person can make a decent living here . . . One of the tragic things about France is the terrible difficulty of making ends meet. And we're still a free country; and the whole McCarthy phenomenon has certainly illustrated that. It is far too complicated a subject to penetrate on this dripping hot day. But the answer is yes, I feel fine about America; and I think I can again present an enthusiastic, but fairly realistic picture . . . probably a truer one in some respects than the last time. You have helped very much, as people like you and B. are our hope, as I see it. God, what a picture I would have had from just California, however!

I like the sound of the Lanyis,[§] please remind us again of them next summer if you think of it. I certainly do like both the Rochlins. There is something very comfortable about them that appeals deeply to both of us, plus their good, active minds. It is wonderful to have a pair like that who fit in gloriously anywhere, and are always fun to be with any old time. Hope we shall hook up with even one pair of good pals like that in Bonn, but they take time to find, and to cultivate.

Your 30th Sept. has just come in, and I'm so glad you like Mrs. Beeton. I've never understood, either, why their cooking is so downright bad when she is the all-out household guide. Her principles are very French, and classical, so why. Oh why. I've never heard about not poaching eggs

* "Culture at Two Bits."
† "Wild West."
‡ Stewart Alsop, a columnist for the *New York Herald Tribune*.
§ George A. Lanyi, a Harvard tutor, and his wife, Susan.

that were less than 36 hrs. old; and don't know where to get a newly laid egg. Wonder if that's true, as I certainly never ran into it in any of my numerous French sources . . . *Larousse* doesn't mention it. Mme. St. Ange* (a remarkably careful source) says they should be just as fresh as eggs chosen for soft-boiling, NOT exceeding 3 days. Hell. Would love to read *Movable Feasts,*† and will return it to you from Germany. MFK Fisher book sounds fun. I shall get it, as she writes well, and must have quite a bit to say. I've been going over Marian Tracy's *Chicken Cookbook,* and find it quite carelessly written; and I feel she gives too long a time for cooking many of the chicken dishes . . .

We haven't run into any anti-German feeling, particularly, here. But I do keep running into anti-French. I think a lot of that comes from the alleged anti-American feeling in Europe, built up in our newspapers to make startling reading. I run into some people who took a two-weeks tour in France, and who say my, how much they noticed it. They didn't speak French, of course, but just sensed it. Very likely, having read about it, they were nervous, stand-offish, suspicious . . . so got what they gave. I keep remarking that after 6 yrs. in France, going every where, villages, hamlets, farmlands, etc. etc., talking, knowing, asking, chewing the fat . . . we never ran into any of it at all. But this has no effect, because THEY ran into it. But we love the French, and obviously showed our affection. (Can we do that in Germany? We can do <u>some</u>, if we can speak their language.) I know that these ANTI stories are mostly global generalizations, depending on who talks to whom and what the prejudices of each side are.

I, too, am depressed about Formosa.‡ How can the mainland Chinese feel safe when Formosa exists as an enemy, and we threaten to "unleash them." I was thinking the other day, also, reading an article about Chiang's son, who is the accepted successor and existing second man in the government . . . I hadn't realized that he lived for 16 years in Russia, and is thoroughly Russian trained. Is that not an anomaly, that a government violently supported by our evidently anti-Communist wing, should have an almost Russian in that position? Would we or the know-nothings, accept Dandy Dick§ as VP with such a history? Or is it, again, a

---

* The author of *La bonne cuisine de Madame E. Saint-Ange,* on the *cuisine bourgeoise* of the early twentieth century, published in 1927.

† *A Moveable Feast,* Hemingway's memoir of his time in Paris.

‡ Taiwan, where the U.S.-supported Chiang Kai-shek relocated his government after being defeated by Mao Zedong in 1949.

§ Richard Nixon.

question of the double standard . . . OK if the China Lobby* does it, but not for anyone else. I don't get it.

We're not eating anything particularly remarkable, not much room to cook much here, and only about 2 pots and a fry pan to cook on, but I'm going to do a duck tonight, despite the weather. And when we go out people seem to be afraid to cook anything for me (horrible reputation) except a barbecued steak. Little do they know.

Paul has suddenly developed an awful pain in the hip and leg and has barely been able to move the last two days. He had his usual reaction to his typhoid shot, of rash, fever, and swollen joints, and then this suddenly developed, and no telling whether or not it has any connection. He has just been to the doc, who said it could be osteomyelitis or TB of the bone or gout. X-rays showed it was neither of the first two, so it may be gout, which appears to be calcarious deposits on the something or other; and can be helped by some kind of drastic medicine developed in England. Presumably he will know this afternoon, when they telephone the results of some other tests. But it is hell for him, and he is depressed. But, better to have it happen here and now than in krautland. Certainly hope it can be fixed up quick.

Two more Pearson's† enclosed, and we'll clip them for you while here. This one on the Alaska oil fields. I think it would be more effective, for me anyway, if he pointed out what the alternatives were. I naturally assume that McKay and the Repub. Program of public lands is wrong; but what should be done about those oil lands. Why shouldn't some one develop them, they don't have to sit around there filled with oil and nobody drilling them, do they. In other words, I wouldn't have no ammunition when attacked by an oil baron.

Must make us some lunch . . . Wish we were in Cambridge with you, drinking iced coffee at the moment! Please think seriously about New York.

Most fond love to all,
Julia

It is now after lunch (canned shad roe and bacon, that is good, never would have thought of it but for you). P. is feeling better. I neglected

* A group of wealthy Chinese Nationalists whose goal was to influence U.S.-Chinese relations in favor of Chiang Kai-shek.

† Columns written by Drew Pearson, a well-known investigative journalist. The column described here criticized Secretary of the Interior Douglas McKay's plan to lease Alaskan oil fields to private oil companies.

to mention possible allergy reaction to the typhoid shots . . . for which Doc. also gave him some big red allergy pills. So maybe that's it. Weird. Doc just telephoned, too, saying urine, blood etc. all OK. He also advised never for Paul to take another typhoid shot, as some people are just allergic. Well, thank heaven.

I neglected to say that we are having a good time here. We're awfully busy during the day, as P. has to attend all sorts of talks, and meet types.* But we do have some very good friends here, like some tonight whom we are going to drop in on with a duck to cook. (I haven't tried a US one yet, and saw a nice duckling today.) Washington is fun. But P. dislikes the climate intensely . . . He just wilts in hot weather, and curses and sweats. He sends you his most fond love and says he wishes he had a pen pal, too.

3 Steubenring, Apt. 5
Plittersdorferaue
Bad Godesberg, Germany

OCTOBER 27, 1954

Dear Avis:
We feel as though we were in the moon, somehow. We arrived Sunday afternoon into Germany, had a fine lunch in a German restaurant, getting beer, meat and potatoes in our faultless German, then found our way to the American sector. It is a great big housing project, set in a sort of plain not too far from the Rhine, in the midst of a lot of German housing projects. They all consist of 6-family apartments, 4 stories, white stucco and red trim with brown tile roofs and radio antennas. We said to ourselves, Gawd, how did we ever get ourselves into this pickle. One of the buildings has been turned into a guest house, so we stayed there, feeling nightmarish, somehow. Next day Paul was taken into the big HICOG† headquarters, about 5 miles away up river. It is a vast rabbit warren, seven stories high, with half as many wings as the Pentagon, and a map for each visitor. I wandered, lonely like a cloud, about the housing development, then up a path along the Rhine, which was lovely. I thought I'd never see Paul again, but he appeared at about 3, saying we were to find an apartment immediately. So we looked at 9 apartments, each furnished almost exactly the same, except for one which had light wood rather than dark wood. So we took that. It is very much like a Statler hotel, perfectly

---

* State Department officials.
† High Commission for Occupied Germany.

good, solid, middle-class furniture. Plumbing is perfect, windows open and shut with ease, heat works like a blast furnace, ice box, stove, perfect electricity, perfectly comfortable. So we moved in yesterday and here we are. Perhaps, when our stuff comes up from Marseille, we shall be able to impose a bit of ourselves upon its rigidity. I have at least taken all the doors off all the cupboards in the kitchen, which has made me feel better.

The Information Operation is colossal here in Germany, they have a budget of some 9 or 12 million, I forget which (France had about 80 thousand). It is just done on a tremendous scale, and they appear to have some very high-powered personnel. Paul says he is scared to death. He is to be in charge of the Exhibits Program for the whole show, but hasn't an idea yet of what it all involves . . . but it appears to be plenty. He says he is going to be terribly hampered by not being able to speak German. Evidently the people here keep screaming to Washington, asking them to give people adequate language-training before sending them out, but to no avail. While a great many people do speak English, all the propaganda is naturally in German. So he will just have to struggle along, learning as he goes, and hoping that in a year he will be reasonably fluent . . . I suppose all these stupid lacks in training of foreign service personnel (not only in language, but in the political and sociological background of the various countries to which they are assigned) may eventually be solved, if we can ever develop some kind of serious national approach to international affairs . . . if we survive that long. I keep thinking of the know-nothing Republicans, and their contempt for foreigners, however. (Shall not pursue this line any further.)

Had a good time in Paris for two days. Rained all the time, but was lovely. Had a full day with Simca and Louisette. Simca has arranged a cooking class with herself, Louisette and 2 or 3 Americans once a week with two kinds of chefs. Very good idea, and they can experiment on our recipes. Louisette, who has never been a very active member of our trio, as her home-life does not permit her to put in a 40-hour week, is probably going to relinquish her full third of rights in the cookbook and will remain on a consultant basis. This is much more sensible; but haven't quite worked out the details.

Gee whizz, I can hardly put words to machine, I feel too weird and uprooted. The address we gave everyone to write to us to doesn't seem to be too good. They say we shall eventually receive our mail so addressed, but god knows when. So we haven't heard a word from anyone at all. Hoping everyone is still alive everywhere. Do, please, write us and tell how you all are. You seem very far away. Any word of Gordon? How is Benny's book, and what else. Has Mark begun to shave?

Your own Gefiltefisch,
Julia

Address for regular letters:

3 Steubenring, Apt 5
Plittersdorferaue
Bad Godesberg
Germany

APO address:

APO 80
Box 955
Care of P.M., New York, NY

NOVEMBER 1, 1954

Dear Julie:
It will take me a long time to get used to your new address. What a language. How I do sympathize with you both, new place, big job, no communication on account new language. Now you know how I felt in Paris. I suppose you will both have to work German lessons into your very full schedule. Before I forget, I have here in the house a complete set of tapes, plus books to go with, for the new system of learning French. These came from the Middlebury [College] French School, where, as you know, students get fired for using even a word of English—they were loaned to me by Mark's French teacher. He is wildly enthusiastic about the method, but says it is very hard to use in a classroom. However I think he uses as much of it as he can. Mark by the way is taking third year French, after only one year of it, and doing all right. I borrowed these because I thought I'd like to try them out with Mark, even if it only could be done half an hour a day. But woe, these tapes do not fit our recording machine, and I got an estimate and it would cost over sixty bucks to have them transferred to tapes to fit our machine. So I am going to return them, and keep an eye out for records.

Germany from your first few impressions sounds pretty grim, but I dare say by now the adaptable Childs are over their first shock and are beginning to seduce the Krauts. Cheer up. You are not too far from Paris, if the going gets too rough. Your apartment sounds horrible, frankly. Esp. electric stove, which I loathe. Hope your furniture arrives promptly, hope you get a cat, hope you find German wines attractive. Well, you adapt, or you die. And I never saw anybody adaptabler than you two.

Blush when I recall how I tore out of Pierre's, saying farewell to nobody.* I had plumb forgotten the time, until Charley [Child] yelled down the table telling me I had better step on it. I was a bit tight, and having such a good time. Still, no manners, no aplomb. And turned out I had plenty of time, as no traffic at that hour, and hung around LaGuardia in blazing heat, oh dear. It was so wonderful seeing you and having that last two days. But I wish I'd said goodbye properly, however confusing and disrupting with so many at the table.

My child is home from Korea, and out of the Army. He got to Camp Kilmer about ten days ago, and they have a policy that if there is less than three months to serve, they give them their discharges. So he called up and surprised the life out of us. B. was in New York, en route to the Great Smokies for a Parks Board meeting and then Washington. So Bunny† spent Saturday with him and then came home, and has been here ever since.

And I have been so sad, so distressed, so worried that I hardly know how to talk about it, which is why I haven't written. He is a very strange and I think a very sick boy, though in all fairness I must say that he seems to have survived the army and didn't get into any trouble. We have failed him terribly, and he still has a very long way to go before he gets on his feet and is able to make a life for himself. If he ever does.

. . .

We've consulted with Gregory [Rochlin] about all this, and he has known Bunny for years and understands his problems pretty well, and he says leave it all to the VA.‡ Bunny has had God's quantity of psychiatric help, ever since he was about eight or nine—five perfectly good psychiatrists have worked with him for varying lengths of time. No soap. They couldn't reach him. He fails in school. He fails at jobs. He gets fired. He can't live with us. He can't live without us. He is unable to do anything except his own peculiar way, and can't and won't follow instructions, from us or anyone else. What the psychiatrists call "a hostile depen-

---

* Avis had gone to New York City to say farewell to the Childs at a luncheon at the Pierre Hotel, then had to dash out to make her flight back to Boston.

† Childhood nickname for her son Gordon.

‡ Veterans Administration.

dent." He just breaks my heart, and I can't eat, or sleep without dope, and I feel as if I'm carrying around a thirty-pound weight in my stomach. I think Benny is too, but he puts up a good front.

This kid is good at a few things. Words, usage, grammar, spelling. He's better than any of us at that. So perhaps a job as proof-reader? But he maddens people he works with. He knows more about guns than a lot of experts, can take them apart, repair them, and so on. Job with a gunsmith? Again, the getting along with people. He's mad about racing cars—so adolescent. He knows all about how they work, their engines and so on. But he's no good with his hands (except for guns) and though he would love to work with car design, so perfectly certain that he couldn't get the engineering degree. He's talking about finishing up his interrupted second year at Boston University, this is the school of General Education. He has gone in this afternoon to see whether they will take him back, and I think they probably will, and should. But I just don't see him sticking with the work that is necessary to finish out the year. I am not such a fool as to even hint this to him. But I know his work habits, all too well. He is also talking about going to the Eastman School of Photography at Rochester. He bought a Retina [camera] in Japan and is mad about it. His pictures are on the way so I haven't seen them. But you and I know how hard it is to earn a living by photography, and he has nothing whatever of the artist in him. He's talking about technical photography, medical stuff and so on. Well, maybe. Again, I wonder if he would stick through the course. His pattern, all his life, when confronted with a job, is gradually to retreat into his own private life of listening to records, reading cheap fiction, etc. Doing crossword puzzles. My God. And any of our attempts to rouse him out of this passivity is worse than useless.

. . .

I don't feel desperate. I just feel sad. I don't even blame us, B. and me, for what we did to him. It's easy to feel guilty. But actually, we did the best we could, all the time. We really did. I wasn't a very good parent, being far too snarled up in myself until I got analyzed. Another child would have survived this. Bunny just happened to be much more in need of a feeling of security than most children, and he didn't get it because I didn't grow up until several years after he was born.

Well, I shouldn't unburden all this on you, except that I haven't spoken or written freely to anyone about it, and it has been piling up. If I don't seem to be reaching out in all directions for the next few years, you'll know why. And don't worry about us. This is a risk of parenthood, and it is some faint consolation to me to observe what other parents we know are going through with their own young. I wonder how I would feel if he were a homosexual?

I just talked to Mabel Souvaine over the telephone, and she says she loves her knives, and keeps them in her New York apartment so her Stamford cook won't ruin them. The staining bothers her, but apparently she hasn't formed the habit of wiping them off immediately after using them. She asked about you and I told her. She thought you were absolutely charming, and hopes to get some articles when you get around to it.

The photographs were here when I returned from New York, and though I look like hell in them, I am having the big one of Paul and me framed for the wall, because it's the only one of dear Paul I have and I do so appreciate the look in his eye, the old goat. I'm also delighted with the Louisburg Square* picture, and so is everybody I show it to. I will now look at the Square with fresh eyes. Has anything further been heard about the proposed Chicago show? I would almost go out to see it. Except I don't see how I can go anywhere at all until Bunny is set in a job, or in school. That's my first responsibility, for as long as it is necessary.

I think Jim Beard's fish cookbook† is very good, and I am going to try some of the recipes when I can. Bunny wants steaks and chops and pasties and spaghetti and lime pies, all of which I duly provide. I haven't even got round to the madeleines yet, though the pans are staring me in the eye. Give me a little advice about the proper sized pastry bag and tube for piping potatoes around a dish. I bought one at the hardware store today, but it is too small, appears to be for cakes and such. Bazaar Français catalogue doesn't seem to offer one big enough either. I want to master that heavenly shrimp and scallop thing I had at Pierre's, with other fish too. How do you get the delicate brown on the piped potatoes? Egg yolk?

Tomorrow is Election Day. Rochlins and Kennedys [Sargent and Elizabeth] coming over to sit it out with us on television and radio, suitably fortified with drink. B. was in Washington and had dinner with Elmer Davis—he is not in good shape, one leg very bad and his speech somewhat thickened,‡ though he is pulling no punches over the radio. Also spent an afternoon with the Dem. Nat. Committee, and sat in on

---

* In Boston's Beacon Hill section.
† *James Beard's Fish Cookery.*
‡ Davis suffered a series of strokes in the 1950s.

Mitchell's press conference,[*] which he says M. handled very well. Phil[†] still wouldn't predict — says he does all his predicting the day after. Everybody else says news from around the country is so good it scares them. They are counting on about 30 House seats, three to four Senate seats. Definitely counting New York (everybody says Harriman is in, the Ives-Dewey vilification campaign has boomeranged),[‡] New Jersey, Kentucky, Wyoming, Illinois. They are not counting Oregon but I definitely am — all that work, all that money I raised, he's[§] got to win. Nixon has been absolutely incredible. You just wouldn't believe the depths he has sunk to. Adlai made the most wonderful speech on TV Saturday night. He was never better in his life. Very angry, very stern, pulled no punches but never hit below the belt. He took out after Eisenhower in no uncertain terms. Likewise Nixon, and Brownell,[¶] and Dewey. Everything he said was perfectly true. We hung on his words, and cheered at the end. He was just plain magnificent. The papers have carried very little of it. Nixon says he is going to answer it tonight. I can't wait.

Massachusetts is such a mess, politically, that I just can't vote for a Democratic Senator or Governor. You must know how hard this is on me. I was going to hold my nose and vote a straight Democratic ticket, for the first time in my life. Jack Kennedy turned his back on [Foster] Furcolo about three weeks ago — there has been bad blood between them since they were in the House together, and Furcolo demanded that Kennedy go on the air with him and denounce Saltonstall, which Kennedy refused to do.[**] Saltonstall is yellow — on the McCarthy issue, and so are all the other candidates — but he has after all worked very well with Kennedy on purely Massachusetts issues. Now it appears that Furcolo, thinking that Saltonstall was going to ask McCarthy into the state to speak for him, which turned out to be a misapprehension, asked Cohn to speak for him. Cohn refused, saying he wouldn't offend McCarthy. This was printed in *Newsweek*. Arthur Schlesinger printed it in his *NY Post* column. It has been widely talked about. It has not been denied by Mr. Furcolo.

---

* Stephen A. Mitchell, the chairman of the Democratic National Committee, predicted that the Democrats would trounce the Republicans in the 1954 House and Senate races.

† Philip L. Graham, the publisher of the *Washington Post*.

‡ In the race to succeed the New York governor Thomas Dewey, Democrat W. Averell Harriman triumphed over the Republican senator Irving M. Ives despite a smear campaign by Dewey and Ives.

§ Avis is referring to the Democrat Richard L. Neuberger, who did win the Senate election in Oregon.

¶ Herbert Brownell, the U.S. attorney general.

** Senator Leverett Saltonstall, the Republican incumbent who was running against Furcolo, was a friend of Kennedy's. Furcolo lost the election.

None of the four candidates for Senator and Governor has even peeped on the McCarthy issue—too afraid of losing Irish Catholic votes. The hell with them all. Furcolo is a liar and a hypocrite—at various times he has put himself down as a Baptist and a Presbyterian—now he's a Catholic. He went on a radio program with our friend Rod Macleish* and denied that he'd ever called the ADA a bunch of traitors, when Rod had the transcript of the speech. I just can't swallow him. B. says in that case I have to vote for Saltonstall, no matter how I feel about him, because it is necessary to defeat Furcolo and that's the only way my vote will count. I can't do it. I'm going to vote Farmer-Labor,† and hope there will be enough votes for the fringe parties to slap both the major parties. Herter‡ has been a fairly good governor, so I guess I have to vote for him, because Murphy is a feeb and it would be just the same old State House gang again if he got in. This is just killing me. And maybe when I get to the polls tomorrow I will have changed my mind. But if anybody is feeling happy about the political climate in Massachusetts, I haven't met him.

How do you like the enclosed Herblock§ my ma sent me from St. Petersburg?

What does Beard mean about the <u>intestines</u> of the lobster, mentioned in several recipes? Not the tomalley, or coral, or stomach, which he differentiates clearly—so leaves only the intestinal vein, far as I can see, and what on earth would that accomplish? What does Escoffier mean page 49 on Mayonnaise—put into a bowl the yolks of six raw eggs, after having removed the <u>cores</u>. Suppose he means that little colorless knot that clings to the yolk after you have dripped away the white? Curious locution. I have time to read cookbooks these days, since I am not secretarying—Barbara [Bernard's new secretary] is turning out quite well—and am not gadding about at all. I find them soothing, even if I do not have much of an appetite. I have planted two trees—a fifteen foot sourwood, a six foot silver bell. I am going to rip out all the old lilacs and put in beautiful new ones, double dark-purple and double white. I planted about a

---

* Roderick MacLeish, the nephew of Archibald MacLeish and news director at WBZ radio in Boston.

† A party representing labor and agrarian interests that existed on local levels, later absorbed into the Democratic Party.

‡ Christian A. Herter, the Republican governor of Massachusetts, who defeated the Democrat Robert F. Murphy, the majority leader of the Massachusetts House.

§ The Pulitzer Prize–winning political cartoonist, who lampooned both McCarthy and Nixon. The letter's date suggests that Avis is probably referring to his caricature of Nixon, who was shown climbing out of the sewer at a campaign rally.

million bulbs and it has been so warm that lots of them are up, but we are going to have snow in a day or two, or at least a frost. I must stop, and I send you both my very fondest love. I am so glad you came and conquered. I will send a couple books in a day or two.

Hugs and kisses. Everybody sends love, esp. Mark. He is in wonderful shape—not shaving but any day now.

Avis

Thanks also very much for pâtés and pepper grinders, which I certainly can use. Will give one grinder to the Rochlins as a remembrance from you, as they only have one and asked me to find a really good extra one. Don't forget, I'll send anything you want or need.

NOVEMBER 10, 1954

Dear Avis:
We're just back from Paris for the day, and then off to Berlin this afternoon for 2 days, arriving back Saturday in the early morning in time to receive all our household goods from Marseille. What a life! But at last we shall be settled.

Thank heaven you are feeling better about things. It does seem to do a lot of good to blow off to some one. I suppose it is the effect of externalizing, rather than letting things stew around in ones gastric juices. I do hope the thirty pound ball has left your stomach! I am sure that inner stewing is hell on everything inside one, making the glands function awry . . . Well, of course it is. Wonderful that Gordon really wants help. I guess it is only then that anything can be done. And wonderful that Mark knows all, and is so willing to help. He is a very perceptive guy.

I certainly wish we had been with you during the election. We feel we have quite missed out on the low-down. And it just happened that we fell in between newspapers. Our *NY Times* is two days late here, one day late in Paris; and there the *Trib.* comes on the day. So we somehow missed the important days when things were summed up. Hell. But I want to know what Benny thinks about everything, including the future, and how the Dems should act now, etc. etc. We ran into the Walter Lippmanns, who were staying at our little hotel in Paris. (We keep running into them, it seems.) She is a good Democrat, and was delighted about everything. Didn't have a chance to probe him, as he was deeply concerned with French politics, and we saw them only for a drink. They are both pro–Mendès-France.* There seems to be a Washington clique who is against him (M-F), which worries both Mrs. L and me. This clique insists that M-F has Communist leanings, or Communist types surrounding him. I had fun, myself, during our stay, asking all the French people we saw, which wasn't more than a dozen, what they thought of M-F. About as many answers, naturally, as people questioned. Like asking Pegler, Knowland, Truman and DeVoto what they each thought of [Dean] Acheson.† One friend of ours felt there was some suspicion of Communist sympathy; as he is a Jew, and for some queer reason many Jewish intellectuals have been pro-Commie, or sympathetic. Another, a fairly old but canny man, just shrugged his shoulders in that French way. No one else mentioned it. So I know little more than I did before, except that there are some French who might feel there is something in the Washington clique's feeling. It also showed me how very little I know about anything, particularly in the matter of evaluating my sources, as French political thought is so damned complicated.

Now you will have two potato squeezers, as I stopped by my dear old hotel supply store and sent you two bags and 4 nozzles, all for the sum of a buck. Recipe for "Pommes Duchesse" enclosed. I haven't done any experimenting on these yet, but have a working recipe.

Tomato Purée. There does seem to be confusion about this. And I have been aware of it for some time myself. I think, though I shall check with Simca before we make our final decision that we shall adopt three [*sic*] different terms:

> French Tomato Purée, made from fresh tomatoes.
> Tomato paste, from a can, which can substitute for the above.
> Tomato Sauce, made by oneself, and that does contain a thickening.
> Tomato Fondue, made from fresh tomatoes, and in which the pulp is only cooked until the juice has evaporated.

I think that in various books Tomato Purée means Tomato Paste. The canned tomato sauce is only a rather miserable substitute for the good

* Pierre Mendès-France, the French prime minister, who ended French involvement in Indochina by negotiating an armistice with the Vietnamese Communist leader Ho Chi Minh in 1954.
† Westbrook Pegler was a conservative columnist. William F. Knowland, the Republican Senate majority leader, tried to appeal to both Democrats and Republicans. The four men referenced would have had very different views of Truman's secretary of state.

stuff you can make yourself. I noticed, too, making a Tomato Sauce the other night, using green peppers, onions and a bit of bacon, that, before it had simmered for about ½ hour, it had a very American taste; but that after it had simmered and blended, it had a really French taste.

Had a wonderful two days with Simca, and feel we have really renewed contact. She is an exceptionally fine old girl, and a real workhorse. We spent some time drafting an agreement concerning Louisette who, sweet as she is, is just unable and incapable of doing any serious work, and never has. She is the complete amateur. But she is a good natural promoter, and she does have nice little peripheral ideas. I'll send you a copy of the agreement, once we have agreed on it. Simca and I, after many months of jockeying around, have finally reached what seems to be a good manner of working together. I keep forgetting that in the European tradition women are not used to taking on authority themselves. So it has taken a little practice for Simca to realize that she does not necessarily have to believe and take for TRUTH all that she reads from the great masters; that she can make some findings herself, which are perfectly valid. She is also inclined to let me be the authority; which is not as it should be. And we both are schooling ourselves in the scientific method. So I am very encouraged about our work, though we certainly have a long way to go before the end. And I, myself, am going to have, at last, peace and time. Paul doesn't come home for lunch, and I shall have almost the whole day to work in, except for marketing and my hour-a-day German. Thank heaven! (Husbands take a lot of attention.)

Best love, ever, and we keep thinking of you continually, all of you,

Julia

NOVEMBER 19, 1954

Dear Julie:

I've got a terrible restless fit, so though it is 8:30 and dinner is out of the way, I can't settle down with a book. I want to talk, and I ought to dig myself out of this empty house and go talk to someone, but got no car. Benny is due home tomorrow night and so is somewhere between here and New Orleans. Mark has gone off to a dinner party and a dance, looking most elegant in his new charcoal gray flannel, black shoes, restrained tie and all that. Gordon is home on a week-end pass from the [Veterans'] hospital and has gone off to dig up his friends. It's the night before the Yale game and the whole place is bulging with visitors and thank God I have nothing to do with the whole business except to go to an after-game cocktail party at the Katz's.[*] And it's foggy and dripping and tomorrow it will pour. If you were only here we could talk all night.

I wrote to Saltonstall this afternoon and the effort left me very shaky. He is definitely playing footsie with Nixon and [William] Jenner and Dirksen[†] and that scum, object is to water down the McCarthy censure business. So I told him I hadn't voted for him, and I hadn't voted for Furcolo, because I couldn't vote for a politician who refused to come out against McCarthy. I told him I was sick to the depths of my soul over his lack of moral courage. I pointed out that he is now safe for six years, and what has he got to lose by denouncing McCarthy. And how can he live with his conscience if he continues to evade and weasel. And more of the same. Perfectly dignified, and not too long, and not a bit hysterical. But very stern and biting. I hope. And then I sent twenty-five dollars to the Liberal Citizens of Massachusetts so they can write to a lot more people and urge them to try to stiffen Saltonstall's spine. If he's got one, which I doubt.

McCarthy's elbow would be funny, if it weren't so cynical, and if it weren't for the fact that he will probably get away with it.[‡] Wayne Morse[§] pointed out that he got himself to the Senate for one vote, in a wheelchair and with his jaw wired after being thrown from a horse. I think the whole censure business is going to fritter out in postponements and evasions and Republican gutlessness. I suppose you saw that list of military, naval and other lights who are out to get ten million signatures saying Joe is saving us from Communism. Jesus. I want to burn down the White House. I am sick, sick, sick. Hope to God when B. gets home he can cheer me up a bit about all this. Though how he can cheer me up after Knowland's[¶] speech, and the Davies firing,[**] God only knows.

Change of subject before I throw up.

Dec. issue of *Woman's Day,* just received, has a quite good article

* Milton Katz, a professor of international law at Harvard who had directed the administration of the Marshall Plan, and his wife, Vivian, the DeVotos' next-door neighbors.

† Everett Dirkson, a Republican senator from Illinois and a sometimes reluctant ally of McCarthy.

‡ During the hearings, McCarthy entered the hospital with an elbow injury, and the Senate recessed for ten days until he could be present.

§ The maverick Oregon senator, who had earlier switched from the Republican to the Democratic Party, returned to the Senate to vote for censure.

¶ The Senate majority leader called the resolution to censure McCarthy "unfair and unprecedented."

** John Paton Davies Jr., a State Department China expert and pragmatist who advocated America's backing of Mao Zedong rather than Chiang Kai-shek, was fired by John Foster Dulles after McCarthy attacked him.

about cooking potatoes in 48 different ways, all French. By Miss Olga Keverkoff, who it says is a Paris magazine editor. Ever heard of her? In the ordinary course of events I would rip this article out and send it to you, but this time I expect to use it, and there is no place nearby where I can buy another copy of the mag. So I wrote to Mabel and said hooray, because I am so tired about 108 ways to do hamburg with margarine and second-hand eggs, and all those fancy desserts, gah. And I told her to send the magazine to you regularly at your new address, and to be absolutely sure you got this issue. I suppose it will go to you ship not air, but let me know in a reasonable time if you aren't getting it. Wonder how much they paid for this article?

Took Mark to Durgin Park* the other night, since we were alone, and it was really great fun. I had lobster sautéed in butter, very good indeed, and he had swordfish as usual. But enormous helpings. Across the table from us were two guys who looked exactly like Mormon missionaries, you know that awfully good, awfully dull look, but they turned out to be from Iowa, and they had never been East before but they had heard about D and P. They had roast beef and it was a whole rib apiece, weighing it looked like three pounds, very rare and wonderful. They both turned quite green and just nibbled around the edges. Mark had Indian pudding and I had apple pandowdy, warm and terribly good really. A lovely old guy in a market apron and a battered straw hat, stuttering like mad, took us all over the kitchens which are upstairs over the restaurant. Into all the various cold rooms—such cuts of meat you never saw. Poked our noses into the stock kettles, watched the deep-frying, and the man opening clams and oysters invited me to help myself and I found I had room for half dozen of each. Really the only way to eat them. They hang and cut meat the way it used to be done in my youth. From the blackened and moldy object down to a steak you could cut with a fork before it was cooked. We must go there en famille when we are all together again.

That Véfour† menu [you sent] nearly killed me. All that lovely food. And not too expensive. Ah me. Sent you one from Durgin Park—ah the difference.

The VA is really at work on Gordon at last. There was bound to be a lot of delay and red tape at first. You get snarled in the organization

* A Boston institution noted for traditional New England fare.

† Le Grand Véfour, a renowned Parisian restaurant established during the reign of Louis XVI.

and just have to let it roll over you. This is the first pass he has had, but we have talked on the phone every day and I spent Saturday afternoon with him. It's a brand new hospital, a vast and luxurious place. He has had a whole series of psychological tests and is now finished with them and I suppose they are being evaluated before he starts seeing Dr. Murphy regularly. His skin is being worked on, green soap scrubs, sulphur lotion that smells like rotten eggs, anti-biotics by mouth, and Monday he starts ultra-violet treatments. A nurse assists with all this so he does it right. I think he looks some better already, and he seems very determined to get to the cause of it all. And he has had his teeth x-rayed, so I suppose they will make him some new ones soon. He likes his fellow-patients, he has taken quite a shine to one of the nurses, and as they have a fine library he is doing some reading. So, we shall see. I am hopeful.

Dorothy has asked us to Beverly Sunday, to meet Stanley and Laura,* but I hate to buck all that traffic for a cocktail party, particularly when my child has to be back at the hospital at nine o'clock. So I think we will by-pass that one, though I'm regretful. She is sailing Tuesday, I think. I wrote down your new address and sent it in to her, just in case you hadn't, so that she can get in touch with you when she arrives. And if and when she does, tell me all about it.

I am nothing but domestic this year, so except for a few cocktail parties I haven't been anywhere or seen anybody. I have been working like a dog in the yard, rooting out tired old bushes which I will start replacing in the spring. I am beginning to evolve a basic plan, which in my mind's eye is a dream of beauty. In one corner three white lilacs with a blue tree wisteria in front of them, and so on and so on. Then I pick things out in the catalogues and make lists and when I add it all up of course I come down to earth with a bump. However, as I mean to do it all myself, except for planting trees, perhaps by the next time you come I will have reached the half-way mark on my ambitious design. All I need is energy, which I have plenty of right now. When it stops drizzling, I'm going to whitewash that brick wall under the library windows, and that is going to be so enticing, with the brick color bleeding through, that I will just damn well have to plant a lot of roses and delphinium and whatnot, won't I? But it will take years to grow things to screen off the garage, and the back fence, and the dear Katz's, and give us a private green place that smells heavenly. And I hope that Mark gets over his present violent urge for violent reds

* Dorothy's son and his wife.

and oranges, because I want it mostly blue and purple and white. Well it is all very good for the figure and the complexion, and I am bursting with rude health.

There isn't any low-down on the election. I never saw the pundits floundering around so. The really rotten right-wing Republicans seem more obnoxious and more vocal than ever. I cannot think why Dulles felt he had to throw Davies to the wolves right now. He was in a very good position to put his foot down and show some courage. But no, the appeasement goes on. Well, watch for DeVoto's post-election Easy Chair, which I think will delight you. And tell Paul to watch for the Nov. 19 issue of *U.S. News and World Report*—yes, you heard me—which has a very good interview with Wayne Morse. Also a restrained and reasonably fair piece about [John Carter] Vincent, Davies etc. Reasonable for them. I read this sheet regularly to find out how the ultra-conservatives are thinking. They certainly were unable to manipulate Morse. He is going to be the next object of my financial affections, and I am warning all my friends that in two years I will be collecting to get him re-elected. My political stock is very high these days and it is the strangest feeling to be thanked for having wrenched ten bucks from somebody. And B. of course didn't think that Neuberger had a chance. But I did, except late Tuesday night when I thought all was lost.

Very interesting what you say about Lippmann, and Mendès-France. To tell you the truth I haven't been following his progress about our great and fair country too carefully, on account of being ensnarled in McCarthy again. How I wish he would die, so that I could think about something else. And of course I feel just as fierce about Nixon. A document has just come in for B., from a Montana friend that I wish I could send you. If I get another copy, I will. It is an expensive brochure trying to prove that Murray* worked hand in glove with the Communists, and it is a lulu. Everything violently wrenched out of context, important dates missing and so on. Unbelievable. The Nixon line, of course. Well, it didn't work.

I think I wrote you that Herb Scheinberg† reported to B. that Kennedy was dying of cancer. I'm sure I didn't write it to anyone else, so it must have been you. Now I've got another report. And I don't know which side is up. Dean Clark, who is executive head of the Mass. General, wrote to Kennedy's surgeon (this at my request), and called up the

* Paul Murray, an aide to the Secretary of the Army Robert Stevens, who was accused by McCarthy of covering up spy activities in the army.

† I. Herbert Scheinberg was Bernard DeVoto's personal physician and friend.

other day and read me Dr. Wilson's reply. And Wilson says Kennedy has not got cancer, that there isn't a word of truth in all the rumors, that he operated on Kennedy for a ruptured disk, that he is making [a] good recovery, and will be out of the hospital in three or four weeks. Dean Clark says loudly and cheerfully that the rumors just are false. He's a terrific Democrat. So, what? Who is right? Time will have to tell. Meanwhile most of the Irish in Mass. say Kennedy is hiding in the hospital, afraid to come out because he repudiated Furcolo. Or, that he is hiding in the hospital so he won't have to vote on the McCarthy censure. It's maddening, the whole business. And of course if he does die, it's a tie vote in the Senate with [Vice President] Nixon holding the balance of power. Which is scarcely to be endured.

Thank you darling for the potato squeezers though of course they haven't arrived yet. Mine is more damn fun and I have an intense desire to decorate everything. Will have to practice restraint. I am astonished at the number of egg yolks in your Pommes Duchesse. Should think it would be mostly egg, and bright yellow. But will try, just the same, when I have enough people to cook for to make it worth while. But what happens to your fish or sweetbreads or whatever, masked in a sauce, if you have to brown your potato border at such high heat?

I'll say there is confusion about tomato puree, and I am awaiting a decision from on high.

Poor Louisette sounds just like me. A peripheral character. And of course I think you are quite right drawing up a new agreement, and I hope all can be arranged so you and Simca, and Louisette on the edges, get what is coming to you in the way of kudos and cash, and no feelings hurt. I am so glad you are arranging things to get solid slabs of work done, with no interruptions except dashes here and there. What does Paul eat on his lonely lunches? Is there nothing to be had in Deutschland except sauerkraut and sausage? I never heard anything so depressing.

My goodness, it is ten o'clock. Nothing like correspondence to make the evening fly away. I'm for bed, to await Mark's giddy return at eleven thirty.

Much, much love,
Avis

DECEMBER 8, 1954

Dear Avis:

I've got just an hour before I've got to get to my German study, and hope to read the papers as well . . . report in the *Trib.* we got today about McCarthy's speech denouncing Eisenhower. Lord, what a weird situation that is in the GOP. Although we all have seen it right along, none of the "Young" Republicans I happened to talk to this summer seemed to feel there was a split at all . . . Look at the S[Southern] Democrats, they would say.* They were just plain Freudian about it. I did read that slush, David Lawrence,† who seemed to have disappeared from the *Trib.* for a good month, and I thought they had turned over a new leaf. My, what a loathsome creature McCarthy is right down into the smudge between his toes. It is all fascinating, and horrifying and frightening, too. I thought a recent editorial in the *NY Times,* "Knowlandism" was extremely pertinent in describing the right-wing mind, as being guided purely by emotional reactions, and being riddled with such glaring inconsistencies as only the childlike right-wing mind could tolerate. And I was very much moved by Walter Lippmann's "Bonds of Affection" piece, on the "necessity of maintaining those mystic chords of memory which makes it possible for men to be free, and to differ, and yet to be one people" . . . all of which the right-wingers are tearing down. A fine article, and one that I should like to [have] graven on my image, somewhere in a permanent place. (Except I cannot regard the Republicans as people, somehow, only as monsters, fools, beasts, and foul excrement. Must I turn a new leaf, or another cheek?) Well. I wish Benny would write a column. Why doesn't he? I, for one, would very much like to know what he thinks about this and that, and not only once a month, but two or three times a week. Have no direct word on where my old Pop stands, but have a pretty good idea! Right down the far end of the farthest right line, straight as a bee or an Arbuthnot.‡

German. It ain't easy. Now I am in adjectives, and do they have the same ending as the other words in each of the four cases for each of the four genders? No. And do they have one ending for one case and gender when preceded by all words? No. Or do they have another set of endings when used alone, but referring to a noun in another place? Yes. However

* The southern Democrats were more conservative than their northern counterparts.

† A conservative political columnist for the *New York Herald Tribune.*

‡ John Arbuthnot, a Scottish-born eighteenth-century satirist, was a Tory, unlike his brothers, who were Jacobites.

I impressed the nice woman in the post office twice, after having carefully practiced two sentences each time, each containing two datives, a possessive genitive, and 4 scattered adjectives. And I also called up a German store on the telephone and ordered 2 dozen oysters, which impressed me immensely. Suddenly my eyes are going bad, I can hardly see to thread a needle, and I can no longer see clearly to read at 8 inches. WOE to the 12th power that this should happen to me. But, taking courage in palm, I telephoned a German oculist and have made an appointment for tomorrow afternoon (I believe!).

Our life is little but work and German. Paul gets home at six, does some violent exercise, takes a shower, and studies. I finish my desk work at 7, we have dinner by 8; an hour or so of frittering, like now, and then back to the German. We have met one or two people that we may like, which is nice. But have done powerfully little in the social way. I have been doing quite a few ducks lately, so we have had various people over to eat them, and have a pair coming tomorrow for a pheasant. Saturday we go up to Dortmund, which is about an hour or so away, for the opening of a big American art show at the museum there. Mrs. Conant* is going to give a talk on art; it appears she is an art historian? First time we shall have seen her; and we all go up in their private train. Everyone here seems to think she is a fine woman. She takes her job seriously, and is very popular. I doubt if I shall meet her, as I am just going along for the ride with another wife/baggage type. But I look forward to getting a look at her.

We'll be very much interested to hear how you liked our friend, Cora. We are, as you know, extremely fond of her, and admire her immensely. We got to know her awfully well when we were in Ceylon, and she was marvelous, in a very top-notch position, the only woman in a swarm of men and brass, to hold any executive position at that level. And she got along famously, and was much liked and admired by everyone. I think she has had a very tough time, emotionally, and has been much hurt; but has managed to survive and blossom; and to adjust herself very well to her special kind of life.† The fact that she was chosen for this evidently much-coveted chair is a great thing for her psychologically; as well as being a real recognition of her worth, which is considerable. We both feel that

* Grace T. Conant, who had studied art history at Boston's Museum of Fine Arts. Her husband, James B. Conant, formerly the president of Harvard, was serving as high commissioner to West Germany.

† Cora Du Bois was a lesbian.

Julia making bread in the Child family log cabin on Mount Desert Island, Maine.

people in those devious sociological positions very much need the support and affection of those not so afflicted . . . And god knows, with any of those problems, they could be me, hair lip, beard, club foot, or whatever. Paul loves an evening with her, as she has such a clear, well-stocked and professional mind. She has a wonderful sense of humor, too.

All your garden plans sound fine; though knowing nothing about gardening at all, I cannot rise to manure with such enthusiasm as you. However I do remember with pleasure all the beautiful vegetables raised in China because they used human manure, which seems to be the best of all. Have you ever tried that, I wonder? Now, if you were talking about the arrangements of a cow's stomach, and which are the best parts for tripe, I would rise to the bait like a flash. (Fine picture of same in *Webster,* under ruminant.)

Speaking of *Webster,* under Egg, is a drawing which will explain the "core" you were asking about. I finally thought of looking up "core" in the French Escoffier, and in French it is *Germe*. Germ is the Cicatricle, figure H in the drawing (*Webster's* recent *Collegiate*), though it looks as though it were in the middle of the yolk. Those white knobs are those ropey things, figure F. I suppose if you were being fancy you could strain the egg yolk through a fine sieve, and get rid of both. Phooey, but interesting.

POMMES DUCHESSES. Excuse me for not answering your original question. Yes, browning the duchesses for 4 to 5 minutes under the broiler would be just right for also gratinéeing the cream sauced stuff inside the border. I think that the consistency of the potatoes is dependent on what you want to do with them. If you want to make separate little patties on a pastry sheet, squeezed out from the bag, you must have them so they will hold their shape enough for them to be removed with a spatula, so you can place them around a roast, or whatever. However, if you put them in a big pile altogether in one platter, they need much less stiffening element. This would also go for a border, as the border doesn't have to be moved once it is browned. After quite a number of experiments, I have concluded, that for patties, the correct proportions as follows:

Per 1 cup cooked mashed potatoes:
1⅓ TB butter
1½ yolks (the US "large" egg, 1 TB per yolk)
1 TB cream (or milk, which is less rich!)
salt, pepper, etc.

I found 1 yolk per cup was not quite enough for patties, but would be sufficient for borders, etc. I don't find it too eggy; but I don't find it as good as plain, mashed, yolkless, creamy potatoes.

For browning: Sprinkled a bit of grated Swiss cheese on top, and dotted with butter. Can stand that way for hours, ready to pop into broiler.

Idea from Simca, making pancakes of Pommes Duchess, patted into shape on a floured board (disks, really), about 3 inches in diameter and ⅜ inches high. In middle of one, place any kind of a filling, such as mushroom duxelles mixed with a bit of Swiss cheese, lay a thin slice of ham on top of that, then moisten the outside edge and place on top the other potato disk and press the edges down. Paint with melted butter, and put in oven or under broiler to brown. She said it was very good.

CELERY RAVE is Celeriac, celery root, those big bulby things. I've seen them in Washington, on our last trip there. Maybe in some foreign markets in Cambridge. KOHL-RABI, according to André Simon's *Encyclopedia of Gastronomy,* and a most useful reference book, is a hybrid of the cabbage family which must be grown very quickly to be fit to eat. Its base stem is swollen and nutty in flavor and tender. Quite popular on the Continent, but not in England or the US. I ain't never seen this baby, but will keep an eye out.

I am surprised your potato soufflé didn't turn out, as I thought all of *Woman's Day* recipes were foolproof. But I do seem to remember that the oven thermostat on your stove is off . . . so maybe it was never at 350. How much potatoes for ¼ lb. of butter? Certainly isn't much more than a Duchess plus egg-whites, but seems like lots of butter unless for over 2 lb. potatoes.

MUSHROOMS. I don't think you get your butter and your pan hot enough before you start in . . . Remember doing those scallops, how hot the pan had to get so the juice wouldn't run out? If juice runs out, you have failed, actually. As water content evaporates, they begin to brown, but should do so only lightly, so as not to get a burny taste.

This little business with Louisette is turning out to be something of a problem. My fault, as I don't see why we should have all the responsibility and do all the work, and she come along for the ride. And poor old Simca has been the work horse all these years. After having discussed it with Louisette in Paris, and having said I would send her a letter, I have now heard nothing from her at all, after our most carefully thought-out letter, batted back and forth between Simca and me. As she puts in probably less than 4 or 5 hours of work a week, and has produced just about nothing except a long chapter on game, mostly copied directly out of a book, I suggested that she consider herself Consultant, rather than Co-Author, and then she would have no twinges of conscience that S & I were putting in at least 40 hrs; and that we have a 10-45-45 split. A bald statement of the facts, I must say; and I really don't think she knows that she is not working, as she doesn't know how. Beside, the book has gotten beyond her anyway. What she should really do is to write up smart, chic little recipes for *Vogue,* or something, that would be charming, and you would give them to your cook and let her worry about the details. Too bad, too bad. And we need her as Consultant, as she does have some very nice little ideas, and gentle suggestions, and she does have the Fed. of Women's Clubs. So I have written her another letter, asking her please to come across with her opinions, and saying that we were dopes not to have thought out this problem when we first began (we thought we were thinking of everything, of course). I also said that we should, when we had reached a decision, write up an agreement that would take care of the incapacity and/or demise of any or all of us, and have it legalized. And I also waved the big stick of saying if we could not reach an amicable agreement we had perhaps better just cancel our contract with HM (did I put my foot in it, I don't know, but have sent copies of all of this to our NY lawyer). Both Paul and Simca think she will not answer at all, and will wait until she gets to NY for Christmas and talk with HER lawyer. Her husband, a big canny bull of a man, self-made, and rather crude, has perhaps told her to put nothing in writing. My, the problems of tri-partite cookery. I can't make much out of our HM contract, as legal lingo throws me. But, as it says on it that the ms. should be submitted on or before June 1954, I should think it might be invalid now anyway, as it has never been renewed . . . and I don't want it renewed until we get the book finished; it don't seem fitting.

Heavens I must get to my German.

How is everything? Are you doing your detective reviewing? How is the new secretary shaping up as office mother?

Much love from us both,
Julia

JANUARY 5, 1955

Dear Julie:
Coincidence. Your first draft was waiting for me when I got home from the Square this morning, bringing with me a four pound roaster to try out in the rotisserie. If I had read it before I stuffed the chicken, I probably wouldn't have stuffed it, but I might at that. Stuffed or not, I would have run into trouble with your rotisserie cooking times. The <u>book</u> that came with the machine advised an hour to an hour and a quarter, medium heat, for roasting a four pound bird, not mentioning stuffing. No mention either of starting on high heat. My machine (the very latest, Roto-Broiler model 400) has three heats—high, medium, low. So I thought in my dim mind, Julia says page 15, in general estimate 20 to 30 minutes more than

oven roasting time, so I added on half an hour to an hour and a quarter, and thought maybe little extra for stuffing—and allowed two hours, thinking probably done some time before that but could turn down to low heat, and even let sit a bit while cooking the carrots Vichy. But I started it on high, for fifteen minutes, as per your instructions for fast beginning. I made that chicken liver and cream cheese stuffing, quite dry on purpose so wouldn't ooze out. And I trussed very well indeed, and sewed up vent tight, and put six strips blanched bacon (tied over breast).

Fun watching the spit turn, watching the skin brown. But after about an hour and a quarter noticed stuffing beginning to ooze out, and in another fifteen minutes it was all out, the rear walls of the vent had collapsed and the damn thing was falling apart. This on medium, with fifteen minutes of high. So I don't see how you can recommend more time for rotisserie. They must vary enormously as to actual heat given out. I did think last week when I did that steak that it took longer than it would have under the gas broiler. It was a second cut of the rump, about one and three quarters inches thick. Took about twenty-five minutes and was fairly rare—did it close up to the electric element, high heat. Seems to me meat timing is the very devil.

The chicken looked rather revolting—I leave it to your imagination as to effect when stuffing kept rolling out. Stuffing then took on quite a lot of drippings, so it wasn't very good, though it didn't burn. I scooped it all up and put it back in the cavity. Carving—breast was moist enough, since I didn't remove the bacon until about ten minutes before I took the whole business out, in imminent danger of collapse. Leg bones came right out in my fingers. But nothing dried out. Skin quite nice. But didn't salt it until I took it out of the machine. I keep burning myself on that beast—will get some padded gloves tomorrow. Just not used to it yet. And the timer is useless—can only tell from look and feel. If I ever again stuff a chicken for the rotisserie, will use it quite dry, and besides usual trussing, will tie cavity very firmly to legs after the spit has gone through. The side skin just gave way.

I will finish this tomorrow, when I have had a chance to read through your ms. again. Things that occurred to me today, on first hasty reading. Page 5. I have a strong impression that very few American butchers these days know how to pull tendons, but I will check with Tom tomorrow. Page 5 also, where you say lower leg (have lower leg left on) maybe you clear this up in illustrations, but don't you mean ankle or whatever? Except for that confusion, heartily approve all your notes about tendons, trimmings, etc. Not at all sure they will appeal to a limited audience only. Very valuable, I say. Will try out the tendon pulling in kitchen drawer next time—sounds easy and fascinating.

For me—don't use "Frenchie" in the book. OK in letters, or conversation—not somehow right in a book. Don't know why I find this offensive in print, but I do.

Page 10. Cooking time usually based on Dressed Weight! Well! Now does this mean in Child? Or in Mrs. Joy, Fanny Farmer, any of the tomes that give tables? I am astonished. Remember my confusion at Christmas, wondering what weight to time on—butcher's scales, my own stuffed or unstuffed. Think this should be made quite clear. If it's the way you are going to recommend timing in <u>your</u> book, say so.

Page 12. So that's why drippings in Christmas pan got so dark—too big pan. Had to have it that long, because bird was long—but lots of waste spaces on the sides. Next time, too, will use a mixture of fats, to prevent so much darkening.

Page 16. "Light but optional whiff of garlic" doesn't seem to make sense. If it's optional and they leave it out, it ain't a light whiff.

I very much like your résumé system, then going on in considerable detail. Hope everybody likes it as much as I do. Gives you a nice general idea, and then the details seem logical and easy. None of these will look as long in print, anyway. Yes I do think the length is necessary, and any serious cook will find it so, and thank you for it. Nothing more maddening and frustrating than to tackle say a *Gourmet* recipe, and find time stretching out because they skipped so lightly over things. Makes you feel feeble-minded and no good at your job. Also delighted to have you get down to brass tacks on quality of poultry, freezing, bad storage care and so on. High time somebody said all this.

May have more to say when I re-read. Oh gosh I wish Paul were here. Quite apart from his obvious charms, think he would be able to help me understand that damn machine. Also having trouble with my wonderful cutting block, laminations on which are separating. Find this is fairly common trouble. Had carpenter fill cracks with glue, and sandpaper. On his advice, am now treating daily with hot olive oil, well rubbed in. This to be followed, when Dickson gets me some, with many applications of beeswax. But if it doesn't work, will order a new one next year, and this time have carpenter run long steel rod through laminations, with tightening device. Then I'd have something permanent.

The night Henry[*] left we took our dinner on the spur of the moment over to the Walter Edmonds, who wanted H. to see their really superb kitchen. I was making Blanquette de Veau and had enough for five, and it turned out beautifully. Did everything but the sauce here, and we carried it over in the pots, looking very silly no doubt but it was dark. Still we seemed to meet an awful lot of people in one short block, under the street lights. Had about three and a half pounds leg of veal, when it was all trimmed up, and I must say I do think the dish is much better with good cut of veal instead of breast or whatever. The pieces cook almost tender in about forty-five minutes, and sitting in the stock keeps them moist. Thank goodness I have licked this one, and everybody very enthusiastic. Edmonds donated two fine bottles of Romanée Contée,[†] from their boundless cellar. Also did your marinated pork for Henry and he was wild about it. B. also very enthusiastic about this one.

More.
Avis

JANUARY 13, 1955

Julie:
Bits and pieces this time. Gotta get dressed pretty soon and go to a cocktail party. B. in N.Y. Snapshots from Xmas and New Years . . . You recognize Helen [Rochlin]—the other gal in both pictures is our English friend Mollie Brazier[‡]—very sorry you didn't meet her. New Years' Eve she arrived first, to find me buzzing round the kitchen. She rarely talks about her work, because I suppose she knows none of us know anything about it. But that night she sort of exploded. That very day she had got final results on a piece of research she's been doing at MIT—large amounts of dough furnished by MIT, on project that was pure hunch of hers—and mathematics done by Norbert Weiner.[§] Something to do with brainwaves, which is her specialty, and Gregory [Rochlin] who comprehends a bit of it says it's so important he wouldn't be surprised if she got the Nobel Prize. Quite a payoff, after all these years. She is divorced from

---

* Henry Dart Reck, formerly Bernard DeVoto's researcher and a close family friend.

† Domaine de la Romanée-Conti, an estate in Burgundy, which produces some of the world's finest wines.

‡ Mary A. B. "Mollie" Brazier, a world-renowned neuroscientist who worked at Massachusetts General Hospital, Harvard, MIT, and UCLA.

§ Norbert Wiener, an esteemed mathematician who received a Ph.D. from Harvard at age eighteen and taught math at MIT.

an Englishman who is a great authority on cables and stuff, did all the barrage balloons during the war. Has a son [Oliver] at Harvard, and I suppose is my closest woman friend. Probably because she finds professional women pretty hard to take, and I'm no threat to her and we can be females together. Strange, wonderful and aggressive woman she is. We always celebrate our birthdays together, she being four days older than I, and she and Oliver have spent Christmas with us since 1941. How odd if she does get the prize.

Louisette came to call last Saturday morning, and is she ever a cutie. She telephoned out of the blue the day before, catching me flat-footed as Gordon and a friend were to be here for the week-end and I always try to keep the decks clear so he can sleep as much as he wants to, and eat when he likes. She was pretty well booked up too, and only in these parts for two days, so she arranged to stop by on her way to Beverly. Too bad such a short call, as I could cheerfully have talked to her for hours. She asked over the phone who she could see at HM—could only tell her that Dorothy was abroad, and Lovell, only other person who knows anything about the book, probably unavailable at such short notice of a Friday afternoon. If we'd had more time, would have put the bee on him to take us both to lunch. However, we have at least met, and taken one another's measure, and liked what we saw, and now I know two of you. I am very glad to have met her, and touched that she took the trouble, for trouble it was in a strange neighborhood on a wretched cold day. She was brought by her hostess, a Mrs. Davenport—elderly and very charming. Louisette is sweet, and amusing, and warm, and I liked her thoroughly. B. didn't lay an eye on her, as it was morning and he was beating his typewriter and she got away before he knew she was in the house.

I have not yet fully digested your chicken material, and your stuffings. But I have talked to Tom Mahoney, my butcher, and to Sam Collins at Sages,* about chicken and turkey ages. Both agree substantially. Both say flexible breastbone test not much good—butchers in the old days, when women knew about it, sometimes broke the breastbone on purpose. Peculiar. Both test by digging fingers into upper part of breast, near wing-bone. If soft, a young chicken. Don't know if I could count on this, without practice. Both say practically no chance of getting old birds in a market. U.S. chicken business so much a beltline production, turnover so great, that birds go to market when young. None of the old business of

* Sage's quality grocery store in Harvard Square.

running into a rooster. Broilers go to market six to seven months—roasting chickens seven to eight months—a fowl will be a year or more. Both say, to my surprise, very little frozen chicken sold in markets—except the regular frozen packaged stuff, of which they have as low an opinion as we have. Sam, an old man, takes a very dim view of icing chickens, even as you and I, says it spoils flavor. It's turkeys that are routinely frozen, I suppose it being more of a seasonable business. Tom says that's what he meant when he said this was approaching when I wouldn't be able to get a bird that wasn't frozen. Turkeys, not chickens. He says a twelve to fourteen-pound turkey would be about a year old, a five to ten pound turkey around nine months, a turkey broiler seven to eight months. All depends on the feeding, and you must have that dope from the Agriculture* bulletins. Just keep in mind though that it's really big business, and pretty well standardized. And huge turnover.

Both say the housewife-shopper is apt to be a dope. Both told gruesome tales of housewives or cooks pointing to a bird and saying I'll take that, then kicking because they roasted a fowl. Or, over and over again, a telephone order for "a fowl" when they really wanted a roaster or a fryer. Like the bride who ordered a leg of beef, Tom said, if they point and say I'll take that, I don't stop to ask what they want to do with it.

In both places I tried breastbones of a couple of roasting chickens and found the flexible breast-bone, all right. Tried fowls, too, quite different. But both men insist that softness of upper breast, near wing, is the way to tell, and I couldn't take up more time (theirs not mine) to find the difference between a dozen birds.

I don't know if this is any help whatever. However, yours to command, to the best of my poor ability.

Think I will try a goose this week-end, if I can get one. Friday evening, when B. will still be away, and Gordon home.

Knives now on sale. Prices—$1.25 for paring knife, $1.75 for fish filleter, don't know what on the enormous one yet. It's sheer weight of enormous ones that ran all-over prices up—Roy Dickson says they wound up paying nearly sixty bucks transportation charges. Awful. But these prices perfectly fair, I think, and actually slightly less than for similar sized American made stainless steel knives of good make, such as Flint and Robeson. The steels he is selling at $3.50 and are they ever worth it! If I weren't so damn broke this month I'd buy four or five and set them aside

* U.S. Department of Agriculture.

for wedding presents. Adore mine. Beautifully made and absolutely solid worth. Should think he would sell out the small knives almost at once, but it will be quite some while before the enormous ones are moved—he is very wisely circulating word of them through the clubs and restaurants, for chefs. And he talks about his next order, so he is planning to go on with it. Hooray.

Covered with beeswax—why didn't anybody tell me it was such damnable stuff to handle? Got a big block of it, melted some in a bottle over hot water, poured on back of my block—it hardens instantly into something you can't possibly rub or polish. Scraped most of it off, and will hunt up an expert before I go further. I am not a do-it-yourselfer, except over the stove. Maybe needs an electric polisher, which I haven't got, only the floor-waxer equipment. Think also it ought to be mixed with some kind of solvent, which is quite beyond me.

Your goose stuffing sounds wonderful, but I probably won't attempt it for only three. If I got a really small goose, might be fun to do it in the rotisserie except fat-overflow would be a problem. Take it off with the baster, maybe. Haven't done a goose in years and years. Slight slackening of my interest in own cooking, slight slackening of appetite also, and high time if I am to lose enough weight to get into summer clothes in Fla. That ten day trip is probably going to cost us 600 smacks, damn it. Can't share a room with Mark any more, too big and too curious. Full fare on train, car-hire down there, etc. Too far to drive from here, too exhausting with no relief driver. Too boring as well. Ah me. Fondest love to you both, cheers over the way you are turning out the work, and hope Paul's cold is long forgotten.

Avis

Plittersdorf

JANUARY 16, 1955

Dear Avis:

It is certainly about time I sat down to a good letter to my pen-pal. What with Christmas, New Years, ducks, geese, turkeys, pineapples and what not life has somehow been wild, and two trips included. I feel like a steam shovel, but have little to show for my diggings.

Christmas was a fizzle, as Paul was in bed with a bad cold; after having dragged around at the office all week he finally collapsed on The Day. So we did nothing and just treated it like another day. But for New Year's weekend we went up to Nuremberg to visit a chap who runs the Amerika Haus there (Amerika Hauses are US Information centers, 22 of them in [West] Germany, which have libraries, movies, lectures, book-mobiles, press sections, student-exchanges, etc. etc.). The ones near the East Zone have special rooms for East Zone tourists into the West, so they can be hidden from the populace, and soak up the news of the outside world, etc. I don't see how the East Zoners dare to go into them, but they do, hundreds per day in the Berlin house, for instance. This last weekend we visited two more houses, one in Hamburg, one in Bremen. These are the first trips in Germany we have taken, and Paul hopes to do a trip a month until he has covered all the houses, as that is part of his job as Exhibits Officer. And now we are having more of a sense of what Germany is like, for around here it is just small townery. And now I am beginning to see why such importance has been put on Germany. It is incredible to realize to what an extent they were battered and flattened, yet one does not have the impression, now, of a war-torn country. They are building like mad, so things don't seem to have been devastated, somehow. One's impression is of immense vitality, vigor, bustling activity, prosperity. What a difference to my dear old France, where, "Poor me; France is through 3 wars in a generation and the flower of our youth killed." . . . Same happened here, of course, whoever's fault it was; but I certainly don't sense any "poor me" here. The trains are beautiful, the automobiles are sleek, trucks are solid and fine, "things" well-made, everything seems to work, buildings are solid, and there seems to be a great surge of creativity in the decorative arts, such as textile design, ceramics, furniture (compared to the paucity of creative activity in France).

It is impressive, and thought-provoking. What's going to happen in the next few years, and they shall continue to grow and prosper and produce. Whom are they going to sell it all to? Where are they going to get their raw materials? Where are they going to expand? How long can they stay divided? And, the big question, what happens when Adenauer* dies, which can happen tomorrow. It is horrifying to think of them going over to the Russians.

We shall not have any valid judgments about Germany for months, or about Germans. I pick up everything people say, just as a matter of interest. A chap named McCormack,† of NBC, who has been here a year and has covered almost every foot of Germany at a snail's pace, just hates the

* Konrad Adenauer, the first chancellor of West Germany.
† Robert McCormick, an NBC news reporter.

Germans. He says he is ashamed of himself, but he does. He loathes the Ruhr industrialists who treat servants and underlings like dogs, etc. McDonald, of Nuremberg, doesn't like Germans either, and has been here 5 years. Of Hamburg, Bernard* says, it depends, and there are still a lot of the old guard who are rigid and Nazi in their points of view; but he has quite a few good friends, having been here 6 years. Somebody else says they are still half-savage, and Christianity and the Christian concept of people as individuals is only skin deep. Others say they are so used to being led, most of them, that they can not take any creative responsibility. Etc. Etc. I still have met no one who so loved Germany he wanted to live here forever. But I think all this is somewhat beside the point . . . which is, we've got to keep them on our side. So far we have found everyone to be extremely pleasant and friendly; but we have little but shopkeepers and people in Paul's office to go on. In sum, I am not unhappy here in Germany, and feel we are damned lucky to be here and see it and feel it for ourselves . . . and I'll be glad when we can get this damned language under our belts . . . so we can find out more things for ourselves.

I meant to tell you about a train ride to Dortmund (3 hrs away) we had with Mrs. Conant, in the Conant private train. I had supposed loads of people were going, but there were only about 10 of us, so I got a good opportunity to look at her carefully, listen, talk, and think, both going and coming. An interesting phenomenon, and my, what a pure-type Bostonian she is. She is very nice, natural and totally without pomposity, certainly. Earnest and hard-working, and works just as hard as her husband, I guess. Speaks German fluently and nicely, and is very friendly and natural with Germans. Looking carefully at that face, it still looks like that of an innocent student-government president, from Boston. Is it innocent? How canny and worldly is she? I will probably never know. It is said that she has warmed up considerably in their few years here. Anyway, I liked her, and admired the way she is going at this job. She seems to me to be wonderful for Germany, where things are on the serious side, and the rather plain and sturdy side too. A fancy pants would never do.

Wonder how your roto-broiler is working out?

I imagine you breathed a sigh when Congress convened with no Democratic holes! The GOP roster of bigwigs is awful. How could they put that horrible Bridges in a position such as he has . . . It is incredible. We

* McDonald and Bernard presumably were news reporters whom Julia and Paul met during their travels.

read with interest *Times* article on Morse and Neuberger.* We expected it to be one of their most snide, and it wasn't. Though Morse certainly came off better than Neuberger. How did it affect DeVoto and you?

Hope all goes swimmingly in Cambridge. How is Gordon doing?

Much love to you all,
Julia

FEBRUARY 1, 1955

Sweetie:
This is mainly to send you the enclosed, which I consider priceless and which I have shown far and wide. God how I do love Herblock. Both on paper and in person.

I'm in a bit of a rush—late afternoon and gotta cook presently. Our first real snow today and very welcome at least to me. The horticulturist got up in arms about the continued drought.

Do you get *Gourmet*? Occurs to me that if [Louis] Diat makes a book out of his present series, it is going to be your only competitor. Just finished reading him on the cutting of vegetables, consommé and marmite, etc. Very, very good. But he will certainly not produce your kind of book—he is being far chattier and more discursive, perfectly delightful of course but not nearly so basic and detailed. Fine to have both books together, I think. And in spite of him, I think you should go right ahead and tell how to cut up vegetables. It just cannot be said too often. He really is the only good thing in *Gourmet* which in general is a lot of hooey.

Also been thinking about something Louisette lighted on during the short time she was here. She wondered if Americans would bother to do cooking that meant getting every pot and pan in the kitchen dirty. Wish I'd had time to go into it with her. Because I am deeply convinced that it just is not necessary to let everything pile up to be washed. I suppose it is a sort of fixation of mine. I certainly had it drummed into me thoroughly by my old ma. And I wish you would write something about it. It is so easy to wash up as you go along—absolutely no soap needed. Everybody who reads your book will have a kitchen where the water is continually hot. All that is needed is plenty hot water coming out of the faucet, and a brush. The nylon ones stand up better, but ordinary Fuller Brush sink

* Wayne Morse, the senior senator from Oregon, and Richard Neuberger, the junior senator, had a long-standing personal feud that began in 1931 when Morse was dean of the University of Oregon law school and Neuberger was a student. The animosity was played out in a series of angry letters they exchanged while in the Senate.

brushes do very well. Finish with a pan, take ONE MINUTE to stick it under the hot water faucet and brush it out. Turn it upside down to drain and it will be dry in a few minutes. No soap. I just never use soap on utensils, except the detergent that goes into the dishwasher. And it works on the very greasiest of pans, roasting pans and everything, if you do it at once. If you are dishing up, and hurrying to get things hot to the table, have a sinkful of very hot water and put your bulb baster, meat rack, thermometer, skewers and the like in and let them soak. After dinner, use the brush and the running hot water and they are done. I realize this is very hard to knock into people. My last maid was a dream, and a wonderful cook, but she would let the potato pan and the ricer and the strainers sit around and dry hard every time, and I suppose it never entered her dear little head that she spent half an hour extra in the kitchen every night as a result. Let alone wear and tear on pans. I suppose you noticed the way I snatched things from you last summer and washed them up and I hope it didn't get on your nerves. I just cannot bear to have things pile up. I've only seen one article saying all this, and it was in *Gourmet* sometime back and written by a man who felt as strongly about it as I do, bless him.

I simply can't wait to hear what you've got to say about "The Investigator."* Write me quick, quick when you get it. I'm going to Fla. Feb. 19, gone ten days. Wish it was over. I'll write a proper letter one of these days.

Lashings of love to you both,
Avis

FEBRUARY 6, 1955

Dear Avis:

Just went out to do a little weekend marketing across the street (what luxury), and your 1 Feb. was in the mailbox, with the wonderful Herblock. What a man. We shall attach it to the "Investigator," which will have its 2nd gala hearing tonight, with Joe Phillips (our big boss), and 2 other kindreds. I am anxious to hear it again. We were quite bowled over the first time. Shall report later, at end of this letter, as I doubt if I get this done before they all come. Have to make some cocktail snacks (doing Simca's

---

* A CBC radio satire of HUAC and McCarthy by the Canadian scriptwriter Reuben Ship. More than 100,000 recordings of the play were sold between 1954 and 1955.

bastard puff-paste . . . just an ordinary pie-crust paste, 1 part flour to 1 part butter, but rolled out and turned over onto itself 4 times, and the 5th with cheese).

We have just gotten our Jan. *Gourmet,* and I noticed the new Diat series on "lessons in classic cuisine" . . . Certainly just what we are doing. After a hasty reading-over, it seems very good indeed to me. We shall just have to do better. This will undoubtedly be turned into a book at the end, and will most definitely be serious competition. Wonder if they will have technical illustrations . . . as description, no matter how good, is never as effective for technical matters, as the comic-book technique. (I was thinking of veg. cutting up, and kindred items.) Ah well. There will be so many things to come out ahead of us, I refuse to worry, but I want very much to study everything that does appear, so we can try to better it . . . which I think we can in many instances. But, as Simca says, we weren't born into the trade, more's the pity. Had we started in at 12, apprenticed to a good master, we would be far ahead of where we are now. But, we also have the advantage in being housewives, which gives a different approach.

Was interested, too, in your reactions about "something Louisette said." She will return filled with generalities, I am sure, about what THEY do and do not do, want and do not want. She has talked to such people as Ida Bailey Allen* (*NY Times,* perhaps?), who says the book should sell for not more than $4; should not be complicated, etc. etc. etc. etc. But this is talk of American-type cooking, and a mass market . . . to which our book does not apply. I feel ours is addressed to an audience who likes to cook, and wants to learn how, and is interested in good French cooking . . . and that is not a mass audience. I think if we listened to the Louisette's and Allens we would end up with a bastard dish, which is just what we don't want. Louisette is a born promoter, and as you probably observed from the excerpts of Simca's letters, the financial element is important to her. Money and mass sales are of no importance to Simca and me at all, except in the matter that if the book doesn't sell, we have failed to communicate. So I don't intend to pay much attention to what she has to say. But when we finally get this elephantine project finished, she will be of quite some value, I think.

Couldn't agree more on your method of washing up as going along;

* A syndicated newspaper columnist, the host of a radio cooking show, and the author of more than fifty cookbooks, who was a household name thanks to the recipe booklets she wrote for products such as Pillsbury flour and Coca-Cola.

and I think we found we both did work about in the same manner in this respect. I think it must be touched upon, tellingly, in the introduction, as an explanation of French technique, and how it can be handled. Diat mentions it, too, that good cooking is not easy, and there are not many shortcuts. But you can train yourself to work quickly and efficiently; and you must, if you want to be a COOK, in the fine sense of that word. Too many people consider it a chore, rather than an immense pleasure and a true creative outlet. However the very fact that *Gourmet* is having such a success is an indication that cooking is becoming something of an art in the US.

Did another turkey the other night, as they have US frozen ones here the year around. This was a "Turkey broiler," 6 lbs. drawn, which I estimate would have been 8 to 9 pounds undrawn. It did have a small amount of flexibility at the end of the breast-bone tip, about ½ inch. I defroze it in the icebox, taking 3 days! It disgorged a cup of juice. Feeling its breast where it met the wing, it was hard and stringy! (Or so it seemed to me, as I am so prejudiced about frozen birds that we get here in our US grocery store.) As it had a somewhat old smell, I washed it thoroughly, then rubbed it inside and out with lemon juice, which I let dry on it, but don't think it did much good. I decided I would do everything to it I could think of to give it flavor. So I stuffed it with herbal mushroom Duxelles, plus its liver, onions, etc, and some Madeira. Then I cooked down some frozen mirepoix with Madeira and thyme, about 1½ cups of it. And first I browned the turkey in the oven, for 30 minutes at 400, turning it (too big to brown in a casserole on top of the stove). Then I salted it, slathered it with butter, and spread the Madeira-ized mirepoix all over it, and wrapped it in a cheesecloth. I then cooked it, covered, in the oven, basting every 15 minutes. Unfortunately my timing got off a bit somehow, probably not accounting for the previous browning in the oven, as I had let it cool off completely, as I was doing it ahead. The 40 plus 7 minutes theory would have had it done in under two hours (not counting previous browning). Damn! Anyway, it had taken on quite a bit of flavor, though the white meat was pretty dry. (Have noticed, trying out some of the packaged frozen chicken here, that the dark is not too bad, but that the white is dry and stringy . . . I usually like white meat, myself, but only of supreme and juicy quality.) Made the sauce of reduced turkey stock, plus the mirepoix. Sauce was delicious. Paul and our guests thought the turkey was very good. I felt it was interesting, but I could taste a suggestion of old, rancid fat on the skin . . . (I would!) It is indeed horrible stuff, this badly frozen produce. However, the experiment was useful, as I think it would be useful for a frozen turkey bought in the US, where it would have not been given such bad treatment.

Tried this same method again, of the mirepoix (this time, cut in julienne matchsticks), again deglazed with Madeira, on a duck. Cooked in the same way, but stuffed with canned sausages first browned, then cooked with vermouth and consommé, then cooled and crushed, and mixed with canned liver paste (instead of foie gras), and raisins. This was superb, and the resulting sauce very good. So it is definitely a good method. This will be along to you shortly, when I finish up the "Covered Roasting" of chickens, now on its last legs and blue-pencil stage, for preliminary first draft.

Wow, so much to do. But I feel I am at last understanding duck and goose, and about time. What a lot of duck, turkey, goose and chicken stock we have in the house, too, which is certainly a help in sauce-making. I have just now finished boiling up the carcass of last night's duck with some frozen chicken necks and frozen–cooked–re-frozen turkey white meat from last December. Freezing is indeed a godsend, particularly when one is but two, and Paul is not home for lunch anymore.

And this is now, also, AFTER THE INVESTIGATOR. A mad success, everyone was enthralled, esp. Mr. Phillips, who listened to everything on TV during the hearings, while they were in Washington. He thought it was fine, and said he wanted to borrow it. Good. He's an attractive chap, I think, evidently lived in Paris during the 20s and helped Julian Street* write one of his books, *Dining out in Paris,* or some such name. Loves food and wines, which shows he's of the right sort. I had just been thinking this morning, as that record does bring up again the day-dream of "what would I do, how would I answer." . . . He's (McC) got to be attacked. And Phillips said, out of the blue, "You've got to accuse him . . . How do I know, for instance, that you are not a Communist Agent. You are disregarding our laws and our constitution, etc." ATTACK. Those are his (McC's) tactics, surely. Well, he was attacked in the end, and he was beaten, at least temporarily.

In a second hearing, many more subtleties pop up . . . such as "The

---

* A novelist, playwright, and short-story writer who contributed to *Harper's* and the *Saturday Evening Post.*

Paul Child, painting in the Plittersdorf apartment.

Chief never interferes," etc. And the great quotations are most moving and heroic. Well, it's a great piece of work. How good of you to send it!

Wonder how your dog battle has turned out. I don't understand Vivian doing nothing about it.*

* In an earlier letter, Avis had complained about the marauding Dalmatian of her neighbors Milton and Vivian Katz.

Sad that John Carter Vincent needs work so badly, poor thing. I think he always was a thin man, but he does indeed look frail now. But thank heaven they are in Cambridge, in a friendly atmosphere.

Loved the Mrs. Haskell story.*

Hope Benny is feeling better. Men, it seems to me, are far more complicated than women, on the whole. Paul and I have often discussed the matter. The problem of men being breadwinners, having to be aggressive, masculine, successful, as well, in the case of B, creative. It's an awful strain. Women like you and me, with no social pecking order to maintain (I think of us in contrast to army wives, or social-ladder types, or big corporation wives) . . . we really have none of those stresses. We have our work, our dinners to cook, etc, but we are not competing about anything, or struggling to be great, or to keep abreast, etc. Thank heaven he is not the ulcer type, anyway, or he would have to give up martinis.

Our *Reporter* subscription has finally caught up with us, and we are both impressed with the mag. Seems to me it has some of the most lively and literate writing I have read lately . . . even to and including *Harpers*. "A Negative thought on John Vincent Peal,† and his power of positive thinking," for instance. A beautiful and profound piece, really. And their foreign stuff is interesting. Has Benny ever thought of writing for them? I should think it would be fun, and rather his meat.

Fine that you are involved in another knife fight, and shall be interested in the results. I should think there would be no debate at all once the two are tried out side by side. I notice the *New Fannie Farmer,* 1951 edition, says "Stainless steel has the advantage of being easy to clean but is difficult to sharpen. Old-fashioned or industrial steel makes the most practical blade." Hooray!

Paul is fine. I notice a distinct improvement in morale the past two weeks. Well, he is getting more familiar with the people, papers, and operation. Good health, too, but watching weight carefully because of how very good this beer is.

Much, much love. Wish you could come to Paris with us over W[ashington]'s birthday instead of going to Fla.

Julia

* Avis had related a witty anecdote about another neighbor's reaction when Queen Elizabeth's chaplain visited Boston.

† Julia meant Norman Vincent Peale, the Protestant preacher and author of *The Power of Positive Thinking.*

Paris

APRIL [N.D.], 1955

Ma Plusque Chère:

I certainly hope that this will not be finding you in the hospital with your neck being pulled apart.* Oh dear, oh dear, I hope, I hope. Funny, I just felt something must be wrong with you; always hearing from you at fairly regular intervals, you have become part of my family life . . . and I would have written, but we got so bogged down in our own mire. I am quite short on details myself, as to what has been going on with Paul and why. But I imagine he must have gotten in touch with you by now so I shall not go into what scanty details I know about his investigation; except to say that it seems very queer. Why would they spend all that money to send him back there, when they could, I should think, investigate him here. I have just been wondering today if someone is out to "get him." I happened to remember today two reasons which might make him a target, which, we realized very well as we did them, might have consequences.

1) Sent $25 to the "I believe" fund for investigating McCarthy.
2) Wrote a letter to a Mrs. Buckley† who attacked Smith College for "harboring Communists on the faculty." (She a sister of *God and Man at Yale*.) Think I sent you a copy of the letter, which we very much enjoyed writing.

It is more than likely that lists of people subscribing to these two enterprises have been handed to the McCarthy-Bridges gang, with instructions that anyone who can be "got" should be gone after. Knowing the temper of this present placating administration, and the influence the Bridges element has on selection and retention of public servants . . . I would not at all dismiss this suspicion. I suppose we shall never know.

But, in a way, I am very glad for Paul to be back in Washington and for him to get to know some of those big wigs. He is not the kind of man who ever pushes himself, and he is not "ambitious." And I just feel that more of these people ought to know him. He has wonderful judgment, is so deeply human in his approach, and is damned realistic . . . and is a man of such integrity. (I think that is really one of the things that drew

---

* In a previous letter, Avis had complained of neck pain.

† Aloise Buckley Heath, the sister of William F. Buckley Jr. The latter was the author of *God and Man at Yale* and the founder of the conservative magazine *National Review*.

me to him more than anything.) They are just very lucky to have someone like Paul in this crazy USIS set-up. He has been out of Washington for so long that no one there really knows him; and this will give them an opportunity. So I think, for all the agony (if it is over), it will be a good thing.

As for me, I feel human again, being home in Paris. Really, widowhood in Plittersdorf was too awful. I have the most darling little room here, on the 6th floor, with a view over all the rooftops; I can see Montmartre, the two ends of the Louvre, a bit of the river, the top of the Grand Palais, and the heavens over Paris. It is the kind of little eyrie one dreams about in Paris, and sees in the movies . . . And gee whizz, what a city. Why live anywhere else, is the eternal question. I just feel myself expanding in all directions, I am so happy to be here . . . even in my widowhood. And we have good friends here, thank god.

I have been so busy with cookery-bookery I have had no time to wander about, but am going to take a three-day leave tomorrow, and poke about the old streets, and the river banks, etc. I have had three fine sessions with Simca. We are really working like a team, now, and it is most satisfactory and inspiring. She is now on lamb, and is really going at it with the profundity that we need. It has taken us a long time to jockey around to a joint conception, but I think we have finally gotten it. We both want each chapter to be a "monument" in quintessence, which means, by gum, that we really have to know the subject from one end to the other. We have both noticed that this is not done in the books of most of our competitors . . . and feel that there is where we can make our contribution. For instance, the *Gourmet* "Cuisine Classique" articles really had me worried until I came on their section devoted to eggs . . . It ain't much. They really reveal nothing on poaching, scrambling, omelettes . . . same old stuff you read anywhere, and it does not clarify THE EGG at all. Lucky for us, and I only hope they keep up the bad work, just like the egg. Have you run into a recently-published book from Little Brown, ANYBODY CAN COOK, by Gwen French. Very interesting, and she has worked like a dog, and knows her stuff; but I do not think she has presented it in a readable and workable enough fashion. It is slightly too text-booky; sends you back to too many "main recipes" . . . and, I think, will not really fill most people's needs for a companion cook-book. These are purely faults in presentation and communication . . . technical conception. So I can see, when we have this elephant of ours finally finished, the whole thing must be looked at, as a whole, with a cold and clear eye. Because it must communicate, and must be "the cook's best friend." Quite a little problem. But how stupid, to spend one's life juice on something, and have it a failure just because of a faulty technical conception.

Had a lovely afternoon today with my old chef Bugnard, now 74. Every Thursday he teaches a class of American ladies at the American Cathedral. They come at 6, and he comes at 3 to get everything ready. So the two of us spend a happy 3 hours cutting up veg, cleaning chickens, pre-cooking things that needed it. Then he demonstrates, and they eat. I was acting as "apprentice," but stayed only until all was ready. Such fun, so useful; and, working with a professional, you just pick up a new trick or two every session. I am so glad that I spent the hours I did learning how to use the knife professionally. As, if you can't do fast knife-work, no chef will take you seriously. I can hold my own with most anyone, now, but it took me a good 5 years to get my hand in. He has a fine trick for splitting a big flat whole fish down the backbone. You force a big larding needle down the center of the vertebrae; then with a great big knife, while you are holding the fish by the tail (and easier of you have a son or assistant to hold one half of the tail while you hold the other half) you hack down the back bone or either side of the needle. Beautiful . . . but I never would have thought of it.

Tuesday there was a luncheon of the "Club des Gourmettes," at which Bugnard was also the chef. Poor old club is pretty sad. All the original members are now so old, in the 70's, and were of the generation who never really did any cooking anyway, just eating. So here were 20 of these old dames, chattering away, drifting in and out, and nobody doing any cooking except Simca and me. Fine for us, as we had a free private lesson. Thing is, with a club like that, you have to have a president who makes that her life work, is full of energy, has a wide acquaintance, brings in new blood to refresh the old. Etc. Old Mme. Ettlinger was that, *dans le temps,** but is now about petered out with age, illness, etc. So the club is just dangling. Too bad, it used to be quite a thing, and held a position of respect among all the gastronomical events. And it will just die off, probably.

Keep wishing you were around, as you would love all of this; and

* "In her prime."

would love working with these old chefs. Such fun to discuss the famous old chefs, Escoffier, etc. etc. You'd love it.

Really, what a city to live in. If one is interested in the creative arts, there is so much so much; you can never touch a tenth of it, a hundredth . . . but it is all there if you want it. And, to me, it is all so beautiful, so cozy, so lovable. I think we shall have to buy a little pied-à-terre here, for when we are 60 and 70; and spend the winters here, from Jan. 10 to May 12th, every year. I am half French, and that is all there is to it.

Now I must study my German. I am now fairly good in the subjunctive and have to get the passive into my being. Just in the last two or three weeks I am beginning to be able to speak it, and feel a bit at home in it. I find it a fascinating language. And, as I keep saying, I am liking the Germans (if we only lived with them!). And one thing that is exciting, and fraught with hope, is that it is really a new Germany. Really for the first time they are a real democracy. But, my god, we have got to realize it, and help them. They have the awful problem of teaching the whole population how to be a democracy . . . that the government is "me" not "they." If this damned fool business administration loses sight of that, and I am not sure they ever grasped it anyway, Germany can perfectly easily lose this . . . which is still only a tenuous thing. They can be wonderful, and are the hope of Europe . . . the only one, I think. They have far more "nobility of thought" than the froggies, to my mind; deep and turbulent and questioning as it is rumored to be. (I hope to know more about it in a year or so.) Well, as that great man, Mr. B. DeVoto, has remarked, if only the Republican Party had a few egg-heads. Maybe Germany can develop a lastingly free and democratic and responsible government all by itself. I certainly hope so, as I don't see much real understanding emanating from us at this point.

Eh bien, my very dear, I certainly hope you are all alright!

*Gros panniers d'amour fleuri.**

Julia

PS: I shall just stay right here until Paul returns, so do write me longly about everything.

Hotel Pont Royal, 7 Rue Montalembert, Paris 7

* "Big baskets of flowery love."

APRIL 25, 1955

Julia dear:

I will have to compress, much more than I like, but must get a short letter off to you. I am on the mend, though still not bouncing, and I sidetracked that acute neck so I didn't go to the hospital, but B. has had a relapse and is back in bed for more of the same. Temperature, and feels achy and wobbly. Second round for him. He felt increasingly well on his travels and felt wonderful Saturday, which was Neuberger day.* Sunday, foul. Mark went back to school today, after ten days sick and then a week's vacation. If it weren't pouring rain and cold besides I would have to be outdoors transplanting six boxes of annuals I bought Friday. This all would happen when I am swamped with garden work. After your penciled letter, I put in a call for Paul and finally got him.† Husband-like, he seemed a bit disapproving that you had written me—exactly like B. in same circumstances. But if you can't explode to one of your best friends, well what? He didn't give me too many details—they told him to keep mum—but I think I have the general picture. Which is far from fatal but so damn small and messy and idiotic that you must both feel defiled. I do not really think this will hurt either of you. But it leaves a mark on <u>you</u> which you won't forget easily. When I was feeling my lowest last week or maybe the week before I read a wonderful article in the *New Republic* about State and security and wanted desperately to send it to you, not knowing whether you see the mag. which I think is absolutely wonderful. Sent B's secretary to the Square to pick up an extra copy, but it was sold out and I had to keep our copy for B. and then things got thicker and Paul was in Washington and I hoped he would see it. If you missed it, let me know and I can now send it because B. has seen. You should subscribe to *N.R.*—or would that brand you as hopelessly red? Or do you give a damn at the brand any more? I found after McC. took out after B. that I just didn't care a bit, but your situation is somewhat more delicate. I would so hate to see Paul resign in disgust. Hope and pray he fights it out and gives as good as he gets. Overjoyed that he [Paul] told the committee—I still don't know which—B. thinks Immigration, I think McClellan‡—they

---

* Avis hosted a dinner for the Cambridge supporters of Senator Neuberger who had contributed to his campaign at her behest. Neuberger was in the area to give a lecture.

† After Paul was recalled to Washington, Julia had confided to Avis her fear that Paul might resign in disgust over the intrusive interviews he was enduring.

‡ John Little McClellan, a Democratic senator from Arkansas, was head of the Permanent Subcommittee on Investigations, which looked into problems within the government.

were mishandling things. So I sent you a cable, first I ever sent in my life and hope they didn't garble it, and now feel slightly silly that I got so het up. But why not? I have told only the Vincents about Paul, and only barest details, and will tell no others. Paul said everything hushed up though of course it is bound to come out gradually, I hope you know that. Just one more item to add to our record of infamy. I too wondered if somebody had fingered you. God help me I even wondered if your father had talked in the wrong places, but of course he's too innocent politically and he loves you even if he doesn't love Paul. B. was spellbound when I told him, Saturday morning when he returned from his travels. To think he was in Washington when Paul was and I didn't know—but undoubtedly both too snarled up in bureaucracy to get together. It was so wonderful to hear his voice, warm and loving and Paulish. So it wasn't primarily Paul who was being investigated, but I suppose the bastards know about your donation to that anti-McC. Fund, and the Smith* letter, and also being so pally with the DeVotos. I simply cannot see why after the Corsi fiasco† and the damage it has done the party they don't just pull in their horns for a while. Idiots. Beasts.

I have one million things I wish I could write reams about but I can't sit too long at the typewriter. Our heart to heart must wait until you come back, I think. Anyhow you know how I stand basically and I know how you think and feel, and that's all that matters. I feel I can communicate more readily and freely with you than almost anyone in the world. Having you at a distance is in an odd way a help in this communication. If we saw each other every day there would be that curtain that says you can't transfer your troubles to your near and dear ones because it just adds to their burdens and friendship can't be handled that way.

Medical picture. We all had what is apparently Virus A which about once every seven years is a real stinker. The docs know very little about it, don't even know if it is contagious. We certainly infected each other, but Barbara was here five days a week and Mary three days a week and neither got it, thank God! I got it three weeks ago yesterday, B. two days later, and Mark the next day. Gordon never got it on weekends and he pitched in nobly and cut grass and emptied wastebaskets and cleaned up the

* Julia's letter to Aloise Buckley Heath concerning her comment about Smith College.

† Edward Corsi, a liberal Republican, was appointed head of the State Department's immigration program by Secretary of State Dulles. He ruffled feathers by trying to streamline the immigration process and was accused by a congressman of belonging to a Communist organization. Dulles forced him to resign, provoking a public outcry.

kitchen etc. And cheered me immeasurably that hideous long weekend when I couldn't move because of my neck, and his gay [happy] friends came in and made me laugh, otherwise I think I would have lost my mind because B. was away and I was going stir-crazy. I couldn't get an acute neck because of Neuberger coming up. So spent four days alternately on electric pad or in padded collar which I simply loathe. So gradually my neck relaxed. Orthopod says I have an arthritic spine—x-rays showed deposits last year—and I must learn to live with it. No more hard labor. I can garden, all right, but only a little at a time and no heavy stuff. Very hard to hit just the right balance between much needed activity and exercise, and over-straining. Will have basic talk with him when I can get in town and get all the poop, diet, rest, exercise, etc. And have talked to my beloved family and said look, you will all have to be more responsible and helpful because I'm not able any more to crawl under the bed for that discarded sock, or carry in cartons of groceries, or mop up the bathroom floor when you haven't had the shower curtain inside the tub. They are very good about it but they have a long way to go, having taken these duties for granted all their lives. B. is going to be the hardest because he is always eight fathoms deep in his work and has for fifty-seven years dropped and flung things where convenient.

The Neuberger evening was a howling success. I had a simple dinner, but just the same it took me the better part of four hours to prepare, plus all the best china and glass and shoving furniture around and concealing litter in drawers. Schlessinger Jrs. and Jack Fine. People came in at nine and stayed until after twelve. Had two Harvard boys in white coats to tend to the drinks and wash glasses etc. People came and went, most we had at once was 26. Dick [Neuberger] was at his best—absolutely no sign that his new eminence had gone to his head. Working most earnestly at his job and extremely receptive to ideas as to how to be more effective. Long discussion at dinner, for instance, with Jack who is a very eminent surgeon on vital need for govt. to clamp rigid restrictions on distribution of Salk vaccine—which govt. is not going to do so we will have a fine black market.* Jack for instance who does nothing with children except surgery can get all the vaccine he wants from pharmacy at Beth Israel [Hospital]. He won't of course, but any doc could. All wrong. Then long and wonderful discussions all evening—conversation mostly general—be-

* That month, the government had granted permission for Jonas Salk's polio vaccine to be given to U.S. children.

tween Dick and [John Kenneth] Galbraith, Seymour Harris (economist), John Gaus (city planner), Schlessinger Sr. and Jr., Louie Lyons (curator of Nieman Foundation—journalists), John Carter Vincent, Ed Mason (top man in US on resources—head of [Harvard] Economics Dept.), Arthur Maass ([Harvard] Govt. Dept.—knows all there is to know about water resources) and several others. All these guys ardent Democrats and absolutely top-flight political thinkers. And reason for all this talk is general agreement that what the party needs for '56 is some red-hot domestic issues, since Republicans have got us more or less boxed in on foreign policy. Golly I wish you had been here—right up your alley.

Now I am feeling my back—tight spot has moved about half-way down and heat is the best thing—so back to my sofa. Ghastly having to be so careful, having to learn to walk slowly instead of run, as it were. Deeply interested in everything you have to say about cookery in Paris and about your thinking about the Germans, but dare not go into either because if I have a set-back I just don't know what will happen to this family. Did I tell you I spoke again to V. Katz and Milt about the dog and they couldn't have been nicer or more understanding, and they think he should live in the country but meanwhile are taking him out on the lead and you wouldn't know this neighborhood, it is so peaceful and tidy. Dunno why I waited so long, knowing nobody else would say a word. They were just darling. Oceans of love and one of these days I hope to be able really to carry on a correspondence again. Haven't even written my family more than a line in three weeks.

Avis

Plittersdorf on the Rhine

EARLY OCTOBER 1955

Ma Belle Amie:

Sat. the 15th we take off for two weeks in *La Belle France,* that irresistible magnet. Shall drift southwards, through the chicken country of Bresse, then to Les Baux, then to Marseille, and then through a SW district around Coreze [Corrèze] and the Dordogne to Bordeaux, then through a Romanesque area and, oddly enough, there is Paris, blocking the way back to the Rhine, where we shall pause for 3 days. (Dangling clausal structure, which would be quite clear in German, because of gender designation; but must be clear, from context, in English . . . as who would stop for 3 days at the Rhine when Paris was in between?)

My, Plittersdorf is a strange half life. It is difficult to wrap up an explanation as to why. Today we had a really civilized couple from Düsseldorf (American cultural officer there) for lunch, and things seemed as they should be, in a human way. Last night we went to an American-sponsored concert in Cologne, and to a reception afterwards, and it was all rather unreal and bureaucratic. But, of course, it was "real"; it was happening. But we were both unable to participate in it fully, possibly because too many Plittersdorf people were there, and they brought along with them the half-life feeling we have here in Plittersdorf. But in Berlin, on our own, we felt human again, met some people we liked, and some Germans, and the blood seemed to run again in us. Very odd and tenuous. Undoubtedly one of the reasons this is not our teacup, is that we do not want it to be, and refuse to drink from it. If one were seriously ambitious, in a bureaucratic way, one would conform to the brew. Anyway, it seems that Paul's stock has risen considerably since Berlin. Whether we shall see any tangible results I have no idea. But at least, the boys on top seem to realize they have a good man . . . I think.

Will you be wanting any knives in Paris? Or anything else? Let me know. You can reach us as follows:

Oct. 19–20. c/o American Express
La Canabière, Marseille (B du Rh)

Oct. 27–29. Hotel Pont Royal
7 Rue Montalembert, Paris 7

Am about through with my cut-up chickens, and may get it typed up this week, I hope to God. And then comes the amalgamation of chickenry. It is certainly GD long, at present, with numerous over-lappings. I really feel I have an understanding of the chicken, whose complexity I had not suspected before. The problem is to present this complexity in smoothed-out, logical, and cookable form, in a lovable and compelling way, etc. A very interesting problem, but I find myself pursued by guilt-feelings, because it is taking so long. It seems to include almost the whole gamut of cookery, as most all the sauces are represented here. One we have been hassling around with is the Voluté [Velouté] with egg yolks that can be boiled (essential for gratinéed dishes, made-ahead fricassee, Blanquette de Veau, etc.): Escoffier p. 20, "Allemande"; St. Ange, p. 94 and 87. St. Ange, actually has given the clue. A great many of the recent masters say it should not be boiled . . . But if it cannot be boiled, it is of no use for the kind of cooking you and I do. I enclose a preliminary work sheet, which might interest you. Another is the brown sauce, for brown fricassees, of which Coq au Vin is an example. When we have licked these, we shall have contributed something valuable to the life of the home cook, I think. I am certainly anxious to see how [Dione] Lucas has handled these problems. Actually, neither Lucas nor *Gourmet* have me worried. My biggest concern is that we be able to give a readable and compelling presentation.

My, I do wish we could be in Cambridge with you and Benny for even an hour, to hear what you think about the domestic political scene. I should imagine that Benny is in the thick of things, and may even at this moment be with Stevenson.* I do wish, in some way, that the Democratic Party could also be considered the friend of business publicly. We gotta have business. I can imagine the coming campaign is going to be horrible, and the GOP is going to try and make "socialism" stick onto the Democrats, plus Communism, and everything else they can think of.

Much love to you all, and write every day. Special love, as always, to Mark. We saw the most tremendous display of glittering organ pipes last night at the new city auditorium in Cologne, and I wished he had been there to explain it to me, as it looked very impressive.

Hoping all is fine with you!
Julia

OCTOBER 26, 1955

Julia Dear:

. . .

Our big party in NY† last week took place at the wine room at Pierre's, small world I always say. Did you see this room? Through the kitchen and upstairs. They had tables set up in the form of a letter O and the only drawback was that with about 20 people there just was no space to mill around and talk. But on the other hand it was cozy for dinner. Delegation from Houghton Mifflin, Henry Laughlin (president), the Thompsons, Paul Brooks, Dorothy and Anne Barrett—from *Harper's,* all the people named in the dedication of the book. (I mailed you the book this morning.) (Note lovely blooper on the back of the jacket, quoting Jus-

* Adlai Stevenson was planning to run for president again in 1956.

† *Harper's* and Houghton Mifflin feted Bernard at a party in New York City to celebrate the twentieth anniversary of his "Easy Chair" column and the publication of his book *The Easy Chair,* a collection of his articles.

tice Douglas from the *New Republican!*)[*] (Note also nasty note hurled at J. Edgar Hoover, in the back of the book, I love it.) Also Irita Van Doren and Belle Rosenbaum from the *Herald Trib,* and Francis Browne[†] from the *Times* book section, and Herb Scheinberg, and Mabel Souvaine from *Woman's Day,* and Garrett Mattingly (B's closest and oldest friend)[‡] and I guess a few others but can't remember. B. was in an absolute twit, knowing he would have to make a little speech, and damn if it didn't turn out he had to make the speech of the evening, but he did OK, thank God. And Henry Laughlin damn him called on me and I about died and don't know what I said but apparently it was all right though scarcely audible, not to say quite un-dazzling. Dinner was pretty good, but just pretty good. Little overbalanced, I thought. Started with their foie gras, which is the light pureed kind but stinking rich, full of truffles and on a bed of aspic, and a great lot of it. Followed by lobster bisque which was delicious. Followed by guinea hen with a little wine sauce, and wild rice and string beans. Not mad about guinea hen. Dessert was a rather elaborate bombe—a very thin chocolate cake crust, then chocolate ice cream, vanilla, and raspberry ice, with chocolate sauce. Two wines and the red was a very good Beaujolais but I don't know what the white one was. I was in sort of a daze as you can imagine. Two fixed points of the evening were long loving conversations about gardening with Jack Fischer and Francis Browne. I wore the most beautiful of my two very beautiful new dresses—both Harvey Berin.[§] It's a ruby red wool broadcloth like velvet, three quarter tight sleeves, deep scoop neckline, empire waistline and tight sheathe skirt, loops of red satin ribbon at each side of the neckline and a piece of the ribbon hooked around the back. I also have a thin dark blue wool one cut very much the same, only the neckline front and back (rather low) is wide and V, and all is edged with dark blue braid stuff. Very dashing and I have not felt so fashionable in many years. Find I can wear Empire waistlines with the proper brassiere and girdle, and thank God my hips are flat. I was overcome when you said you had bought a girdle. I somehow never thought you didn't have one. Knew you weren't wearing one while you were here but thought it was because it was summer. I flop in the summer,

* The Supreme Court justice William O. Douglas, an ardent environmentalist. The publication was the *New Republic.*

† Francis Brown, the editor of the *New York Times Book Review.*

‡ A professor of Renaissance history at Columbia University.

§ A New York City manufacturer who adapted styles from Paris couture to the American market and a favorite of several First Ladies.

in hot weather, all over the place, but don't see how you can live without a girdle when you are really dressed. And they are so damn comfortable, keeping you in one place gently. Naturally no bones or armor plate, just nylon net. For heavens sake forget stomach exercises; you will never have those girlish muscles again, dear. Unless you ride horses all day, God forbid. It is not so awful to begin to sag, age comes to us all, though must admit that I will be very dispirited when my jaw begins to sag.

What in God's name has Louisette done? Or is this the little Ripperger job you originally sent me? If it is something new, how come? Is she allowed to nibble away at the edge of our book? Say it isn't so.

Blanquette de veau. I just do not find that using leg of veal means the meat tends to be dry. Sitting there quietly covered with stock it just absorbs juices and everybody remarks on how tender and juicy it is. Sure you will find this is so. Haven't noticed (in blanquette) acid taste in onions, but haven't got the taster you have, and anyway I used (last time) part onions I had cooked myself, which undoubtedly helped. Velouté. Cannot—of course—lay hand on your letter where you went rather deeply into the question of an enriched velouté that could be cooked further, as under the broiler, though remember very well raising this question in a letter last year, in regard to filet of sole bonne femme. (Desk looks hit by hurricane, am so far in arrears.) Will be damn glad to get a final word on this vexing question. Proportions of flour to butter—just find I am happiest, in an ornery way, if there's enough butter so I can give the roux a good bubbling before I start adding liquid, and if there isn't enough butter, it won't bubble. Making the kind of thickened gravy, as with roast beef, which my family demands, I never measure anything any more—throw in flour until the roux is the thickness I like and let it cook and darken thoroughly before I start adding stock. About cooking that thickened velouté for the veal—I do think it improved it. If however you made a veal stock as Lucas says in her book with a good chunk of veal in it, you certainly wouldn't need to reduce the velouté. But there wasn't any real veal in my stock, so in spite of much cooking it was not particularly rich. Didn't add any meat glaze, but might next time. And you must be right about covering the pots. I really know this, particularly about potatoes, but have never happened to have a stock turn sour.

Haven't tried marinating chicken or lobster for salad first, but have meant to, to add a little zip, and will whenever I do one, which probably won't be until warm weather. Only thing I am really firm about in chicken salad, which is one of our hot weather standbys, is about twice as much chicken as celery and no other fillers.

Your new clothes sound wonderful and I wish I could see you in same. Absolutely love those big woolen shawls and can't wear them as overwhelming in one my size, though I wear stoles a lot, but light weight. Should be wonderful with your height. I have told you (or Paul) before and tell you again, I just love the way you move—so utterly graceful and feminine, like a dancer.

Gotta quit. Wore myself out on a letter to our dear old friend Henry, told you about him and all his homosexual troubles, poor lamb. Got a letter from him that simply bowled us over (B and I and the Rochlins, who also love Henry). He's had years and years and years of analysis. He's 43. Now he is engaged and my God I really think he is in love. Sounds like a man in love, unmistakable. Helen thinks he is living with this gal. Who is an MD, getting training for being a psychoanalyst and so she ought to know what Henry is all about. He burbles on like a man in love—she's no beauty, but warm, feminine, graceful, funny, intelligent etc. etc.—and we all think maybe he's living with her. Oh I said that before. He admits to a few misgivings, and wants to come up and see his old psychoanalyst here before he takes the plunge. Which also seems awfully normal to me. Anybody has misgivings. But we have all got such a lift from this, it just sounds right, and we are cheered and immensely hopeful. I am closer to him than anybody in the world; after all I have known him since he was 8—and several years ago he was muttering into his beer about matrimony but it just sounded like a counsel of desperation to me and when he talked to me about it I advised strongly against it and never felt the least bit guilty about having done so. I felt in my bones it would be disastrous. This time I feel good about it. Oh my, life is exciting, isn't it? Ansel Adams just this minute called up and is coming to dinner tomorrow night. B. has a fixed idea of massive proportions that I don't ever entertain. I am making a little list. If there is one single solitary visiting Elk who doesn't come here to dinner he must have us on his black books. They average two a week. And there are those Sunday cocktail parties. Mean to flay DeVoto with this list the next time he opens his trap.

I love you,
Avis

. . .

Plittersdorf on the Rhine

NOVEMBER 2, 1955

Dearest Friend:

It was lovely to get a letter from you on our return, made us feel very much at home. We did have a lovely trip, and find we clocked up just over 2,000 miles! We started from here and went South quickly on the Autobahn (much like our speedway from NY to New Haven), but rather scary as people go so fast, and quite murderous when there is fog or sleet . . . No speed limit at all, and the great Mercedes pass us like we were doing 20, when we are doing 70. Spent the first night in Colmar, then did a wander about the Jura area where all the grapes were being pressed and everything smelled nice and winey. Then on to Les Baux, stopping in at Tain L'Hermitage on the way to stock up on some red wine. Les Baux was heavenly, dry and hot, and the sunny black hills filled with herbal smells. Marseilles was wonderful, and quite its old self. We called on most of our old friends, concierges, etc, and regretted that we still didn't live there. Then we took a series of smallish roads along the Cévennes mountains, following various river valleys, and that was lovely. It is most medieval, with stone villages, lost churches, looking as though little had changed for centuries, except for the happy advent of electricity. I was chatting with an old biddy in a most romantic little village, saying how beautiful it all was. She said, well it was certainly picturesque, but, Madame, to live here . . . *hein!*[*] Quite true, I imagine. They are so remote, you wonder how anyone ever could get out if one were born there of peasant stock. And Paris. I had a fine whole morning with Simca, which was very useful. And we got hold of our attractive *Vogue* friend, Susan Train[†] who had a *Vogue* assistant editor in tow, and took them to lunch with the Fischbachers.[‡] That makes a nice little connection, business-wise as well (if the time will ever come to do business!). We also saw quite a number of our good friends, which was a joy after living in this desert. A fine two weeks, and Paul, for once, admitted that he was about photographed out.

We also saw our friend Jane Foster and her husband.[§] She finally wrote us a letter what all the trouble was about, and it is a harrowing story. I

* "What!" An exclamation indicating exasperation.

† *Vogue*'s Paris bureau chief.

‡ Simca Beck and her husband, Jean Fischbacher.

§ Jane Foster Zlatovski, a well-educated American who worked for the OSS in Indonesia in the 1940s, and her husband, George Zlatovski. She was indicted as a Soviet spy in 1957 but could not be extradited from France.

am making a copy of it and will send it on to you. It is all very sad and maddening, and downright frightening to see what has happened in our country. I don't know what is to be done, but something must be if we are to survive as the Home of the Brave . . . which we are far from being at present.

You do indeed sound busy, but at least you are leading a full and mid-stream life, which is something. The great NY luncheon sounded like great fun, lunch rather on the heavy side. Rather surprised to hear they would serve a Beaujolais at such an elaborate affair, but it seems to be a popular wine in the States. Your own new dresses sound heavenly. Wish we could see you in them. Who is Harvey Berin? Wonderful feeling to feel really fashionable, does a great deal for the morale.

As for girdles, I was surprised myself that I hadn't thought of that as a solution. I wore one when I was 17, as everyone did, and then gave it up later, as I really didn't need one; and quite forgot about them. Paul doesn't like them as he says a girdle buttocks is no fun to look at; but sees the point in regard to pots.

That little book by [Louisette] Bertholle is the same little old Ripperger job. Putnam sold the rights to England, I believe; and there was also a French edition put out by a bookstore. Jean Fisch[bacher] was furious when he saw "Bertholle" on the English blurb, and said he never would have settled for that 18% when he arranged our affairs last year, if he had known that would happen. Simca refused to let her name appear on the French edition, and Louisette's didn't, either. I don't know what happened with this English one, and have kept strictly out of the business, as it is none of my affair. But she don't have no nibbles at <u>our</u> book.

Blanquette de Veau. I won't have anything more to say on this until we have gotten into it, and heaven knows when that will be, as beef is next, then lamb, and THEN veal. (By that time I hope we shall have a Democratic administration!)

Very interesting about your friend, Henry. I do hope he really is in love, and that it will work out. They must both certainly know what things are all about, particularly the girl, at her age. We know a very nice English pansy slightly, a friend of our Cambridge Bicknells, who, after years with a boy friend, suddenly up and married a girl. It seems to work out very well, but we know nothing about his background, really.

We are not at all surprised that Mark received 99 in an aptitude test. What is his IQ (and if so, what does it mean . . . I forget what the percentages signify). To me, one of the wonderful things about Mark is that with all his brains and interests, he is so damned full of life and so devastatingly charming. It is an incredible combination. How did you ever do it? I found the two-part profile in the *NYer* about Menuhin* extremely interesting; did you read it?

The Dione Lucas has come (check for $4.20 enclosed, and many thanks). I find I am a bit disappointed in it, but also that I am at this point not much of a judge of other people's work. Her technique is certainly not classical French, more personal Lucas . . . which is certainly OK, and OK for the USA . . . But I somehow would not put nearly as much confidence in this book as I would in dear old Mrs. Joy, whom I somehow really believe. Interesting to find that Lucas doesn't make a cooked flour/butter roux for most of her sauces (I have read mostly chickens); that she puts in salt pork or bacon without blanching it first (smokey bacon flavor overpowers taste of other things, according to froggies); doesn't always say "taste for seasoning." Frankly, I find it a bit sloppy, even though I am trying to look at it with an unprejudiced eye. You had me scared, and I was dreading to open it; now I have bounced back to my pre-Lucas confidence. However both Simca and I find the Diat sauce article in the Oct. *Gourmet* extremely good, and giving serious food for thoughts. Lucas has quite a face, doesn't she, most determined and cool. I don't think I'd like her, but who can tell, from a photo (but you and Benny I liked immediately). The other woman [in the photo] looks rather nice. Certainly don't want any photos of us on our book, we'll be too old by that time anyway, and besides I don't think it helps the appetite and might hurt the sales.

Much love to all, and from Paul also,

Julia

* Yehudi Menuhin, the violinist.

PART THREE — 1955–1959

Black News: No, Neg, Non, Nein

Julia and Avis Grow Closer in Sorrow; Revisions and Rejection at Houghton Mifflin

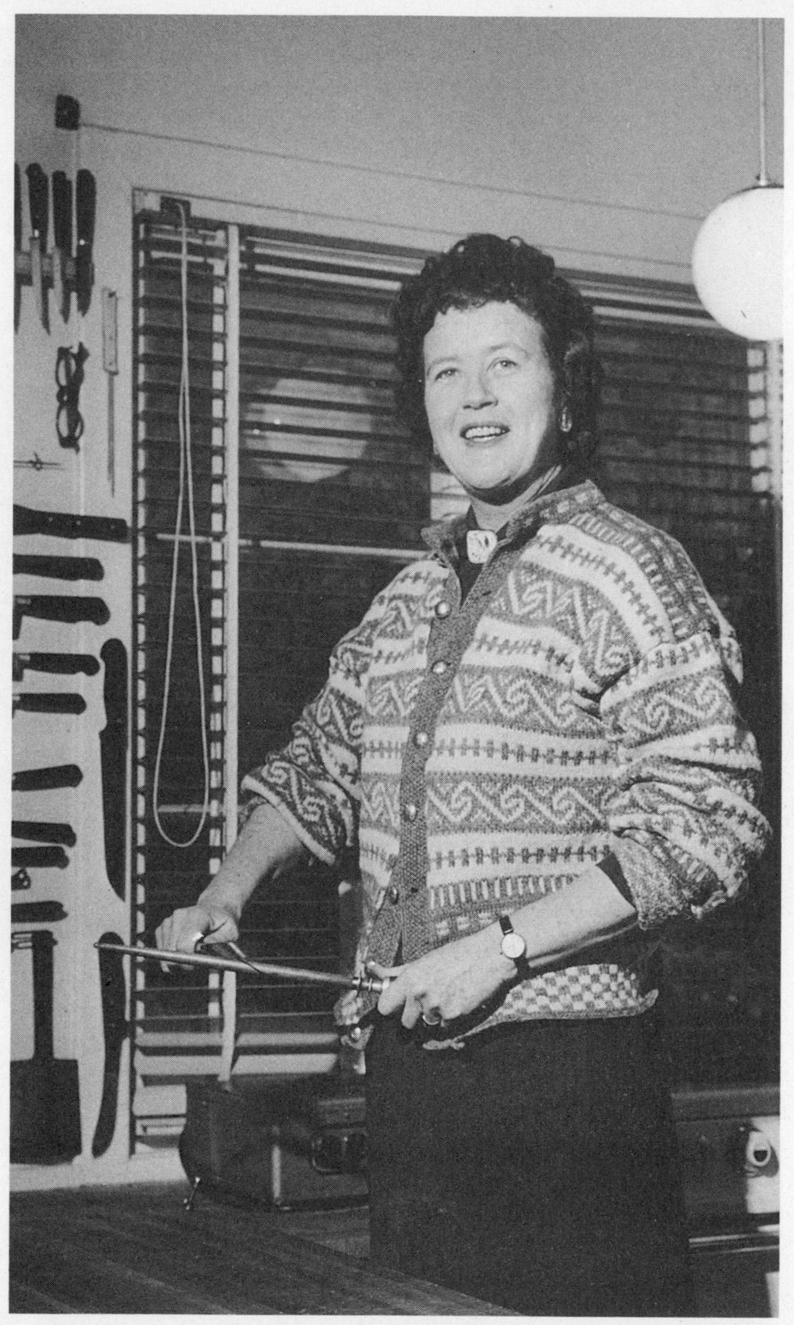

So I am deeply depressed, gnawed by doubts, and feel that all our work may just lay a big rotten egg. I can see that such detail cannot go into a magazine article, and for them we just have to forget our book-methods. But I would hate to take what I consider to be the guts out of our book. I suppose the only thing to do is to go along as we have been, and then when the MS is done, to have a careful and detailed pow-wow with HM; and do it over if necessary.

JULIA CHILD to AVIS DEVOTO, January 12, 1958

Julia in Oslo, where Paul was transferred in 1958.
PREVIOUS PAGE: Avis and Julia in Rouen, France, 1954.

On November 13, 1955, while in New York City on a business trip, Bernard DeVoto had a heart attack in his hotel room and died shortly after arrival at the hospital. Avis and their two sons in Cambridge could scarcely comprehend their loss. Family, friends, and colleagues gathered for a memorial service in Cambridge, and his ashes were scattered into the white-water Lochsa River in Idaho by his friends at the ranger patrol station there. For the next several months, Avis shared her grief and loss with her many friends, including Julia and Paul.

For their part, the Childs expressed their sympathy by sending Avis airfare for a three-week trip to London, Paris, and Germany that would stretch from mid-May to the beginning of June. The trip, well documented by the photos that Paul took of their travels, brought the Childs and Avis together again and became an occasion to meet each other's friends in both England and Paris. For Avis, it was an opportunity to become acquainted with Simca, attend classes at L'École des Trois Gourmandes taught by Chefs Max Bugnard and Claude Thillmont, have lunch with Le Cercle des Gourmettes, and see Germany for the first time. While they were in Plittersdorf, a district of Bonn, Julia asked Avis to read the first draft of the poultry section. It was well over one hundred pages long, and reflected more than a year of working with fresh and frozen chickens, ducks, geese, and turkeys.

About four months later, Paul was transferred to Washington, D.C. Because their house in Georgetown was rented, the Childs moved about the country, from Charlie and Freddie Child's home in Pennsylvania, to Avis's home in Cambridge, to Chicago, to southern and northern California, and then back east again. During her two visits to Cambridge, Julia conferred with Dorothy de Santillana about the possibility of expanding the scope of the book they had under contract into a multivolume cookbook, tentatively titled *French Cooking for the American Kitchen*. The first volume, as she planned it, included recipes for sauces and poultry, probably because both sections were near completion. Initially, de Santillana seemed to agree to the plan.

In the late fall of 1956, when the Childs were able to reclaim their Georgetown house, they spent more than two months renovating it and outfitting the ground-floor kitchen with a professional stove and modern equipment. In the supermarkets that dotted the area, Julia discovered products she had never seen in the Plittersdorf commissary. She made multiple changes in her manuscript, shared the convenience of things such as electric frying pans and Uncle Ben's Converted Rice with Simca, and entertained Washington friends and associates as well as visitors from farther away, including Avis.

Less than a year before the Childs returned to the United States, Avis had taken a job as a literary scout for the prestigious New York publishing house Alfred A. Knopf, which she was able to carry out from Cambridge. Although traveling and interviewing prospective authors kept Avis from testing Julia's recipes as much as she would have liked, she expanded her contacts, already considerable, in the publishing world, and her expertise in editing cookbooks grew when Knopf passed a few her way, including Elizabeth David's *Italian Food*.

For her part, Julia "entertained like mad," conducted cooking classes, revised recipes, and looked forward to finishing the first volume and welcoming Simca and her husband, Jean Fischbacher, to the United States early in 1958. After completing the manuscript during January and early February, Julia and Simca traveled to Boston in an icy snowstorm to hand deliver the manuscript on sauces and poultry to Houghton Mifflin. It was more than seven hundred pages long.

Deciding that it was encyclopedic in scope, much too expensive to produce, and extremely limited in its appeal to the American market, de Santillana wrote to Julia on March 21:

> Our most careful group eye has been brought to bear on the fruit of what is self-evidently the most careful labor of love i.e. the Sauce and Poultry division of *French Cooking for the American Kitchen*, and the problem presented is complex. With the greatest respect for what you have done (for the labor involved is gargantuan), we must state forthwith that this is not the book we contracted for, which was to be a single volume book which would tell the American housewife how to cook in French.
>
> From here we must talk publishing, not cooking. It would be impossible to break in an audience on two sections of a series, and two sections which do not have any strong reason for being issued together except the fact of their simultaneous completion. What we could envisage as saleable (now that the question of one volume is out) is perhaps a series of small books devoted to particular portions of the meal. Such a series should have a logical sequence of presentation following in a way the natural sequence of the meal, such as soups, sauces, eggs, entrées, etc. We also feel that such a series should meet a rigorous standard of simplicity and compactness, certainly less elaborate than your present volumes, which, although we are sure are foolproof, are undeniably demanding in the time and focus of the cook who is so apt to be mother, nurse, chauffeur, and cleaner as well.
>
> I know this reaction will be a disappointment to you, but I wonder if this isn't the time for you to do some re-thinking yourself on the project which has been . . . so busy just growing that it has grown into something much more complex and difficult to handle than the original book. I should like very much to have your own ideas.

Julia responded:

> Both Mme. Beck and I are sorry that your collective eye cannot use our monstrous masterpiece in its present form. But we are not unduly surprised as it is indeed not the book you contracted for.

> We well realize that the continuing trend in this country is toward speed and the elimination of work, and that our treatise on French Sauces and French poultry furthers neither aspect of this American dream. We have therefore decided to shelve our own dream for the time being and propose to prepare you a short and snappy book directed to the somewhat sophisticated housewife/chauffeur. This would be about 300 pages or so of authentic French recipes including a dozen soups, hints on the improvement of canned and frozen soups, a few sauces, a good handful of chickens, meats, fish, eggs, soufflés, unusual vegetable dishes including the pepping up of canned and frozen vegetables, a few cold dishes and some good, easy desserts. Everything would be of the simpler sort, but nothing humdrum. The recipes would look short, and emphasis would always be on how to prepare ahead, and how to reheat. We might even manage to insert a note of gaiety and a certain quiet chic, which would be a pleasant change.
>
> All the recipes we need are at hand and fully experimented upon, and within six months or less we could have a completed manuscript in your hands.
>
> Please let me know how this appeals to you. I am sending this special delivery as Mme. Beck will be here only a few days longer. In the meantime, would you be so kind as to have Gargantua and the Photostats returned to us?

There was no disguising Julia's great disappointment, and in a letter to Simca in the summer of 1958, she wrote, "HELL AND DAMNATION, is all I can say. WHY DID WE EVER DECIDE TO DO THIS ANYWAY? But I can't think of doing anything else, can you?" For the next two years, all of the entertaining that Julia did included testing recipes for the book that she and Simca were revising. And she continued to teach cooking classes to determine whether their recipes were clear and doable. Meanwhile, Houghton Mifflin concentrated its publishing expertise on a Texas cookbook by Helen Corbitt, the director of the Neiman Marcus restaurants.

During the fall of 1958, Paul learned that he was to be appointed to the post of cultural affairs officer in Oslo, Norway. It posed another language problem, and he requested at least six months of language training before accepting the posting. Julia was pleased about the timing, believing that their single-volume book could be delivered to Houghton before they left the United States in the spring.

During previous visits to Cambridge, the Childs had looked for a suitable home to purchase, with an eye to Paul's retirement. They were told that a desirable property seldom came on the market. Nevertheless, they continued to hope. A call from Avis during a snowstorm in early December 1958 made it happen. They immediately boarded a train bound for Boston, and accompanied by Avis and a real estate agent, they drove to 103 Irving Street. The historic home, located between Beacon Street and Harvard Divinity School, was in an enclave of past and present famous Cantabrigians—the philosopher Josiah Royce built it in 1889, the author Gertrude Stein lived nearby at 123 in 1893, the Galbraiths and Schlesingers were current neighbors. The big white clapboard house appealed to them in every way. After Paul inspected the foundation and the three floors above it, the Childs purchased the house and rented it to the owner until she could find a new home. Before setting sail for Europe in April, they also rented their Washington house.

Their first stop was Paris and then a trip to the South of France and a nostalgic return to Plittersdorf. On May 17, they landed in Oslo and began to search for a suitable home. Between her hostess duties and embassy lunches, Julia completed the shortened book and sent it to de Santillana, who acknowledged receipt of the 785-page manuscript on November 6. Two weeks later, Paul Brooks sent the Houghton Mifflin executives' response.

> After the first project grew to encyclopedic size you agreed with us that the book we had originally talked of together was to be a much smaller, simpler book. In a letter of March, 1958, you yourself spoke of the revised project as a "short simple book directed to the housewife chauffeur." The present book could never be called this. It is a big, expensive cookbook of elaborate information and might well prove formidable to the American housewife. She might easily clip one of these recipes out of a magazine but be frightened by the book as a whole.
>
> I am aware that this reaction will be a disappointment and wish

I could send you other news. I suggest that you try the book immediately on some other publisher.

To be fair, de Santillana did recognize the merits of the book and wrote to both Julia and Avis, "This present manuscript is a superb cookbook. It is better than any I know of. But I could not argue with the men as to its suitability for a housewife-chauffeur. It remains time consuming."

Neither Julia nor Avis intended to give up on the book. Although Avis was no longer employed by Knopf and had taken a position as the secretary of one of Harvard's undergraduate houses, she wrote a "Dear Alfred" letter on November 11, 1959: "I want the firm of Knopf to have a look at a unique book which can be a classic if it is promoted right and gets some breaks. I want Bill [William Koshland, vice president at Knopf] to take it home and play around with it and do some of the recipes."

NOVEMBER 27, 1955

Dearest Julia:
Small detail but distressing to me. I had a lovely letter [of condolence] from Simca and it lay on my big coffee table and it was passed from hand to hand as so many others are. Mark (who just got special commendation in French!) was going to help me write an answer. Mollie Brazier wanted to copy it because she has to write many such letters to her French friends and wanted to note down the proper lovely formal style. Well it has just disappeared—maybe temporarily—but I have not got her name and address anywhere that I can lay my hand on. I am drowned in letters and clippings and now the articles from the weeklies coming in. I hope someone has sent you Alistair Cooke's* superb piece in the *Manchester Guardian*. I hear there is a fine one in the *New Statesman*,† and—this is

* Cooke's article concerned the case of J. Robert Oppenheimer, the director of the Manhattan Project, which developed the atomic bomb during World War II. Oppenheimer was accused of being a Soviet spy and his security clearance was revoked. Cooke noted that the hearings would destroy Oppenheimer's reputation, whether or not he was guilty.

† The British left-wing political magazine was critical of the decision to revoke Oppenheimer's security clearance.

priceless—the *Daily Worker.*[*] How B. would have roared at that. (I feel safe—Charlie [Charles] Curtis my executor has just published a book on the Oppenheimer case.)[†]

Send me Simca's name and address and do I call her Fischbacher-Beck?

Your letters help so. They all do. Haven't had time to feel hollow yet, still half-expect to see him coming in the door. The boys are both just wonderful—matured over night. I cling to Gordon—this will get him moving and he is full of sense and we have never been so close. All his friends come in, big army types, and peel potatoes and sharpen knives and empty trash and massage my back which of course has tightened up from nerves. Tummy also giving discomfort, pure nerves—can't eat much and drink tea all the time. Cooking a farce, steak and baked potatoes, hamburgers. Cooked Thanksgiving dinner for 8—Greg [Rochlin] and daughter Martha (Helen and little Greg had to go up to the farm in Maine to her parents), Mollie, Julie Jeppson,[‡] an army friend of G's. Never did a better turkey. It's going to be like this for months—infinite details, letters, papers—I haven't even gone through his personal things yet, no time. Then bang—emptiness. Wonder what happens to me then? One thing certain—proof-read the cookbook like mad. Hope I still have the house when you come, so we can cook in a proper kitchen. Will hang on here as long as I can, loving the place so damn much, but silly when I am really alone. Wish you could come share it with me, when you leave Europe, but know this seldom works out. Families should be by themselves. But so many women in my spot, alone one way or another, perhaps I can double up with one—I have so many, many friends, so many women I love and men too. So damn good to me I can't get over it. HM working hard to get me all possible money and all sorts of things opening out. I don't know what is happening in the world—two weeks just dropped out of my life and can't read newspapers yet, or much of anything else, though I wrote my little column, faking it because I couldn't really take in those silly books. But sensible to hang on to it and will eventually do some reading for HM. Should make all I can though there is enough for

---

* A newspaper published by the U.S. Communist Party, to which Oppenheimer had subscribed.

† Curtis's book questioned the impartiality of the officials. Avis seems to be saying that she feels "safe" in feeling that Oppenheimer was unfairly charged, since her own lawyer thought the decision was unfair.

‡ Arthur Schlesinger Jr.'s secretary and a friend of Avis.

our safety, bless him he thought for us all the time. Best thing in the Easy Chair book is the sassy note to J. Edgar [Hoover] in the back. I just love it. How is Paul? I think about you two so much. Can't write a proper letter probably for months. Trying to write twenty or so little ones every day.

Best love,
Avis

DECEMBER 1, 1955

My darlings:
Only wrote 16 notes and letters today, with three long calls in the morning and one in the afternoon, so I am exhausted and will go to bed on my electric pad and read a whodunit. Shouldn't be writing to you and will make it telegraphese. Forgot to tell you by all means send me copy chicken chapter, don't know when can read it and report, but probably before I do other things. My book, you know. Terribly glad Paul feeling better — he has more than his share of afflictions by virus. Hope they stay away from this door. Avoided getting neck in traction last week by a hair, with metacortin prescribed by dear family physician and Herb Schineberg over phone terribly disapproving. Hard situation. But I got no side effects and didn't get neck pain. Spine tight all the time, nerves probably. Otherwise not bad, eating a bit more and sleeping with adding much whiskey. Even cooking a little bit; we just (Mark and me) had sort of mixed grill, lamb chops, veal kidneys, stuffed mushrooms, peas — very good and easy. Don't forget send Simca's name and address please.

. . .

I don't need any money, bless your little hearts and thank you very much. Got a nice chunk of cash in bank and substantial insurance. Don't know how much of everything yet but not particularly worried. Think will rent rooms on top floor (advice Betty Vincent* who is very practical and does it herself) to graduate students — she says I can get $140 a month without putting in a kitchen. That will pay taxes and most of my tiny ($51 a month) mortgage payment. Surprised you think this too big to handle, size of house I mean. I crave space, always have. Mary comes three days a week, five hours each time, keeps it spotless. All I do is beds, empty baskets and ashtrays, cook. Truly. Pay her fifteen a week and carfare. I'm no meticulous housekeeper and care only to have kitchen and bathrooms

* The wife of John Carter Vincent.

shining. But of course I must wait on all this until I talk to Charlie Curtis who has been having bursitis and whose court appointment has not gone through yet. He will advise on investments—B. did it all himself and was awfully good at it, but Charlie handles some of the biggest trusts in the country and has written a book called I believe *The Cautious Investor.* B. said he couldn't understand it which caused Charlie to roar. B. said I would be all right if I could hang on to those investments and Charlie will certainly try. Don't know about inheritance tax yet. I just refuse to worry until I have to, Benny would have hated me to, he was the worrier bless him—if I have to get out and scramble, well there are a lot of things I can do. Type, review, publishers' reading, secretary to academic type.

I find that because I wasn't with him, because Greg and the others took absolutely all arrangements off my hands, to spare me shock, that I don't feel shock so much as complete emptiness. May Sarton told Anne Barrett it was as if he had stepped down an open manhole. I feel he just went off into space. And I feel in the most curious way cheated of all that anguish. Can't quite describe what I feel. Almost expect him to walk through the door, back from a long trip. And no funeral, no burial, no marker—ashes to the Lochsa [River]. Best, I know. But I wish I'd had some of the other. Real loneliness has not had a chance to descend on me yet. I am constantly with people. Flocks of callers every day, flowers still coming, more and more mail and all the glowing and wonderful editorials and articles. *Manchester Guardian, New Statesman, New Republic, Time.* When the loneliness really hits me I will probably be climbing the walls. Helen Rochlin, very tense and tired for a long time, went up to her family's farm in Maine for Thanksgiving and on the train was hit with a real panic—fear, cold sweat, the works. First in her life. She is back now, very changed. First time she would ever admit, or talk about, the terrible inadequacies so many of us feel toward the people we love. I've had that and I know all about it, and Benny knew, God how he knew. Better than anybody because he had it all his life, off and on, and never could get the satisfaction out of achievement most people get. Always devalued himself, worried, was tense and depressed so much. I'm tougher than he was. But I dread waking up in the night alone, so the whiskey and the pills.

All this confidential, naturally, but what follows even more so. B. planned to come back from Washington (after New York) and do two things—go out to see [Adlai] Stevenson and tell him he <u>had</u> to have a strong public power man on permanent duty in the Chicago office, and draw up a sort of conservation plank for Finletter* and his advisory group, to which B. belonged though he didn't go often. I knew who he would have recommended, Clay Cochran of the National Rural Electric Cooperative Association—not civil service which knocks out so many of B's Forest Service and REA† friends. (I have removed from the files and burned all the letters he had from them over a good many years—he always protected his sources.) So after three days of brain cudgeling, and consulting one or two people, I wrote a rather long letter to Finletter, who had written me a terribly nice letter, very warm. (Also had a wonderful telegram and letter from Stevenson—affectionately, Adlai. Nice man.) And I explained about the Forest Service and other sources and said why I couldn't give names, and then I did name Cochran. Said B. had it very much on his mind but I was only suggesting, not asking. Said also I had no qualifications whatever to do more than report what he was thinking, wanted no part of public life and so on. Sent it off and quaked. Arthur Schlesinger Jr. called me today and he'd just got back from seeing Finletter and they both wholly agreed, said it was wonderful letter, asked permission to send it to Stevenson, and are setting the wheels going on Cochran. I'm so relieved. Not elated, how could I be? But pleased and relieved. Will see the Finletters some time I think—have met them once and so very, very nice.

Also, Anne Barrett has re-read B's manuscript‡ and is talking very strongly about the absolute necessity of having it finished. Six and a half chapters out of maybe nine are written—many notes, and he talked with several people about it. HM mad to have it done; I have held back not wanting a half-finished job. But we think now perhaps Wally [Wallace] Stegner who after all is a very distinguished writer and knows the West like a book and owed Benny everything, plus Bob Hansen who is here as a Neiman fellow from the *Denver Post* and who talked to B. a lot about these things, might be able to do it properly. Wally just got back from Arabia—had a letter from him from NY this morning. So he hasn't been approached yet. Paul Brooks is reading the ms. now and we shall see. Anne said what is true, that it is absolutely packed with wonderful

* Thomas K. Finletter, a lawyer and statesman who played a leading role in the Democratic Advisory Group, which was composed of politicians and activists who developed platform positions for various candidates.

† Rural Electrification Administration.

‡ *The Western Paradox* remained unfinished at Avis DeVoto's death and was finally published in 2001 by Yale University Press.

material and some of the best writing he ever did. I will <u>not</u> wring his stuff dry for the last penny. I would rather scrub floors. But this book is so needed and perhaps we can get it. Not, sadly, in time for the campaign.* His broad theme was the west against itself, how the west destroys itself, how we cannot separate ourselves from the west without wrecking the country. Stevenson wrote me that he is already wondering what B. would have said on some questions—this book has a lot of the answers. It is Anne's job mainly—she was extremely close to B., because he put her together again after her divorce fifteen years ago and has always advised on the children and all her problems and walked her and driven her around in the middle of the night countless times when she fell apart. She is a perfectly wonderful woman and a rock to me. As so many others are, Rochlins, Edmonds, Murdocks, de Santillanas, all the HM people, *Harpers* people, Carl Brandt, you two. Nobody ever had more wonderful friends. Now I am going to bed.

Avis

FEBRUARY 13, 1956

Dear Julie:

I wrote you a rather long letter a couple of days ago and will make this short, since I owe so many others—to acknowledge your more than welcome report this morning. Golly how glad I am that Paul is really better, and that you had such a weather break in Rome. I yearn to go there, and the food sounds perfectly beautiful. Simple and exquisite. I am passionately fond of *finocchio,* or *finocchi* [fennel] or whatever it is spelled like, and put it in salads whenever I find it which is not often. It is also terribly good eaten raw, like celery, dipped in olive oil and toasted poppy seeds. Bit messy but tastes delicious. Surprised you had never been in Rome. Thought you had been everywhere.

I have only cooked a couple of geese in my life—they are very hard to come by here, have to be ordered specially, and you can't be sure how tender they are going to be. Last one I did remained tough to the last—too old. And I got it through Sages and paid plenty. All right dear, I'll bite—what do you do with all the goose fat? A quart and a half would be an embarrassment of riches for me. Grandma used to think it was dandy for rubbing on the chest and covering with red flannel. The liver, now—that I envy you. And of course the skin has got to be well

* The 1956 presidential campaign.

pricked and [Dione] Lucas, as usual, is off her head. If you didn't prick it all over, should think it would split with the bubbling fat and make a mess. I prick ducks all over, certainly—I love to eat the skin of a duck if all the fat has been drained and it gets nice and crisp.

I don't know that Dorothy is going to write you on cut up chickens—she seems to have turned it all over to me. It isn't that she isn't wildly interested—she is and will continue to be. But she is swamped, and feels the ms. is in safe hands with me. Dear old HM now pays me two bucks an hour for reading and making suggestions on the ms., and the same for trying out a dish and also will pay for ingredients used. This is new and I've only sent them one bill, for $10 which was underestimated, in my usual head in the sand fashion. D. rebuked me so I will charge more next time. In spite of the fact that they are in Boston for the winter months I have seen very little of them—they dropped in unannounced one Sunday afternoon several weeks ago but I was just going out and could only talk to them for fifteen minutes. Their lives, like mine, are just inconceivably complex and busy. I don't know any of the idle women you read about, and believe it to be a fiction.

If I see Dorothy I will mention your question about publishing the book in two volumes, but I think you'd better write her about it in detail. I don't know enough about publishing problems to know how they would view this proposal—I would sort of hate to see two volumes, but then having too fat a book would be as bad, and hard to handle and use. I dunno. We must all think about it and do a lot of talking. Wish to heaven you could be here, when you have decided just what you want, to talk to them directly—always more satisfactory, and you make such a good personal impression, with your air of gentle authority, totally lacking in female aggressiveness. And oh hell, just occurred to me, if you had sauces in one volume, fish and meat in the other, the cook would have to juggle two volumes to prepare a meal, and that's a real snag.

My goodness it would be absolutely wonderful if you got home leave next winter—I kept thinking morosely in terms of another two or three years, when the ms. was reasonably in shape, and then a good long session with you while we worked on the technical part. For all I know, by next winter the villa DeVoto may be crawling with assorted tenants, but there will always be room for you two, if Mark has to sleep on the bathroom floor, which I think he would be happy to do for Julia and Paul. So let's hold the thought. And if I am working, I will be here as much as I can, and weekends, and you certainly cannot be paying guests, but we can split the food costs. This is a luscious carrot to dangle before my nose and will help greatly during the months before I see your dear faces. But Julia—twenty-five pounds! And how will you take it off when you have to be tasting constantly? I am well under normal weight, but then we are not eating very well. Lots of meat, though. Meat is fine for reducing.

Tons of love—in haste—

Avis

FEBRUARY 20, 1956

Avis our dear one:

What a relief to know you are all right, or were as of 13 Feb., and sounding as though you were going strong, so we are not worrying about you as of today. As for me, I am rather bilious as of after lunch yesterday, and have had but 4 cups of tea since, which seems to be taking care of it . . . and will also remove a pound or two, I hope, from my stomach. Too much duck and goose lately, and if I weren't such a pig, this wouldn't keep happening. But biliousness does indeed seem to be my occupational disease. Paul, fortunately, is still eating very carefully so will be immune to these troubles until June at least, when he is off his diet. He continues to be fine, which is miraculous; and seems to be filled with energy and ideas. He has again, poor dear, started in with his German lessons, and really hates it, but must learn it. It means, too, that he must do so much studying, as nothing comes into the noggin at all otherwise, that he has little time for reading. And that is a great deprivation.

Our two duck splurges went off very well. We haven't had anyone in for so long for a meal that we have rather forgotten our systems for serving, etc. Little details such as how the cocktail stuff gets cleared off effortlessly and invisibly; forgetting to put on enough forks, etc. We work as a good team, though, and always go on the theory that no matter whom we have, things should go off as though we were entertaining royalty (though in a simple way), so we will always be in practice. Each of the duck dishes involved a boned and stuffed one. The first was a liver and sausage stuffing, and the duck was braised. The second was a sausage and veal stuffing, the duck was browned, cooled, then encased in a pie crust, decorated with pieces of cut-out pie crust to simulate wings, feathers, legs, etc, in an abstract fashion. Then baked for about 2 and ½ hours, and cooled. Then, before serving, you cut around the sides of the crust, so the top can be lifted off. So when the duck comes in, the server lifts off the top of the crust, and there is a little old duck revealed inside, looking just like a duck. It's a pretty cute dish, and serves 8 to 10. Nothing at all difficult about doing it, but takes a bit of time.

The braised duck was extremely good. I did a very high-class braise, following a recipe of Simca's. First you make a fine brown sauce, using the duck neck, gizzard, carcass bones, and vegetables . . . These are all browned, then you make your brown roux, then you add a good stock and some wine, and simmer it for about 2 and ½ hours. Then you strain it, and you have a slightly thickened sauce, and this you use for braising the duck. Then you degrease your sauce and reduce it, and you have, after flavoring it as necessary, a really elegant deep-brown luscious and wonderful sauce the like of which . . . well, it is about as delicious as anything one could wish for. The simpler braise consists merely of browning your vegetables and carcass bones in the braising pan, adding and browning the flour, then adding the stock, wine and duck; so duck and sauce cook at the same time. Makes a good sauce, but it hasn't quite the quality of the great braise.

I am certainly coming more and more to the conclusion that this book, we should frankly state, is for people who like to cook; and who want to be able to produce the most delicious things it is possible to do. There are plenty of simple things, like a plain roast duck; but when you want a great effect, you can't kid yourself with half measures, as you won't get a superb result.

Having so many duck and goose carcasses on my hands lately, I have been doing a bit of experimentation with the pressure cooker. In two instances I have browned all my bones and vegetables, then divided them exactly in half. One half I did in an open kettle, the other half in the PC [pressure cooker]. First time the PC brew was so horrible I threw it away. It had a horridly PC taste, which I can only describe as that. Ugh. The next time, concluding that it was possibly the vegetables that gave that ghastly flavor, having cooked under pressure for an hour, I put in only the bones, and added the veg. for five minutes at the end. Again it was loathsome so I threw it out. It may be that poultry bones exude something loathsome under pressure; I don't know. I shall do another experiment or two using beef and veal bones, and see what I think. I do find the PC marvelous for dried chick peas and beans. Tried it the other night with some artichokes, but it turned them brown greenish and ugly, though taste was OK.

Goose fat . . . one of the great fats for sautéing, and browning, for chicken, stock, basting roasts and chickens in the oven. Also good for vegetable flavoring . . . cabbage, etc. Sautéing eggs. It is able to reach a high temperature before burning, 428F, according to Escoffier, versus 392F for lard, and 275 for clarified butter. (Escoffier has a good bit on various fats on page 126.)

Mushrooms . . . new idea for sautéing, following suggestions of Simca's . . . and I find it works extremely well. For whole small mushrooms, or caps, or quartered mushrooms (haven't tried it yet for sliced ones). Trim, wash and dry them thoroughly in a towel, then set your skillet over very low heat, dump in the mushrooms, and let them cook, tossing once in a while. No liquid comes out, and they cook through . . . takes about 10 minutes. Then, when they have shrunk a bit, and browned lightly, add your minced shallot and oil and/or butter, raising heat slightly, and stir until shallots are cooked a minute or two. Then salt them, and there you are. I find they work out perfectly well done ahead . . . Do the dry cooking first, and set them aside (shouldn't leave them in an iron skillet, in case they might discolor or pick up a bad taste) . . . Then heat them up before serving in the oil, butter and shallots, and salt just before serving (this takes about a minute or two).

Doing book in two volumes. We'll take this up with Dorothy when we send in the complete poultry chapter. If each recipe is complete enough, giving full enough details on all the essential preparations (cooking onions for garniture, etc.) it might work out. Maybe the first volume would also have to include vegetables. It just seems to me it is going to be so long before we get the whole thing done . . . Meat and fish are going to take several years, I think! Well, we shall see.

END OF FOOD SECTION, except to say how wonderful it is that you are completely in charge of the MS. What luck for us! And that is interesting that HM will also pay for recipe testing; though when, with all you are doing, will you ever have time to do that?

Your trip to New York sounded like good fun; and must have given you the perfect change from the trials and anguishes of Cambridge these last few months. It does do so much good to get away every once in a while; and I hope you can do it fairly often. Which brings up the following proposition. We can't get away in the summer, as that is Paul's busiest time, so we are quite definitely planning to take off two weeks in May. We want to spend a week in England with our dearest Cambridge Bicknells, and see a bit of London; then a week in Paris. Would you accept a round-trip plane ticket from us, and join us? You could meet us in London, then we could return to Paris, and then come on back here, and you could take a plane back from Düsseldorf (2 hours away from here), either directly or again via London. It would be heavenly for us to pal around with you again, and we would adore showing you our Paris, and perhaps make you feel it is just as romantic as England. Our finances are in a particularly lush state this year. We live on Paul's salary, and anything from my income is used for pure squandering. We have set aside a more than comfortable lump for our home leave next year, and for a new car . . . and the rest is gravy. The point of money, we think, after you have taken care of the minimum living essentials, is to spend it. We always live on a strict budget, and pile up as much as we can for squanderings. Please think it over and say yes . . . What fun!

That is a terrible thing about Ellen Stevenson* and her scandalous book about Adlai. What is the matter with her! She must be insane. We were both very much impressed by his statement on the de-segregation question; and also his remarks to the effect that he is not going to make any empty political statements, but will say just what he thinks and feels about various problems, no matter whether or not it is politically wise. That is what I have been wanting to hear from him. Gee whizz . . . it is certainly an untraditional approach, but if he can have the strength to keep it up, it may change the complexion of politics a bit . . . and get us away from the Madison Avenue type of political fakery that is now so prevalent. It is certainly a weird political situation this time! Hate to miss it by being out of the country.

Alfred Knopf's proposition of [your] being a literary agent sounds as though it has good possibilities. Paul says your concern is not in the "justification of taking the money"; if they want to pay it, they pay it, no matter what comes of it. You are so familiar with publishing and magazines and writers, and you know so many people, and your name is known; I should think it would be very good. And the fact that your time would be pretty much your own is a wonderful factor. And I would think you'd be extremely good at it, as you are able, conscientious, and experienced. And it also might work into something more important, like roving editor or something.

Very interesting about the library.† I was wondering what would happen to that, imagining it must be a real gold mine for an institution. It is wonderful that it can be taken over in its entirety, too. Let us know

* Adlai Stevenson's former wife. They were divorced in 1949.

† Avis was looking at various libraries that might house her husband's papers. Later in 1956, on the recommendation of Wallace Stegner, they were acquired by Stanford University.

what happens . . . I hope Brandeis gets it. Golly, what a lot you have to do! Thanks very much indeed for the *Nieman Reports* with the section on Benny . . . a fine tribute; and I shall keep it. I do hope Benny knew how much he was loved and admired . . . He must have to some extent. My, I am so glad we met him; as I never would have known, I don't think, that under all that fire and passion there was such sweetness too.

To think of Mark now going out to dances. How is his love affair going? Do wish we could see him; I imagine he has changed quite a good bit in a year and a half. How is Gordon?

I must stop and get to work, writing up my ducks, and communing with Simca. We are getting along just wonderfully . . . It took a while for us to adjust to each others different approaches to things, but now we are a real team . . . each quite different, but each indispensable to the other. Very lucky. And Paul and Jean like each other very much, too. This winter we are having is perfectly beautiful, about a foot of snow all over, and very New Englandy. But I can't enjoy it fully, as this is not the type of climate for such weather, and it is causing terrible hardships everywhere. Hate to think of those poor Italians, all their crops and trees killed . . . dreadful.

Much, much, much love from us both,
Julia

FEBRUARY 28, 1956

Julie:
Charlie Curtis says I would be the biggest goddam fool in the world not to accept your offer. He says I can't afford not to go. He attempted to tell me how much money I would save by going, but didn't work at it very hard. The Rochlins, the Kennedys [Sargent and Elizabeth], the Fairbanks* and God knows who also have invited Mark to stay with them, and he would prefer to stay here.

I have ordered some more prints of my passport picture and I will go in to Boston to renew the passport as soon as it stops snowing and I get an hour. It has just begun to dawn on me that I will see Germany. Whoops. I think I will try to squeeze out an extra week, after I leave you, for England. I have very close friends in Cambridge, George and Nancy Homans†—he is Mark's other godfather—teaching there this year and

* Benjamin and Ginnie Fairbank, friends of Avis who lived in Cambridge.

† George C. Homans, a former student of Bernard DeVoto and a Harvard sociology professor, and his wife.

they have been imploring me to come over and stay in their big house. He is this generation's direct descendant of the Adams's—his mother whom I adore was Henry Adams's* favorite niece. And in London I can stay with Nancy Balfour of the *Economist* at the drop of a hat, very fashionably in Eaton Square. Let alone my rowdier friends. I will just wrench things at this end so that I can relax and take my time with you. Should go into training for Paris, and will try to lose seven or eight pounds before I take off.

Most wonderful thing about all this is that I am longing, have been longing since Benny died, to see you both. Can hardly believe that it is about to happen, if I live very carefully and don't let anything smite me. I have been oddly healthy this winter—not even a sniffle, and now that I know about aspirin, practically no aches and pains. May it continue.

So let me have dates, or approximately so. You must tell me what to do at this end for plane reservations. I can't believe it!

Having dinner tonight with the de Santillanas and will report what they have to say.

When Benny died I felt driven to make a whole lot of decisions—thought I was carefully figuring out the future for me and mine. Everybody said don't, coast a while, coast a year, don't make any decisions except when absolutely necessary. I know why now. Fate or something takes a hand and things happen that you could not have dreamed of. You might almost begin to believe in God or Santa Claus. New and cruel blows, too—but I do believe things have a way of evening up.

This just a note—the librarians are taking over this week. Houghton Library this afternoon, Athenaeum tomorrow, Brandeis Friday. But Stanford still looks best. I won't even have to make this decision—Charlie [Curtis] and Lyman Butterfield† will make it.

Hastily and with much love,
Avis

JUNE 8, 1956

My loves:
This is the first chance I've had to sit down uninterrupted, I hope, at the typewriter. I'm still a long way from having sifted out impressions and putting my thoughts in some sort of order. But today I have a

* A nineteenth-century American journalist, historian, and novelist.
† An eminent historian, a former student of Bernard DeVoto, and the curator of the Adams papers.

hope. I didn't really stop spinning around in time and space until yesterday—perhaps it would have come sooner if we hadn't had so many hours tacked onto the return flight. We left London, we left Shannon right on time—twelve people on the plane!—but sometime early in the morning we ran into weather, so instead of landing in Boston at 5:30 we proceeded to Philadelphia, and set down at 8:30. There was even talk at one time of going to Miami. The ways of planes are beyond me—I don't see how the weather could have been that bad. The Philadelphia airport is new and luxurious and we had the run of it—rooms to sleep in if possible, free food and drink, etc. Free phone calls home. And oh God, what a let-down the first American food is! Appalling. Especially for me, with the conditioning of the previous four weeks. We didn't get off the ground until 3:15 and by that time were exhausted—something about not getting your clothes off in 36 hours is degrading to the spirit. When I finally got home I hardly knew where I was—and there was old Mark DeVoto, cutting the grass, which looked as if it hadn't been touched all the time I'd been away. But the house in beautiful order and Mark obviously happy and in command of his life, apparently not having missed me in the slightest. Nothing whatever went wrong. Both boys behaved in the most exemplary fashion—a quite unexpected bonus of that glorious trip is the knowledge that they are quite capable of managing their own lives, thank you, and even of getting along peacefully and fruitfully together. This means I will never again be tied down by the young, and can follow my career where it takes me with a calm mind. So when we are counting up my thanks to you, let's include this important item. Not that I can ever thank you adequately, and I will not try to because you must know how I feel about it all.

All has been a sort of ordered confusion since Monday afternoon. Telephones and people coming in and three baskets of mail to be dealt with. No unpleasant shocks or surprises.

. . .

No hour goes by without my thinking about you two and the fun we had together. I haven't had time yet to copy down my notes and try to sift them out, but of course food occupies much of my thoughts. That soufflé at Des Artistes.* The duck at Rouen.† My eternal soles.‡ Ohmyoh-

* A Paris restaurant.

† The restaurant La Couronne in Rouen, where Julia ate her first meal in France and to which she often returned.

‡ Filet of sole.

myohmy. I can taste them yet, and I can tell you it doesn't make my descent to earth any easier. I am going to lick that Normande sauce when I get time to do anything in the kitchen, if it's the last thing I ever accomplish. Though what to eat it on remains a mystery. I must find out if the fish man can get me a whole sole, which would at least be more satisfactory than the filets. Flounder, I mean. Why can't we have decent fish? Though to be sure, Mary and I just had some nice lobsters for lunch, at fifty-nine cents a pound, and very sweet they were too. But I look on the usual display with a jaundiced eye. Also, Paul my love, we've had lovely green asparagus three times, as it won't last much longer, and I do agree with you that it is better than the white—infinitely more flavor. No local strawberries yet, as the season has been cold and late—perhaps there won't be many strawberries. But we'll have a shortcake soon or I'll know the reason why. And I did a *suprême à blanc,*[*] finished off with cream, and Mark was enchanted. Rubbery, I find, <u>is</u> the word to describe how it feels—and it was just right inside. Boning couldn't be more simple. It makes a terribly nice meal for two, and I expect to try all the versions.

As soon as there is time I must write my old parents a blow by blow account of my travels, and this is when I think my impressions will be sifted out and put in order. Dunno when I can manage it though—something always comes up. As Mark having eleven of his school friends to dinner last night—I will admit he told me at once he wanted to have this party, but I couldn't pin him down to the exact number until yesterday. So I fed a round dozen noisy young males—roast beef, risotto, tomato and cucumber salad, and three quarts of vanilla and chocolate ice cream with chocolate sauce. They demolished eight pounds of beef—six quarts of milk. Lord, what appetites. But such pets. They are all around 16 and very courtly, full of graceful speeches and anxious to be helpful.

Rochlins in somewhat improved health and spirits, and I went over there my first night home and submitted to cross-examination—they <u>really</u> wanted to know what we did and ate and where we went, and all the news about you two, and are still chortling over that postcard. By contrast my kids are totally uninterested in what I was up to all that time—perfectly normal I would think. I'm going up to the Vincents later this afternoon to deliver perfume and they will want to know all about you—V. [Vivian] Katz asked dozens of questions also.

I could have brought home the crown jewels without any difficulty

* A boneless chicken breast in butter and lemon sauce.

—my bags went through under bond from Düsseldorf and were totally ignored at customs in Philadelphia, where they were too busy giving the works to an unfortunate Burmese student to pay any attention to the rest of us. Might as well have brought all those knives along.

. . .

Dear England. Half the men in dark blue suits, bowlers, tightly rolled umbrellas, and they all marched together, followed by the ones in unmatched coats and trousers, caps and casual hats. But all wearing rows of medals. A nice little sort of thing in England, and think it is very touching that they cherish their class-consciousness—can you imagine Americans behaving that way? So it was a very pleasant few hours, my last look at England for probably a good many years.

A heavenly trip, start to finish, especially seeing you and relaxing in the companionship the Childs seem able to give me. No strain, nothing hidden, never pulling in different directions. Seeing things I never would have known the existence of with other people. It was awfully good for me and the benefits will increase with time. It gave me the change and breathing spell I needed so terribly.

I am beginning to get things together to send to you and to Simca—it will take a bit of time. Sending four copies of the *New Republic*—I have not read them completely and won't have time to, but they are average issues and I think you will agree that they give a kind of coverage you get nowhere else. Certainly in no other weekly. It's very desirable to have weekly coverage on these political things. I don't know who T.R.B.* (inside front cover) is, but it's the most biting political writing today anywhere and at the same time objective and fair. I know how much reading you try to cover and know, Lord I should, how hard it is to get time for very much of it—the point of the *New Republic* being that you won't have to piece it all together from the daily papers, or wait for the delayed reaction of the monthlies. I just plain couldn't live without it. Must look up your APO address instantly. Time out. Well where it is I do not know. I went through one file drawer of letters, and three address books. Humbly beg you send please and I will this time put it in <u>all</u> address books. Phooey. When it comes will send *NR*'s immediately, air-mail. Will I ever get more organized?

Found, while rummaging for address, two communications from my insurance broker and must retire to sofa to find out what they mean.

* The initials assumed by the writer responsible for the issue's lead editorial, often Richard Strout.

Never can tell whether he is sending me a bill or a credit, and damned if I think it is all my fault. Will finish this tonight.

After dinner, and about to go to see Grace Kelly in *The Swan* with Gordon, so will hurry. Mark telephoned me at the Vincents to tell me about Eisenhower's illness* — sensation. Have since listened to Louis Lyons and whatever else I could get on the radio. Golly. How I wish you were here so we could buzz about it. Vincents elated, and I can't help being somewhat so though I suppose sorry for the man, but I still think he's a dope. Our chances have improved a thousand percent. How I wish B. was here to hear what he had to say about it — as on all occasions of any questions whatever. Mark made a good crack — deadpan — What if they opened him up and found a golf ball? Well, well and well. Can't wait to see the papers tomorrow, can't wait to observe the confusion of the R. [Republican Party].

This is far from the letter I had in mind and many things still untouched on, but will get it into the box and continue later. Send APO number again, please. I miss you dreadfully, every hour on the hour. So much we didn't get around to. And now I must go because I can't remember when my son asked me to the movies last, and I'm touched and pleased.

Loads of love, all kinds sacred and profane,
Avis

JUNE 16, 1956

Darling J:
Here I go again, ignoring all those letters I owe, to write to my Julia. Dunno why it is, but have the most awful reluctance to write a letter, or sit down at the typewriter, except when talking to you. Wonder what the head shrinkers would make of that?

Everybody horribly restless because after four-day frightful heat wave a real humdinger of a storm is toying with us — cold front trying to get through. It darkens up, rains a few drops, vagrant breezes whisk around — and then the sun comes out and it seems as hot as ever. Been going on all afternoon and finally drove me in desperation to the kitchen to make another batch of mayonnaise. In this nasty weather we eat nothing but cold food, on which mayonnaise is welcome — lobster, salmon,

* After severe stomach pains, the president was operated on for a bowel ailment.

chicken, shrimp—I'll be damned if I'll turn on the oven or the broiler until the air is fit to breathe. Made some mayonnaise a few days ago, Mark using the electric beater, me adding the oil—today did it alone. So, if you can bear a few worthless observations. Used your recipe both times—dated 1952. Did you ever send me a new recipe on this? Have vaguely in mind that you did, but went carefully over all cookbook ms. and couldn't find it. So—think you should specify what oil used. You just say oil, and we decided all olive oil is too strong. I used both times one cup Wesson oil to one half cup olive oil and like it much better. First time we did it, Mark kept the beater on high from start to finish—think this is a mistake as mayonnaise gets too stiff for my liking. Today kept it on low, and am very pleased with consistency, which is creamy and light while still holding its shape. (It's so damn hot I decided to put it in the ice-box—I did the other day too and it didn't separate. Left out in this weather am sure we would get ptomaine.) I used red wine vinegar both times—OK. Think I got a bit too much mustard both times and will try only a pinch next time. There's a problem about an oil-dripper—would be very hard to cut a wedge in a cork so it would drip properly, I think. I used an eyedropper, fairly fast—I'm sure "drop by drop" should be modified, as I think we talked about that once, but one person doing it alone, beater in right hand on low speed, eyedropper and cup of oil at left hand, can manage perfectly well. Only with the eyedropper it took fifteen minutes or a little more. Not 7, as in recipe. Also I found, once I began pouring in a thin stream, that the oil absorbed much better if I put the cup down and turned the bowl as I beat, rather than trying to cover the bowl with the beater. That's a clear sentence for you. Anyhow, I'm very pleased with this second batch—delicious.

. . .

My God it is hot, and the sun is now shining brightly. I cannot diet in this weather, and have paid it no mind, but what with living on cold fish and watermelon and mangoes, and sweating gallons a day, it seems to be coming off. Your Calvinist regime fills me with admiration and horror. Don't see how you can do it. Do be human and say you sneaked out to the icebox at midnight and ate something rich and filling.

I have had a little correspondence with A. Knopf, and we have jointly decided to wait a few months before I get to work. This is the deadest season in publishing, and since I'm taking over on a six months trial basis, it seems silly to tackle it now. Lord knows I have plenty to do to keep me occupied this summer, and my finances are not pressing. Checks—insurance, social security, investments, come in driblets, but oftener than I expected them to. I am not aware that we have cut expenses to the bone, but somehow my checking account remains healthy. I can't understand it, and am vastly reassured. If you are worried that I spent too freely while I was with you, relax. I know exactly what I spent of my own money, and it is a good deal less than I expected to. Charlie couldn't believe his ears when I told him. Of course this is the cheapest part of the year for me—no school bills, no heating—big taxes and insurance payments come in the fall. I have a thousand dollars advance on the Cady Hewes book waiting for me at HM, which will take care of the house taxes and part of Mark's school bills. Got a check for $2500 in royalties the other day. I'm almost beginning to think I can live on my income, but suppose that is crazy. And when I start working in September, and renting the top floor, there will be $400 a month right there, though I'll lose my social security. My old man really took care of us, bless him. Perhaps some day I will even learn to enjoy life without him, but it will take time.

Think at this point a long cool shower is indicated, and a clean dress. Among other things that recommend European living is the climate. Much, much love to you both and I miss you terribly but am very glad, for you, that you aren't here right now.

Avis

Plittersdorf in the Rain

JUNE 22, 1956

Dearest Avis:

No time to write, and have had none for weeks, it seems, even to you, my soul mate. I suppose one reason we can write so easily to each other is that, for one, we have established the rhythm, and for another, that our lives are not much involved with each other, actually. Perhaps if we lived next door, we would have developed curtains and veils and various tender heels. Anyway, it is lovely to be perfectly at ease, and to be able to discuss anything at all; and may it remain ever so!

Just can't seem to pull myself together lately, and find myself all over the place, especially from a cooking point of view. I feel I must do some vegetables, while they are in season; I must think about the one or two basic pastries we shall include, and try them out continually, thinking deeply on how in hell to explain them clearly; soups for *W's D* [*Woman's Day*], which I must send next week; and the end of the poultry. Then there is Plittersdorf life and German; trying to keep up on the news; and most important, WIFEDOM. Phoo. I wish I needed only 5 hours sleep, that's what. And it has literally rained most all day every day since you left, and that is going on three weeks . . . except it now seems as though you had left months and months ago. And poor Paul is very much down in the dumps physically, with a protruding hemorrhoid, stiff shoulders . . . As he says, if he could only get good and drunk it might make him feel happier, but his liver won't take it. Presumably this will all pass, but it is pretty discouraging to him, feeling that he is rather quickly breaking apart. And I don't think a good operation for ileitis would make him into a new man, either. (Now I am not sure how to spell that word, which is odd, after having read so much about it.)

Anyway, we read that your heat wave has broken up at last. My, I do hope that will be the last bad one this season for you. Probably a good idea you aren't starting in with Knopf right away, so you can get all the various little things settled. And good that you think you can probably manage on the income. I do hope the BOM club* puts out Benny's histories as a book dividend . . . That would be nifty, wouldn't it!

All those lobsters in mayonnaise sound delicious. They cost about $5 a piece here, so . . .

MAYONNAISE. Shall try your proportions, and use the electric beater. But I do think adding the oil with a dropper is going too far! When you measure it out, add the first half cup or so with a ¼ tsp measuring spoon. Or, if you are steady, you can just pour it in droplets from the cup . . . and keep your eye on the cup, not on the beater. You can perfectly well add it from the bottle if you rest the lip of the bottle on the edge of the bowl, and keep your eyes glued to the bottle, then tilt bottle back, and observe the mixture. Certainly think we must have directions for using the beater, and shall note that low speed is best. Myself, I always use the bowl and wire whip, as I am used to it, and I hate to organize the beater and then wash it off (lazy) . . . But my hands are stronger than yours, which makes quite a difference. (Most of the sauce recipes need doing over anyway, as we have progressed quite a bit since that period.)

Tonight we have to go to some kind of a diplomatic dance, which sounds as though it might be quite dull; and besides we know practically nobody, which will make it even duller . . . but is why we decided to go, in case we might meet even one kindred soul. We have invited 3 rather excruciatingly pedestrian couples here for drinks beforehand; and

* Book-of-the-Month Club.

I have been making puff paste, and croûtes, and a quite nice filling made of canned clams, canned shrimp, fresh mushrooms, minced scallions and a Sauce Parisienne to bind it all together. Tastes very good, I must say.

MAYONNAISE again. Will also put some in ice box. You know, if it does turn, you can put a TB of the turned sauce in a bowl over simmering water, then gradually beat in the rest. Or, sometimes, just putting the whole mess into a bowl over simmering water and beating it will bring it back. Certainly, in our climate, you have to be able to put the stuff in the frig.

My, I was reading the most disheartening article in *W's Day,* contrasting an old Do-Do [dodo] who had time to do things the old fashioned way, and a smart young thing who did everything the "New Modern Way" . . . using cans and boxes, and frozen stuff. After cooking a chicken with boxed dressing, and doing several other smart things, she then served frozen asparagus with the new modern sauce of Hellman's mayonnaise heated up with some lemon juice, instead of that silly old fashioned old Hollandaise sauce. Ugh, and Hollandaise is so quick to make. Maybe we should do a little bit for *WD* on old Do-Do methods, which may take 3 minutes longer to do, but which do have a rather nice and old-fashioned taste. What is the country coming to, I'd like to know.

Now I must get to work again, and must also somehow concoct a letter of condolence to a cousin whom and whose husband I have never liked one little bit at all . . . have rather the feelings you did about that French artist cousin . . . never can do anything right, and so forth. What to say, when if I were to be cold-blooded . . . Well, I won't even express my thoughts.

Ach, this Ike business. I got a most typical letter from my old Pop, "Everyone has been feeling better here everyday as the word of the President's splendid recovery gets better daily. Butler[*] of the Dem. Nat. Com. states these reports are put out for political advantage. My reply to him would be: it reminds me of the meager false reports on FDR, they are no different!" I was wondering how "they" would relate Ike to FDR, and am glad to see. That old double standard again, as usual. I also like the use of "everyone," that great, undefined body of noble people, the people who really have the interests of the country at heart, and who are the true, blue Americans. It certainly will be interesting to see the next Gallup report, on how the great American public feels. My, I hope Adlai gets

* Paul M. Butler, the chairman of the Democratic National Committee.

the nomination, and then leads a great and noble campaign, which can also be simple enough to inspire the all-too-common man.

Do hope you are feeling better mentally wise. I imagine you will know it as a good sign when men cease to seem like paper dolls. And I do think it will be awfully good when you finally start in to work.

Much, much, much, much, much love,
Julia

AUGUST 23, 1956

Dearest J:

Just shows you how our minds run together—your letter crossing mine of day before yesterday. So though I have got to get through a book this afternoon before 6 when I settle down to look at the final production (it can only be called that) at the Cow Palace,* I rush to get a page off to you. Of course what I am most excited about is the possibility that you may be coming back to the USA to live. Wow. Would it be Washington? How come, Democrats in an election year?† My first reaction is the purely selfish one that I would see you both oftener—I have had no end of invitations to come to Washington but so far have not had the slightest urge. But with you there I could cook up plenty of excuses for the trip. And you would have to come up here to see your publishers from time to time, wouldn't you? But how do you feel about the possibility? Must feel a bit like an earthquake, after so long away. But I will try not to dwell on this, or count on it too much, for fear of putting a hex on the proceedings. And it would slow up the cookbook, wouldn't it? We should get an exchange fellowship for Simca, or something of the sort.

And Paul should see a Boston doctor, of that I am convinced. Is he depressed over this long continued discomfort? It just must be specially hard for a man who always was so strong and graceful and coordinated. Should think he would be very glum. I am just delighted about the vest and hope he wears it to shreds—of course now he ought to have a velvet jacket to wear with it, and a linen shirt all fine tucks.

I am all right, as I told you, only so immersed in politics that I am losing weight rapidly. You would not believe the goings on at San Francisco. Anne Barrett comes over in the evening and we watch in fascinated horror. I don't know why I was so naive as to believe that there would be a

* The San Francisco arena where the Republican National Convention was being held.

† Avis seems to be speculating that the Childs are planning a return in anticipation of a GOP victory, which would likely result in Paul's losing his post to a Republican.

floor fight about the vice-presidency. Now that Ike and Dick have been nominated by acclamation I see that I should have expected it to happen all along. It's just like Russia—one name, one ticket, and anybody who had guts enough to raise a hint of dissatisfaction would be committing political suicide. Big Brother is watching you. I can only pray that this country retains enough of its dislike of being pushed around to reject what we have just seen at San Francisco. The wolfish grin, last night, of Tom Dewey.* The death's head expression of Harold Stassen† as he capitulated—he looked brain-washed. As he was. The Team, plus Whitaker and Baxter‡ and Madison Avenue, will now attempt to brain-wash the country. Do you think they will succeed? Isn't there some revolt in those delegates who sat there hour after hour, grinning and bored, whopping up their synthetic excitement when the cheer-leaders sprang into action? Will they be resentful when they think it over? Surely some of them must have some kind of rudimentary intelligence. We simply could not believe our eyes and ears. But we got some small comfort from the open cynicism and the barely veiled disgust of the commentators. Gosh I wish you had been here to see it. And to join me in taking Phenobarbital which I have been doing regularly to keep from upchucking.

I have got to read that damn book now, and it shouldn't happen to a dog on such a supremely beautiful afternoon. The kids are out and I have to cover the telephone and the doorbell, because students are beginning to look for rooms and I am dying to get my top floor settled—in fact I am getting a bit worried as all the ones who have come so far have wanted more cooking facilities, as I would myself, to be sure. But short of putting a whole kitchen up there I don't see how I can let them do more than breakfasts.

Darling if I write to you while you are in Berlin will my letters be forwarded? I am so distracted right now I can't settle down to a real letter but after I go to NY I'll probably have a lot to tell you. Meantime my fondest love and I think of you all the time, especially now that maybe you are going to return finally to your native land.

Avis

---

* Thomas Dewey, the Republican presidential candidate in 1944 and 1948, who persuaded Eisenhower to keep Nixon on the ticket.

† An Eisenhower cabinet member and former presidential candidate who led a movement to "dump Nixon" at the convention.

‡ Clem Whitaker and Leone Baxter, a husband-and-wife political consulting team who handled the publicity for the convention.

In 1955, Julia assisted Chef Max Bugnard in preparing a luncheon for Le Cercle des Gourmettes.

AUGUST 27, 1956

Avis dear:

Thank god you are all right. We were really so worried; we were casting about as what to do, even thought of telephoning you. Next time you get tied up with politics etc, just send us a postcard once a week so we shall know we needn't worry. Please!

My, would we had been with you for the convention. The good old *NY Times* came through with full transcripts of everything, or about everything, but it ain't the same. Interesting that the commentators on TV seemed to view with distaste [the proceedings] at the Cow Palace. Did they really, or did you read your feelings into them? I suppose that is hard to say. I, too, could not believe that they would really railroad Nixon through until I read a [Walter] Lippmann column a few days before they did, who explained to my satisfaction how it had all been planned for weeks and months ahead; and all the power boys like Dewey were rallying around behind him, and jockeying for their eventual position when he takes over. I wonder how much any of this affects the true believers; or those who might have doubts. One of the phenomena we find incredible and disturbing, having now met 4 Republicans from the US who are visiting around here, is their flat-out worship of Ike, and their refusal to think about anything happening to their symbol of safety. "I am willing to take the chance, if there is one. I am his. He is mine." Etc. I don't remember having run into this before, but imagine there was a great deal of it during the Roosevelt days . . . and he and his certainly did a great deal of railroading through conventions. I keep feeling, as you do, that there is a fighting chance for Adlai and Kefaufer* . . . but can they break through that idolatry onto solid ground?

Do keep us informed on any backroom gossip, and wonderful tidbits like Little Bo-Poop† . . . which we loved.

Glad the Betty Allen Reck‡ visit turned out so nicely, and you think things will work out. And Alma Brodie§ sounds nifty. I hope we shall meet her sometime. How does Mark manage to do all he is doing? Please thank him fondly for his letter, which I shall answer when we are in Berlin. Wonderful letter; wonderful boy. Did Gordon find another job in the interim? I hope so. My, to think we shall probably see all of you in 2 or three months. Nothing definite from that stranded whale back in Washington, which presumes to call itself the US Info. Agency. But we are tentatively expecting to leave from Le Havre on or about Nov. 15th. Heaven knows whether we shall be able to get in any leave before Paul starts in the new job, but we are hoping so. But we must always also remember that nothing at all may come of this. Even so, we are getting quite excited about it; and loving the thought of being home again after so long. My, how shall we fit ourselves and our 8 years of accumulation into that little house? All these books, plus all we have there anyway, etc. etc. Certainly shall get a dishwashing machine first thing. But don't think we shall do any fancy fixings up; as it is not a permanent residence anyway. But we shall certainly have air conditioning; and I shall certainly get a new stove, a black one, gas. And it will be so heavenly to have some friends again! I find I am beginning to relax here, and lose interest in trying to make anything out of Plittersdorf. It will be rather a blow if we stay on after all. And we shall certainly have to come up to Cambridge a great deal; and you shall certainly find you will have business to do in DC, I hope.

---

* Adlai Stevenson received the Democratic nomination for president again in 1956. His running mate was Estes Kefauver, a senator from Tennessee.

† The reference is unclear.

‡ A longtime friend of Avis and the wife of Henry Reck.

§ Alma Routsong Brodie, the author of *A Gradual Joy* (1953), a novel admired by Avis. Brodie visited Cambridge that summer.

This will, of course, delay the book, as usual. Our third move in 4 years! However, things are slowly moving along. Ducks are done, and I am starting on geese as of today, which are about the same as duck, so little problem. But I really have to be home for turkeys. In fact this is just about the right time, from point of view of book, to be coming home. Much better to get that last part done, and the final re-writing, in the US, where everything such as ingredients can be accurately checked.

Saturday we drive up to Berlin (Sept. 1st).

> Hotel Steinplataz (Probably for the month of September)
> Uhlandstrasse 197
> Berlin — Charlottenberg 2
> Germany

Very slight improvement in Paul's shoulders. He can now put on his coat without too much pain. We shall certainly see your specialist chaps in Boston . . . or if not, they might have someone in Washington to recommend. He is not unduly depressed; but just hopes to heaven it is not going to be something permanent he will have to live with the rest of his life. It's a shame for him to lose all that good muscle tone, though.

Otherwise I have nothing at all to report as life has been very dull indeed, and it is still mostly raining. We did, in spite of the weather, drive up to "the castle," which we wanted you to see, on Saturday. Managed to have a nice picnic in a meadow facing it, and without rain. It is a beauty, dating from around 900 on up through the Renaissance. One of the very best we know anywhere.

Much, much love and write soon. When do you go to NY? Do hope you see *My Fair Lady.* We heard the record the other day, and loved it.

Write every day, and much, much love to you all,

Julia

SEPTEMBER 17, 1956

My dear:

If you were here, I would be speaking in a squawk. I will write in one, too, very briefly because I am bushed, but feel a great need also to be in communication with my loved ones. My first full day of work for Knopf, and I am loving it, but it is distracting working here. People come in all day, either not knowing I had been away or to welcome me back, and Cambridge is suddenly in full swing, thank God. I had a bellyful of being alone this summer.

I wrote you a note from St. Petersburg and I hope it reached you. My old pa telephoned Sunday midnight last week to say my mother had just died in the hospital. He was all in pieces and wanted me to come down so I flew down next day and took over. It was a rugged five days and I didn't know how tired I was until I got home and slept twelve hours. Well thank God is all I can say that it was so quick. Cancer of the breast, and two more weeks in hospital and then all over. She was lucky. So was Dad and he knows it now. The thing I have always dreaded most is having either of them require long nursing, which is so terrible for everybody and would pretty well put me in the soup financially as well. I had to arrange everything, and funerals in the hinterland are something—viewing of the remains and all that. But it all went off rather smoothly, none of the horrors I had expected, and I cleared everything of hers out of the apartment, but quick. Only thing to do. Incredible woman. She saved things like the Collier brothers.* Eighteen boxes of notepaper and she used to write me on the backs of old Christmas cards. Unopened boxes of stockings I had sent her. Forty years of medical clippings, some of them yellow with age and quite outdated. And yet when she was sick she never wanted anybody to know anything about it until it was over, for which I am deeply grateful. Nobody but Dad knew she was in the hospital this time. I never got along with her, but I will hand her this, and it's a great deal—she never clung, or whined or complained, and she let me live my own life.

My father is 86, thin and reasonably spry and except for bad eyes, in good health. He seems to be the only Democrat in St. Petersburg and we cheered each other up considerably by discussing politics, about which he knows a great deal. He's a cutie—Scotch and deadpan and full of wry humor. And now an old, old man. St. Petersburg is about as non-U[pper class] a place as there is in this country—dreary beyond measure to me, but he is rather used to it now. And by God they do take care of the old people—everybody exceedingly kind and gentle and friendly. Living is incredibly cheap. Cafeterias are just wonderful. For a dollar and a quarter you can get a big piece of good roast beef and everything that goes with it, good vegetables, fine salads, and superlative apple pie. He does this once a day, and picks up his other two meals in his own little kitchen. Not much appetite at that age, so I left him three bottles of rye with or-

* Homer and Langley Collyer, wealthy New York City brothers and compulsive hoarders who died in 1947 amid mountains of trash.

ders to take a little nip in the evening for his appetite's sake. He ate lunch and dinner with me at the hotel while I was there and had a drink each time — the first in years. His landlady is one of these wonderful lower middle class types who never read a book in her life, but is pure gold, full of energy and kindness. And he has many friends close by. So I will try not to fret about him. But he knows and I know that the next call from St. Petersburg will probably be for me to go down and bury him. Such a nice man. And it's just hell to be old and have no function in life.

My small glimpse of the South appalled me. The planes are full of business men — all building the hell out of Puerto Rico or something of the sort. All Republicans. All segregationists. Wow. And yet St. P. has an absolutely dandy Democratic paper with a first class editorial page and all the good columnists. So there must be Dems. somewhere. Heard Stevenson's Thursday night speech at the hotel — he's fighting vigorously and looks good to me. Some wild swings and mistakes — such as the suggestion about the draft.* Didn't like that. But Reston,† or somebody in the *Times,* says these things are trial balloons, and all will be brought together into firm policy before too long. Hope so. Us eggheads miss the high-level Stevenson of '52, that clear voice of unassailable integrity. But us politicians know this kind of campaign will get more votes. Me, I just like him — if I expected to like all of him, everything he says, I'd be a dope.

I had to wire Nancy Balfour not to come and felt terrible about it, but couldn't do anything else as I didn't expect to be back by the time of her arrival, and wasn't. She wanted me to stop off in NY and see her on Monday but I had to be back on the job. Better luck next time. I was just dying to talk to her after she had covered both conventions, and if she wrote that piece from San Francisco in the Aug. 25 *Economist* she got to the heart of the matter as nobody else has, for my money, and we see eye to eye on everything. I will write her in London tomorrow, if I get half an hour, and say so. Wonderful woman.

And Democratic money is showing up, glory be. This very confidential — Lyman Butterfield's father-in-law is Cyrus Eaton of Cleveland — heavy industry, railroads, etc. He wants to make a donation which I gather will be substantial indeed, and he does this through the Butterfields so he won't have to be hassling with his business associates, which

---

* Stevenson proposed ending the military draft.

† James Reston, the Washington correspondent for the *New York Times.*

is OK with me. I have guided some of this to the committee in Washington, and some will come to the Mass. Committee, which is winging into action right after the primaries tomorrow. I have enjoyed this episode mightily.

My three second-year law students are in residence—they turn out to be Prosterman, Ocheltree, and Langston. All darlings. They keep the most curious hours, working in the Law Library all night, but are quiet as mice and I just love them. Unfortunately my hot water system is on the blink and this is embarrassing—but they seem to understand and I hope they will heat shaving water on the electric plate. A piece of the gas heater just vanished, and why I will never know. I am harassing the gas company with tales of horror but it may be a day or two. Woe.

Gordon started in at Wentworth [Institute of Technology] Friday—ninety dollars worth of books and four hours of homework every night. Golly. He seems serious about it all and I pray to the good Lord that he can keep up. Mark starts Wednesday, and at this moment seems to be coming down with a cold. Woe again. But his colds seldom last more than 24 hours. I came back from blistering St. P. to a Cambridge strongly resembling the Rhine Valley with rain and very low temperatures, so we have got the furnace on. In mid-September, for heavens' sake. The garden is drowned, and the grass horribly needs cutting and can't be touched. We'll have a busy weekend.

I can't even remember whether I got round to reporting on my New York adventures—which were good, including two lunches at Baroque. Unfortunately, though I felt perfectly self-possessed and at ease, something must have been stirring in my subconscious because I had a nervous stomach the whole time and took Phenobarbital to calm the flutters. Pity, with that good food. Three sessions at Knopf's—once to lunch with Alfred, next day with a lovely guy named Herb [Herbert] Weinstock* who is to be my immediate mentor—he keeps the office running and is their expert on music (about which he has written four books) and Latin America. He is just a darling. Last session with Alfred going through the file cards on New England contacts, me racing to keep up. I met all the editors. Over the rank of secretary, there is no woman in the organization except Blanche.† So I hope to make myself indispensable. It's a very

* An influential editor at Knopf.
† Blanche Knopf, Alfred's wife.

impressive list and I'm boning up on the catalogue as hard as I can. They are weakest in up-and-coming American authors—the only one of note right now is John Hershey.* So I'm looking principally for fiction and non-fiction by young writers who look as if they have got staying powers. I am not supposed to operate outside of New England, but if I get a tip on someone in New York, or San Francisco, say—I'll get credit for it.

Of all things, today, my very first day of real work—I believe, I hope, I have stumbled on something that Alfred may like very much. No use raising your hopes, or mine, but it has to do with the papers (24,000 pieces!) of Frederick Law Olmsted,† now in the Library of Congress. The woman who had first dibs on them—Laura Roper‡ who is a great friend of Mollie Brazier's and didn't you once say you knew her?—has given up because in middle age she has settled down to having babies. So she suggests Charlie McLaughlin,§ Eleanor Murdock's son, who is just pulling out of that awful polio—I must have told you about him. And the heirs agree. I have just written Alfred all I know about it—no other publishers on the trail yet I hope and believe and asked him for permission to put an option on the book Charlie intends to do. It's right up Alfred's alley and a couple of years ago he published Olmsted's *Cotton Kingdom,* edited by A. Schlesinger Sr. This would be a real plum for me. But there I go, counting my chickens. Charlie has no publisher even in mind yet, and seemed very pleased by the idea of Knopf. So we shall see and I hope to know in a few days. The suspense is terrific. Please keep this very much under your hat for the present.

Well this has turned out longer than I thought and I have not even scratched the surface, and I'm exhausted and going to bed with a nice detective story—I got a batch of real beauties today—and relax my poor old mind with a little fluff. God knows when I can write regularly. Tomorrow I start out to write fifty or so letters to useful people around here telling them what I am up to and asking for leads. What a mercy I can type fast.

The food in this house is unspeakable. Please come and feed me. I

---

* John Hersey, the Pulitzer Prize–winning author of *Hiroshima*.

† The nineteenth-century landscape architect responsible for many of the nation's most famous public parks, including Central Park.

‡ Laura Wood Roper, an author of biographies for young adults.

§ Charles C. McLaughlin would spend decades editing Olmsted's papers.

think about you both twenty times a day, between telephone calls. Tell me more about Paul's triumphs in Berlin, and are your plans to return any firmer?

Much, much love,
Avis

Garrett Mattingly, B's best friend, in NY, is working desultorily on some new words to the tune of God Bless America. Unfortunately I can't lay my hands on what I wrote down—but the last verse starts

God votes Republican,
Ike told him to.

and ends—

God votes Republican
And so—must—you!

I think something can be done with this. If I only had time.

JANUARY 3, 1957

Dearest Avis:
Good to hear from you, and know that you are still alive, and gradually coming out of your depression. That's right, you are an orphan! I can imagine, to some small degree, the feeling of being suddenly cut off from the trunk, or whatever a just simile would be. There you are; and even if you haven't been conscious of it, you have clung in some way to that antecedent. And now you are it! And they will cling to you, and so it goes. A sobering thought. Such good news that you like Mattie,* and that she and Gordon are really in love. That sounds very hopeful, and it is perhaps just what he needs. Please keep us closely posted, as it all sounds wonderfully encouraging. And how is Mark?

The E[lizabeth] David work sounds devilishly difficult. But don't you think, after all the clarification you have done, that people will understand it? It would seem to me that if they know it is done by an Englishwoman, that they will rather enjoy some of the Englishisms and, as long as it isn't a book on pastry where measurements count enormously, it will not make too much diff. if things are in ounces, etc. No, we have not as

* Gordon's girlfriend.

yet run into decimal measurements. I just wish, anyway, that we would change to the metric system, where everything is in 10's and 100's and there is none of this silliness of 4 of something making something, 12, 16, etc., etc. Idiotic.

Fine that you are blowing yourself on a few clothes and housewares. There is nothing like a bit of new raiment to make the new woman. Myself, I bought two fine hats in New York, picked out by Paul, at $2.95 each. Why pay more!

Our Christmas at Lumberville was very nice indeed. The Sat. before, the two girls were down with two attractive boys, and Charlie and Freddie had their annual Christmas Carol party, which was very gay with lots of noise and singing of sorts, and mulled wine, etc. The weather was mild and fine, though we had come up from Washington in a wild rainstorm. I had some more turkey trouble, damn it! I had decided, seeing it is not possible to do 500 turkeys for a test, that I would take the findings of the Poultry and Egg National Board as to timings, and just put them on a chart, giving them the credit. So we have a 25-lb. fresh turkey from a very good turkey farm in Lumberville, and my chart says 7 and ½ to 9 hours at 325, and I want to follow things exactly. So the turkey is done in 5 and ½ hours. Hell and damnation, and what is anyone to believe, and how is one to write a cookbook anyway. So as soon as we get back here, I go out and buy two turkeys, one 14 lbs, which the two of us chewed on last night, and which will be finished up when we have a luncheon party tomorrow. 14 lbs. at 325, 4 and ½ hours (P&E Board say 5 to 6). But I think 325 is too fast for a big turkey. It can start out at that, but has to be lowered to around 300. Phooey. My other turkey weighs 25 and ½ lbs.! I got it on sale at the Safeway, and it is frozen, so I am defreezing it in the refrig, according to best directions on the subject. I think that I shall just go back to the old timings I cut out of the *NY Times,* some years back, and precede the time table with a statement that, as far as we can see, it is impossible to give accurate timings, and that one should count on the longest, just in case, but that if the shorter one is what yours takes, never mind, it will stay hot for an hour. (I timed that and that is true.) Triple Phoo. Whom is anyone going to believe in this business, anyway. And furthermore, I don't like turkey! And furthermore, I have just had the Turkey section typed up. Well.

New York, to get onto a pleasanter subject, was wonderful. We just loved the Dorset, and shall now always use it as OUR hotel. Small, nice, neat, convenient, and just what we have always been looking for. $16.50

for double room and bath, which seems about average, as far as I can see. We didn't do anything remarkable but just poke around. No particularly good food. The best restaurant was the Italian Pavilion, at 20 West 55; but even that wasn't perfect. Had lunch with Marnie Bacon of *Harper's Bazaar* and her colleague or partial superior, I don't know which, named Perenyi (Eleanor),* who was at one time married to a Hungarian count. A rather weird type, big talker, dogmatic, clever, professional in a New Yorky way, but not, it seemed to me, gastronomically . . . but with very definite ideas on everything. Presumably they will come down and do an article with photos sometime in March, when Simca is here. But Perenyi is such an odd character, I wouldn't put too much on it.

What else did we do in New York. Went to numerous museums, including Mod. Art, Morgan Library, Cloisters, 57th Street Galleries. Walked all the way down 6th Avenue to the Bazaar Français where I ordered some omelette pans for my Philadelphia cooking class. Walked up and down 5th, Madison, Park, 3rd, Lexington. Tried to see *My Fair Lady*, but the ticket man just laughed at us. Took our nieces out to a Polynesian restaurant, the Luau, on E. 57th, which was fun for atmosphere but NO for food. Had lunch with Bob and Anita Littell† and large family group dinner with Marya Mannes Clarkson,‡ lunch with Dick and Alice Lee Myers.§ Napped in the afternoons. Paul did a lot of photography and caught a cold, and I gave him some Coricidin cold pills which disagreed with him and he still feels awful, but is still alive. It was, all in all, a very nice little vacation.

And now we get home to mounds and mounds of Christmas cards, bills, income tax blanks, etc. And someone sent us a brace of quails, which must have arrived the day we left. Couldn't make out what that smell of strong cheese was downstairs, until we opened the storeroom door, and there they were, all boxed, and stinking like fury. Too bad.

You were speaking of Mapie de Toulouse-Lautrec, and I found, back in a letter of Jan. of last year, a passage from Simca on the subject. She [de Toulouse-Lautrec] was invited to a Gourmettes luncheon, which turned

---

* An editor and writer at *Harper's Bazaar.*

† Robert Littell, an editor at the *New Republic* and the former drama critic of the *Saturday Evening Post*, and Anita Damrosch Littell, the daughter of the onetime conductor of the New York Symphony Orchestra.

‡ A social critic who wrote for the *Reporter*, *The New Yorker*, and the *New York Times*.

§ Richard Myers, a composer, and his wife, Alice Lee, had been noted members of the expatriate arts community in Paris during the 1920s.

out to be delicious (Soufflé de Barbue et de Langoustes,* Sauce Newburg, Saddle of lamb, Belle Hélène,† Oranges Orientales). She, Mapie, is the sister of Louise de Vilmorin, a literary figure much in vogue in France . . . Mapie writes recipes for the French magazine *Elle*. A very *sympathique*‡ woman, according to Simca, and a bit taller than Simca. Her recipes, S. says, are not particularly interesting. There are never cooking times, and her daring ideas are not in the French tradition (the French are provincial about that). Such ideas she has, as grapefruit or oranges sautéed, and put around lamb chops; or canned boiled corn to garnish a pork chop. Noodles cooked in Port with truffles. In another letter, [S.] says that she bought the book§ and found it was worth little as far as recipes go, and was again horrified at the mixtures she suggested (not French!). She goes on to say, she, Mapie, has never tried out her recipes. Whether or not that is true, I wouldn't know. You and Simca must have had just exactly the same impressions! Odd that Mrs. Knopf should have fallen so hard; but as Mapie is a social figure, I suppose it is a natural thing.

Hope things are going well with you, dear. I'm in for one hell of a month, and you will probably hear but rarely and briefly from me.

Much, much love from us both,
Julia

PS: Interesting your seeing May Sarton and Cora Du Bois dining together. Hope they aren't two-timing their friends.

Avis:
Loads of love! I wish you could have been with us in NYC, just looking, eating, walking, talking. Nothing pushing us around except ourselves, extremely pleasant.

Affectionately—
Paul

JANUARY 17, 1957

Dearest J:
I still have to look your new address up, so you really aren't home yet. When I forget Plittersdorf and remember Olive Ave., you will have arrived and I can really believe it.

---

* Soufflé of brille (a white fish similar to turbot) and lobsters.
† Poached pears with vanilla ice cream and warm chocolate sauce.
‡ "Likable."
§ *La Cuisine de France: The Modern French Cookbook* would be published by Orion Press in 1960.

Are you pretty well settled in? Longing to hear all details, especially kitchen, but of course you are still up to your ears. Seeing people? Getting your lines out in Washington? Tell all.

I just hope the weather is fairly decent. Here, we got winter in vast amounts. Day before yesterday, 19 below zero outside the living room when I came down at seven. Think of it. Haven't seen a winter like this since my old Houghton days,* when I was far better able to take it. Actually not too bad on me, somewhat to my surprise. No arthritis and I can't understand it. Maybe my health pattern is changing. No aches and creaks, but two colds this winter and I almost never have one. However, not bad, and I remain disgustingly healthy. I'm wearing my nose without even a bandage now,† and nobody notices anything, and in a week or so I won't be able to see the scar myself. So glad I went to a proper plastic surgeon—worth the money.

An absolutely wonderful week for my job, when suddenly everything popped at once. Which makes up for those weeks when I wondered what in hell I'd do to keep busy, every clue having collapsed. First—I think I told you about that novel from Stefanie Lauer‡ in Philadelphia, got for me by her pal across the street here, Sue Miles. Weinstock and Alfred fell mad in love with it and we are going to publish it and I am the white-haired girl. And Alfred said over the telephone yesterday why didn't I add Philadelphia to my territory, as it is so close to NY that nobody ever goes down there. So I will, come spring when it is possible to get around again. You know any leads there? Also, this surely means that they mean to keep me on, and when they come out and say so, I shall go down to NY and make a few little demands. Such as attending a few editorial meetings, finding out about the business end, contracts and such, and being provided with new publications so I will know what the hell they are up to. And also working the year round, and attending writers' conferences and such.

The Lauer book really is unusual—written here three years ago when she was at Radcliffe. She's 28 now. German Jewish refugee, very beautiful says Sue, copper hair and green eyes. Viking and Harcourt Brace turned it down—that is, they wanted her to rewrite it more as a novel, and she couldn't and wouldn't. But everybody referred to it as a novel and I pre-

---

* Avis is referring to the town of Houghton, her birthplace in northern Michigan.

† Avis had minor surgery for the removal of a noncancerous mole.

‡ A German-born Jewish writer and painter who was sent to England in the Kindertransport rescue mission during World War II. Her parents died in the concentration camps.

sented it as a novel. Now Alfred and Herbert say it isn't a novel, and must be presented as family reminiscences of the rip-roaring variety. Really a very funny and touching book, gorgeously written, about a Jewish family who didn't know they were Jewish, in Berlin when Hitler was coming to power. No Jewish defensiveness, no German stodginess, a *Constant Nymph** kind of book. And she's writing another one, a proper novel this time, so it's peaches and cream all around.

Second. Charlotte Turgeon, whom I met at Amherst, wrote offering me her projected Spanish cookbook. Little, Brown, her regular publishers, are doing no cookbooks—when Jeanette and Dudley Cloud† were there they did practically nothing but cookbooks, and I guess they've had it. When I saw Turgeon, the plan (there's no ms. yet but she's got all the recipes) was at HM Co., and I thought they would snap it up, so I wasn't after her. But they didn't, waiting for Child I guess. I regard her as a very hot cookbook property, with a lot of future stuff in her, and I think the time is ripe for a Spanish cookbook, since there hasn't been one in this country since the thirties. So I hope very much we will snap her up. She also wants to translate Ali Bab‡ sometime; I don't know if Alfred would go along on this as it would be a big and very expensive job, but I would love to see it done.

Third. I wrote one of my little notes, in December, to a fellow whose name I'd got hold of, Leonard Drohan of Chicopee Falls, Mass. Just an arrow in the air. He wrote back, a month later, saying he was sending me a novel. I know nothing about this guy except that he's 33 and had an earlier version of this book turned down by Holt. It came day before yesterday. I opened it, thinking Ho-hum. Well, toots, I never laughed so hard in my life. A genuine comic talent of the very first order, and God how we need them. Big fat novel, took me about five hours to read, what with interruptions, and I emerged with sore ribs and red eyes, having laughed myself silly. It's called *Come With Me to Macedonia* (translation from a Roman General of 108 B.C.) and is a howling satire of the foolishness and inefficiency and general red-tapery that goes on in an Army Logistics Area, our hero and his pals being civil service employees locked in a death struggle

---

* A best-selling novel by the English writer Margaret Kennedy, considered shocking for its time because of its sexual content.

† The publisher of the Atlantic Monthly Press (an imprint of Little, Brown) and his wife, who worked at the press.

‡ Ali-Bab, the pseudonym of Henri Babinsky, author of *Gastronomie Pratique,* a book on French cooking first published in 1907.

with the Army. It has plot, form, a wonderful love story, suspense, and a belly-laugh on every page. If we don't take it we are out of our minds. If we do take it we have got a runaway best seller,* either a stage comedy or a musical, a movie and eventually a huge paper-back sale. Musical, I hope. The boys make up a wonderful song about nit-picking, for example. It's bawdy in the nicest possible way, that is funny, really funny. Our hero has an Army friend who can be counted upon to get drunk at parties, so he gets drunk at a proper party and as he is being hauled away he sings, thus—"Hooray, hooray for the first of May / Outdoor screwing begins today." God I hope they don't cut that.

Well anyway I sent it down yesterday and now I am sweating it out and if they don't fall for it I will be tempted to seek a job elsewhere. But I called Drohan up, and he sounds like a complete Milquetoast over the telephone, and knows nothing about publishing, and wants me to carry the ball. In the incredible event that Knopf should say no, I will present this triumph to HM and they can jolly well pay me a fat commission. Best of all, this character Drohan (meekly) is finishing revision on another novel, which he will send me very soon. And another wonderful thing, *Macedonia* is the fourth revision, which means that Drohan is that rare bird who can rewrite and stick with a book until it is in shape. I love him.

So, do you see why I feel on top of the world all of a sudden? Well, if I get to Philadelphia I don't see why I can't extend myself to Washington.

Otherwise not much is happening—it is just too damn difficult to get around. I haven't had the car out for three weeks, and mostly people are staying home minding their own business and avoiding falls on the ice. I've been to one or two parties but nothing spectacular, and am feeling a bit house-bound. But this weekend I'm going out to the Cram's in Shirley Center† to a party, and spend the night, which at this point in my life sounds like what Mark calls "a ball."

Having fun with my blender, whipping up a variety of delicious little cream soups—chicken broth, water cress, potato, etc. And my goodness how convenient it is for bread crumbs. I just love it. This is weather for lots of food, including desserts.

Well dear write me the Washington gossip, and tell me how you both

---

* Drohan's novel later made the *New York Times* best-seller list.

† Robert Cram was a scout for textbooks who traveled to various universities and lived in Shirley Center, a town west of Boston.

are, and what's happening to the house, and in general, all. I miss you. Something fierce. Fondest love to both and how is Paul's back? Any real improvement?

Avis

MAY [N.D.], 1957

Dearest Avis:

Again this is only a short note of communication. I am madly plunged into the final bit of my turkeys and am seeing the end of the poultry in sight, thank god. Ready or not, the deadline to HM is to be on Feb. 1st [1958] . . . but it means all MS typed up for pub and including illustrations. Paul spent a few hours last Saturday photographing turkey in various phases of being cut up and trussed; and useful articles of kitchen equipment. I have a duck defrosting in the icebox, and he will take photos of that being boned, stuffed, sewn up, covered in pastry, etc. However, I must say I find the whole business a bit frightening, as it's got to be so good and so correct. I have been in constant correspondence with the Poultry and Egg National Board in Illinois, who are so cooperative and nice I can't get over it. I asked them various questions on flavor, tough though young, etc. etc. It seems they have made a great many discoveries in freezing, such as allowing the bird to hang a day (2 days for turkeys!), to let the muscles and stuff relax before freezing . . . or it will be tough frozen; at what body temp. to freeze so that the juices are outside the tissues rather than inside which makes the crystals burst the tissues, etc. The effect of growth additives, which makes the chickens very big very young, but also in many instances prevents them achieving age enough for flavor. They say that unquestionably the most successful method for defreezing is putting it in the refrig and letting it do slowly and evenly . . . I have tried to get concrete reasons for this but have not been able to . . . Perhaps if the outside meat gets warm too fast and the inside is still cold, you can develop more dripping out of juices? Well, I certainly found Freddie Child's turkey, defrozen in the turkey farm's refrig, was perfectly delicious anyway. I think I shall send them pertinent bits of the MS for their OK, so to be sure there are no errors.

Bought a copy of *Woman's Day* yesterday, and was much interested in an article on James Beard's "Advanced Cooking School." He is one of the best of the current cook bookers, I think. I unfortunately missed the first article on it.

After a broiling hot weekend and Monday, we suddenly have cooled off and today butter wouldn't melt in its mouth. Odd climate, here. I see you have been at oven heat up there in the Bean country, also. Do wish you would install your air conditioners. We have found ours a perfect blessing. Monday, when it was 100 outside, we were 76 in some parts, and not over 80 in others. We have to follow the cold stream upstairs, and put fans where they descend to shoot them into Paul's room, but it works. Our new electric bill is only an increase of $11; this month's will probably be a few dollars more; and that ain't much for the comfort we have.

I am in the midst of Mme. Prunier.* It somehow is a misfire, I think. Obviously it was written with an eye to publicity, as the 2 Pruniers in Paris and one in London are still going . . . and it smacks of publicity/journalism, and a bit of fakery, so far. Saying, for instance, that her grandfather learned "all about wine" in a bistro is nonsense, for instance. But the historical parts are fairly interesting, though not on the whole very profound. Should have gotten a heavier weight ghost to write it, I think.

Had lunch with Mary Moloney† the other day. She is nice, but I can't quite make her out, or is there not a great deal to make out???? I liked [her husband] Charlie, the bit we saw of him, and hope to have them both over in the near future.

Anne Doyle‡ speaks enthusiastically of you. She is a dear, and has just moved into a house she bought around the corner from us.

P. is fine, awfully busy as he is acting chief of the Division; and actually seems to enjoy it very much in spite of his previous groans and sighs. He says it is very much more interesting than his last job; but is entirely administration. Exhibits got a 62% cut, and it is tough trying to mete out the small amount of dough among the various commitments. Seems to me he is doing very well.

Do hope all is well with you, dear, and that Gordon is finding a solution to his life. Write soon, we miss you.

Julia

---

* Mme. Simone B. Prunier and her husband, Émile, were famous Paris restaurateurs in the 1920s. They opened a London restaurant in 1935, and she published *La Maison: The History of Prunier's* in England in 1957.

† Mary Moore Molony, a short-story writer, whom Julia would hire to type her manuscript.

‡ Identity unknown.

OCTOBER 2, 1957

Dearest Avis:

Ach, so good to hear from you, makes me feel we are again back in the life stream which, in spite of the pleasure of vacations, I really prefer. However, vacations are certainly useful I find as I get older; in this one, for instance, it was such a pleasure to write only 3 post cards and do a great deal of reading-for-pure-self-indulgence. After finally finishing Cozens,* I fell upon a copy of *Peyton Place,*† and read that, too, excusing myself in that I should know what had been a best seller for over a year. Those women, stroked in the right places until they quiver like old Stradivarii! Quite enjoyed it, though feeling an underlying abyss of trash. She does have a style, and a manner of creating atmosphere and character. But I felt it weakened very much toward the end, and she dropped one of her most interesting characters. Her next one, on homosexuality in a small college, should be quite a shocker! I wonder if that can be filmed! But I have now finished with indulgence, and have soberly and happily returned to my Goethe, where I shall remain until another bombshell appears.

We have been madly busy, and what with everything else, have had a few house guests and whatnot. Erica Child, our lovely blond niece,‡ came down for the weekend to join her new beau as his family were here to attend the big economic conference. He is a terribly attractive chap, we think, and so does she. But she says she finds it rather difficult, as both families know and like each other, and she feels hidden pressures which she hopes will not influence either of them. They are being extremely sensible, it appears, which is a good thing. Personally, I would never even have thought of anyone so perfect for her, and am just hoping the affair jells into something.

Broiled chicken is what we are having all the time now. I thought that one could be dismissed in a jiffy, but find there is a bit more in it than I had supposed, such as how is the best way to prepare it for broiling, then the heat, the level, etc. etc., and the timing. Oddly enough, it is a rare bird in France and is mostly half broiled, then rolled in seasoned mustard, fresh bread crumbs and then browned until done. But as it is such a popular item here, I feel it should have more emphasis than it would in a purely French book. I shall have about finished with it tomorrow, after

---

* James Gould Cozzens, the author of *By Love Possessed,* the top-selling novel of the year.

† A racy best-seller by Grace Metalious published in 1956.

‡ Charlie and Freddie Child's daughter.

having Anne Doyle to lunch to try out one with a stuffing piled in the center.

Paul is fine, and very busy indeed. His shoulder seems to have cleared up for the time being, which is a relief, and he is now having a long siege with the dentist.

Last night we went with the Moloneys to *The Egghead,* a play by Molly Kazan. The Washington reviews of it were not very enthusiastic, but we all found it extremely interesting, well acted, and timely. The story is about a professor in a small New England College (popular scene!) who has always been an active liberal, but is now aging. He sponsors a speech by a former student, a brilliant Negro who is accused, by the FBI, of being a Communist agitator in a strategic factory. The professor is enraged by the accusation, will listen to no facts or reasoning, and finally finds out at the end that the Negro is, in fact, a Communist. The theory behind the play being that egg heads must be hard rather than soft-boiled. The Moloneys are leaving for a trip to Mexico next week, driving all the way!

The Nancy Hale* visit sounded great fun. And a wonderful thing to hear that you feel, at last, you can give a party! Hooray! I do hope you will give one twice a month . . . You must. I have a feeling your life will take on a quite different shape this year, as you are more settled.

That chicken liver mousse I sent you the recipe for should freeze. I haven't tried it yet, but will. I don't think it would keep more than 5 to 7 days in the frig, at least I wouldn't recommend it. Don't know whether a jelly around it would freeze, too, and defreeze properly. Will try, anyway.

Proportions for jelly: 2 packages gelatin per 3 cups consommé. (The usual one is 1 pkg. per 2 cups, but this must be stiffer.)

We love thinking of Mark all settled in Harvard, and have already quoted your story of him coming home so many times the first day, and gradually decreasing, and finally never letting you know he was there at all! Woe. But a good thing, as you say. It is good to hear that Gordon seems cheerful, and we wonder what will happen to the meat packing deal. That would be fun, I should think; at least, I would like it. What is Nancy Hale's problem boy doing?

You know, we never had time, when in Cambridge, to take a tour of

* A Boston-born novelist and short-story writer for *The New Yorker.*

your garden, which seems insane. Well, next time. Ours has come practically to a standstill, as we have been having some coolish weather. The wisteria, which grew 4 to 5 inches a day in the summer, has hardly moved an inch in the last week. We have replanted our lawn, which has come up, but is doing it very gingerly. We shall have to spread about some lime come November, I think.

Simca is definitely arriving in the USA, around Feb. 1st. We are so excited about it. She said she had sent you a copy of her prune opus, *Le Pruneau devant le Fourneau,** and very impressive it is, I think. I am trying to urge her to get herself into print under her own name, as it will be a good talking point when our book comes out. She has made approaches to several magazines, and will probably make it. No word, by the way, from *Harper's Bazaar,* on that article I sent them. Hope it hasn't laid another egg, is all. We would very much like to come up and stay with you in Feb. or early March, and do some fancy cooking. Do you have any idea, so far ahead, as to when would be a good time? Paul would like to come, too, and will devote the period to reading Cozzens . . . Would you have room for three people?????

Hoping all is well with you, mss. of magnificent types piling in, and all unanswered mail caught up with.

We miss you,
Julia

OCTOBER 10, 1957

Dearest Avis:

This must be short. I've just finished my draft of Broiled Chicken, and am about to type it up. Do you want to go over it (10 pages)???? I always feel better if you do, but with all you have to read, you may have enough without more chickens.

That's fine Mrs. David liked all your notes on her book. My lord, though I never thought you would end up with so many, but I guess that is possible, with all the notes you had for each page already. She has two other good ones, *French Country Cooking* and *Mediterranean Food*. Maybe you should do them all for Knopf? I took a look at excerpts from Harvey's book† in a recent *Vogue,* and thought it seemed very good indeed.

* *The Prune to the Stove,* a booklet of recipes using prunes and prune liqueurs.

† Peggy Harvey's *Season to Taste,* which Knopf published in 1957, was edited by Avis.

Also, your friend C. Turgeon has a section on Hors d'oeuvres (sp.) in the *House and Garden,* which is good.

This is terrible about more stealings in your neighborhood, and about an old lady being assaulted. I don't know what to think about your having a gun. Some people say it is very dangerous, as if you have one, people will shoot you, and you don't know how to be fast on the draw anyway. What does Charlie Curtis think? I forgot to ask Paul, who is now at the office. Well, be careful. And thank heaven you have your students living right over your head. Anyway, you have that siren machine.*

If you aren't to be in Cambridge when Simca comes, we will wait until you will be . . . would be utterly no sense in going there with no Avis, natch. I don't know just what her projects are yet, and Jean, I am afraid will be only a few days in the USA. Simca did the prunes because I have been urging her to get her name in print, and that was the first thing which came along. And it is useful to have something published (wish I had!). I love prunes, myself, and they are lovely, cooked in wine and stock and butter, and served around a goose. Do try it sometime.

We are so happy to hear that Mark is doing well, and being independent. That is fascinating about the Hell's Half Acre,† and must be very exciting to him and George that all their work is bearing fruit. Am dying to hear about the guided tour with the officials. What a boy. Hope Gordon is fine. Did the meat packing job work out?

Bob Kennedy‡ is coming down to Washington next week, and we shall probably get a load of his side of the story. We do remember that Mary Collins, tall and good looking and nice. Well. We shall not mention that. It is good that Gerta [Kennedy] looks well and has a part-time job. What a business.

Last night we went to the opening of Helen Hayes and Susan Strasberg in *Time Remembered,* a play by [Jean] Anouilh. It is beautifully staged and costumed, but an extremely difficult play, really a satire on the French modern aristocracy. It just doesn't quite come off. For one thing, such a satire would be immediately recognizable in Europe, where they've grown up with the nobility, but is quite meaningless to most Americans who don't seize the implications as it is a social order which doesn't exist here. Strike one. Helen Hayes was just not able to convey the atmosphere

* There had been a string of robberies in the neighborhood. Avis had also purchased a burglar alarm.
† A marshy area on the Charles River in Cambridge.
‡ A former student of Paul Child at Shady Hill.

of someone who has descended from a long line of duchesses . . . She was a cozy middle class daffy-dilly; her performance somehow lacked an elegant condescension (sp, too lazy to look it up), hard to put finger on . . . Maybe, as she is supposed to be a good actress, she can remedy this . . . but does she know what it is? We all found Susan Strasberg (*Diary of Anne Frank*) quite utterly lacking in stage charm, and with a most unfortunately flat, hard unmusical voice, sounding strained. The men were all good, particularly the young male lead, Richard Burton . . . but he is English I believe, and gave a most believably upper class atmosphere. Gawd, what a problem to whip these expensive productions into shape before hitting Broadway. This, unless the two women can be renovated, will lay an egg, I think . . . and too bad if it does.

Must off to the bank, and then back to the machine. Much, much love and be careful of that gun!

Julia

Virtually every time Julia returned to Paris during the 1950s, she joined Simca, Louisette, and often Chef Bugnard in teaching classes at L'Ecole des Trois Gourmandes.

JANUARY 12, 1958

Dearest Avis:

Only a noteski, as usual, hoping things are well with you and that the work is progressing as it should, that the boys are well, and above all, that you are too.

We is OK, except that Paul has had a queer bad cold, which is now getting better but has left him with a flock of cold sores on his upper lip and a deep cough . . . but he is about over it. I am frantically plunging along with my MS, and am afraid that it won't be done by Feb. 1st, damn it. Mary Molony suddenly had to leave town to go to the bedside of an old Aunt who has had a stroke in Florida. So I have farmed out the typing to two other people, one a professional steno. who does work at home, and the other a girl in Paul's office. Heaven knows how they will get along with it, but it is going to cost plenty of mazuma. We have just had another refusal from a magazine for an article (*McCall's*), saying, "If my editor saw these recipes she would faint dead away. I recognize myself that they are not complicated, only detailed. However, to the non-cook they certainly look like quite a chore." That seems to be the general opinion from magazines, and we have now had three of them, so I conclude that we cannot get anything published in our usual form. And I suppose not a few potential book buyers are going to feel the same. *Merde!* I shall have to try another tack for magazines, obviously. However, as I keep saying, the whole point of our recipes is in their detail. I am just doing over the Canard à l'Orange* at this moment, and I come to the sauce which is the crux of the matter. How can I leave out the following passage which gives the whole flavor-trick of the sauce, and which is the part one does not find written out anywhere else:

> Remember the sauce must be thick enough to withstand the dilution of the orange juice to come. Boil down to thicken more if necessary. Taste very carefully for seasoning and strength. It will receive some more flavor from the wine and roasting juices later, but should, even so, be almost perfect at this stage.

Etc.

You could say: "Reduce to thicken if necessary and correct flavor." That is fine for a French cook or really practiced American, but my teaching experience confirms me more than ever in the conclusion that most

* Duck with orange sauce.

Americans don't know anything at all, NOTHING, about the techniques of good cooking and that every detail (EVERY ONE) must be thoroughly explained.

So I am deeply depressed, gnawed by doubts, and feel that all our work may just lay a big rotten egg. I can see that such detail cannot go into a magazine article, and for them we just have to forget our book-methods. But I would hate to take what I consider to be the guts out of our book. I suppose the only thing to do is to go along as we have been, and then when the MS is done, to have a careful and detailed pow-wow with HM; and do it over if necessary. If we can strike the right note at the beginning, the succeeding volumes will follow along quite rapidly. But the first one has got to be right. If the recipes are psychologically horrifying, who is going to do them? . . . Or buy the book?

Well. I hope things are going well with you, at least.
Much love and kisses from us both,
Julia

JANUARY 17, 1958

Dearest Old J:
First things first. I have been in NY, got home late last night absolutely exhausted, after working non-stop with June Platt just cutting her behemoth of a cookbook. Literally, Tuesday and Wednesday from ten in the morning until eleven at night with time out only to dash around the corner for a bite, Platts and me, because of course no time to cook anything. And Thursday from nine to noon, when Mrs. P. had to go to Rhode Island. And we got it cut, too—somewhere between a quarter and a third, but it will still be an awfully big book. And not my kind of food either. Whole thing dated, to my mind. But Pat Knopf* is sold, and says it's for an audience who want to make one great big whopping rich dish and then have the neighbors in. Every recipe serves 6–8 which is nonsense when you consider omelets. Whew! She had over seventy chicken recipes. We cut it in half. Took out about thirty soups. Took out all the impossible things like pheasants en plumage, and how to make a beautiful big butter rose. She is stubborn about keeping in how to make eight meringue swans, which will require a cut. Gawd, I don't know. She can't write for sour apples. I got out most of the little cozy chatty things about friends

* Alfred "Pat" Knopf, the vice president of sales and the son of Alfred and Blanche Knopf.

nobody knows the names of. Had to do some rewriting. But have only skimmed the surface so far. Got out all the brand names, which of course is sometimes ruinous from the point of view of the cook, considering brands of chocolate, mustard and so on, but we are not allowed to have any at all. I am <u>not</u> going to do the styling and copy-editing which must be an office job, on which I will keep an eye and be consulted. But I will read proofs, and pray for the best.

Those poor Platts. They led the life of Riley for many years, international style. He is a designer—stage, Hollywood, furniture, fabrics, and what have you including houses. Made a pile in his time and spent it as it came in. She worked along with him, also on the cooking part, and as you know had a bit of success with those three books, now out of print. He had a stroke four years ago—still a handsome man in his 60's, but slowed up, enfeebled, speaking indistinctly. They have hung on to a house in Little Compton, R.I., which they showed me pictures of in *House and Garden,* quite lovely. Rent it most of the time. And in NY, where they have to be to pick up any job from designing Christmas cards to buying furniture for old customers, they live in a basement flat in the east 60's—two rooms opening onto the street. Front room is their workroom, other room their bedroom and sitting-room. Plus of course small kitchen and bath. But I think they are penniless, and so harried, and such nice people—gentle, cultivated, gallant. I don't use that last word easily, but it's the only one here. Quite heart-breaking. Pat hopes to make her five or six thousand bucks out of this cookbook but I am very dubious.

Julie, I wish you would think about taking a quite different tack on magazine articles. I have not had time—literally, what with end of the year documents to deal with—to think this out in any detail. And God, how I wish I were a writer and could write it for you, or help with it. What I am thinking around the edges of is this—an article with no recipes, designed for *Harper's Bazaar* or *Vogue* or one of the chic magazines, many of whose readers have been abroad—about the basic differences between French cooking, acknowledged by all to be the best in the world, and American cooking which lags way down the list. A sort of teaser article. Introducing the avid eater to the concepts of deglazing and reduction and so on, as many things as you like, and the reasons why they are <u>vital</u>—talking about the absolute necessity of buying the best materials and exercising the most loving care in handling them—talking about the basic differences in the preparation of even the simplest dishes—all this not in any frightening way, but logical as all hell, as concentrating flavors and so on. Easy steps for little feet, but without those steps you can't achieve the final perfection. Hold out as a lure that the American housewife can achieve some perfections in cooking if she cares enough. I haven't had time to think about this in enough detail to be really useful to you, but I wish you would mull it over. If you can think out an article like this and get it on paper in some sort of shape and the right length I will go over it and see if I can improve it. Wish to God I wrote better, but I did re-write a few paragraphs for Platt that were a drastic improvement and maybe I can help you. Think, and let me know.

But it is just unthinkable that you should bastardize your book. You are <u>right</u> and you must be stubborn. We must somehow convince the reading and eating and cooking public that there is nothing psychologically horrifying about this presentation, quite the contrary. What you are doing is casting a great light over the mysteries of French cooking. Stand by your guns, my lamb. We will talk it all out when you come up here but meanwhile try not to be too depressed. Don't you find that the women in your cooking class are excited and pleased to have things explained at length?

I'm sick about Mary Moore [Molony] having to tear off like that, and hope very much that the other typists work out all right. Jesus, life is fuller of snags all the time.

I speak from cause. I didn't answer your last long letter because I was so damn depressed I simply could not bring myself to unload it on you. Feeling somewhat better now, due no doubt just to getting away for a few days, though I did absolutely nothing in NY but work like a dog. But at least not in this house.

. . .

So, I am feeling very much that as you grow older, the walls close in around you. I am feeling every day of my age, and sometimes looking it as well. I keep wondering if there is ever going to be any more fun in life. Bound to be, I suppose, and I wish it would hurry up. All I know how to do is work like the devil and hope for the best. And I have indulged myself to the extent of buying several new and rather expensive dresses, and three beautiful and quite expensive chairs from Design Research* and I wish they would hurry up and get them delivered. These are for the living part of the house and I will retire three of my old heaps of sticks to

* A Boston store that sold modernist furniture.

the library. I am going to spend a couple of thousand on the house, inside and out, and get it all tidied up for the future.

Going back to your letter of Jan. 3—I have always thought it was insane to expect a great big enormous turkey to take all those long extra hours. At some point there is a point of no return, if you know what I mean which I do not myself, but there must be some law that covers it. It has been my experience that all given cooking times for turkey are too long—the best is always done before you plan it. Of course it will wait, thank God—otherwise we would be lost. My turkey this Christmas was done three quarters of an hour before I planned, and that only a 12 pound bird, dressed.

I am so sorry that Paul has been ill and I hope that by now the bugs have left him and his strength is returning. One simply cannot be too careful about the bugs this year. G. [Gordon] had a short sharp bout of flu last week with high temperatures for three days and is still a bit wobbly. Believe me, if I get hit I'm taking to my bed to stay there until it is all over. Paul is most lucky to have a wonderful cook-nurse like you to baby him and fuss over him and make him behave.

Most love to you both and now I must get back to taxes.

Avis

FEBRUARY 2, 1958

Dearest Avis:

This isn't a real letter either, but I feel I owe myself a few minutes respite from the MS to write my little old P. Pal. Simca arrived last Monday, and we've been hard at it night and day ever since. She is proofing manuscript, and I am rushing ahead doing over the sauce chapter, which is a considerable job as we have learned so much about techniques since 1952 (!) when the sauce chapter was first written. Looking at it again, I wonder how you ever managed to sell it to HM! I think we shall have the MS all done except for 3 very minor sections in the poultry chapter by the time we get to 8 B. Street. I have written Dorothy that we would like to present her the MS on Monday, Feb. 24th. Jean Fisch[bacher] is arriving the end of this week, and will stay the weekend. Then Jean and Simca will spend a few days in New York before Jean leaves on Feb. 13th. Simca will stay on in New York, and I will meet her for lunch on Feb. 18th, where we will join *Harper's Bazaar,* who plan to do an article on the Child house, kitchen, and dinners, and will come down to photograph us the end of April for an August article. (Goodie!) Then we will spend the night at Gertrude

Almy's,* give our cooking lesson the next morning, and take the train to Back Bay arriving Thurs. morning Feb. 20th. (Hooray!) We have made no plans at all because I think we shall have quite a bit of MS proof reading to do before Monday. I have read none of the finished stuff as yet, and think I will not have time to do so until we get to Cambridge, as I have to finish my sauces, do a section on wine, and the introduction, all before that time! But I believe Simca's main work will be done, and she could manage the dinner for Friday night. I think it should be stated that the dinner will be a good but not at all a fancy one, as she probably will not have the time to go all out . . . and I will not offer to help at all. The MS comes first. All our meals here have been extremely simple ones, taking not more than half an hour to prepare, but they've been good none the less. So I think it would be wonderful to have Dorothy and the Thompsons, if you just make clear to them what the conditions will be. Saturday would be fine too, for anything; and Sunday, and anytime . . . just as long as nothing needs to be luxurious until after Monday.

Do you have an extra well-working typewriter, or should I bring one along?

We have made no plans at all for after Monday, and would like, if it is OK with you, just to stay as long as it is necessary to get our business with HM settled. We intend to take the attack position. That this is the type of series of books we plan to do, and that Volume II will be ready well within a year of the publication of Vol. I; and that Volume III will be ready within about 6 months of Volume II. This is going to mean hard and constant application but we feel it must be done in order to benefit from the publicity of the first volume. We shall also state that this is a unique type of book, an Escoffier for the average home cook, and that nowhere else is such full information on the subject available. ETC. We must also state that unfortunately there are few people with the professional qualifications and experience to evaluate the book in its true worth. ETC. Think of anything else?

My, what fun it will be to be with you again. It would be wonderful to take a little drive into the country. We are very anxious that Simca see something of New England, as it is so unique. And the more good Americans she meets, the better. (We'll see May [Sarton] and Cora [Du Bois] on our own. Don't quite want to mix them, I don't think . . . too special!)

---

* Gertrude Almy Slichter, an economist and historian.

Glad your spirits are somewhat improved. How wonderful that Mattie is there, to help Gordon. My, I hope it works.

Haven't had much time to read the W. Root[*] book, but from what we've seen, we both think it is marvelous. Can't wait for publication date and the reviews. Seems to me it is a real contribution, and a beautifully presented book.

Back to my sauce, and a great deal of love from us all, especially from Simca.

J.

MARCH 14, 1958

Dearest J:

Mixed snow and rain and wind outdoors, so a fine day to work at the desk and I am accomplishing much. Flow of mss. from the office demanding instant attention, and three long shots of my own also came in—two just plain bad, one obviously University Press—my goodness, it's nice to be busy again.

Better not expect me on Sat. the 20th—I will have only part of a day there Wed. when I arrive, and if my mentor takes me to Duke,[†] as she plans to, I should allow some leeway. So I am planning to spend the weekend in N.C. and probably arrive Washington Monday, but don't know yet. It's hell planning these trips—I have to allow for things opening up and lots of people demanding my presence—but much worse is when general apathy reigns, nobody wants to see me or if they do I don't know about them, and I wish I'd stayed home. I will probably telephone you from Chapel Hill when I see how things are shaping up. But of course I would adore to see the Heinls,[‡] so why don't you see if you can tie them down for some day that week, and make any other long-range plans you like. Will probably be going out to dinner a couple of times, but that can be adjusted to your plans.

Bill Raney[§] wrote me too so I think I have the picture. I am just strongly and immovably against any "great number of compromises," or indeed anything much in the way of compromise. Would be disappointed

---

* Waverley Root, the author of *The Food of France*.

† Duke University. As part of her scouting duties, Avis often met with professors and graduate students about publishing possibilities.

‡ Robert Debs Heinl Jr., a decorated colonel of the Marine Corps and military historian, and his wife, Nancy Gordon Heinl, a historian.

§ William Raney, an editor at Rinehart in New York City.

if you made any, but hope to face up to reality. I'm not going to fret until HM says one way or another—haven't heard one whisper from them, except Anne B. [Barrett] said Dorothy was going through the ms., very much interrupted—and D. is not well, looks awful, Anne says, and is going to Italy in a few weeks. Well, on the knee of the gods. But rather than butcher that manuscript, and negate all the work you have done, I would far rather wait until the times are more propitious. I am convinced that the number of people who are willing to learn the disciplines of French cooking is growing, however slowly. All I say is thank God you have got enough money so that you do not have to rip the guts out of this book to keep yourselves in groceries. You can afford to wait until the right publisher comes along, which he will. (Who knows, maybe HM?) It is all a question of understanding why you wrote the book this way, and your position is so completely logical and unassailable that some publisher is bound to agree with you. If you only had quite a bit more money—real dough—I'd say publish it yourself, promote it yourself, and clean up, slowly but inevitably. Well, this is all footless, until we've heard. But don't you compromise. Phooey to agents.

Thanks for dope on butter-fat content, which I wish I'd had when I was tiffing with Blanche. Interesting. Interesting also on walnuts, though the dish as described does not appeal. And I will be more than happy to have final revised version of the chocolate-almond cake. I am dieting like anything and have lost three pounds so I won't be making that one for a while. I also have had two postcards from Simca who seems to be having a whirl, in bad weather. Dying to know about Detroit.

Mary's [Avis's housekeeper] husband is back at work, praise the Lord—she asked me to tell you as she knew you would be wondering. Not full time yet, but something, and the unemployment compensation is coming through at long last. How in hell do they expect these families to get along through lay-offs?* Also, G. applied to a Cambridge taxi company for a job, and they say they can take him on—he is in process of getting his hacker's license, which takes a bit of time. Don't think the taxi business is very good these days, but it's something, and he's just got to get out of the house part of the day. Too easy here for him to sink back into apathy. Also a possibility of a job at Woods Hole [Oceanographic Institute] on one of the research ships, and he is going down Monday to

* The United States was experiencing a recession in 1958, when unemployment reached a postwar high.

look it over. Think this is being very unrealistic, as it would be very difficult to see his psychiatrist which seems to me the most important thing until he is over this hump. Perhaps they can arrange something weekends, assuming he has weekends. His determination to move out of the house is crystallizing — we are all strongly in favor of that, as the first step to independence, but he has not got the vaguest idea how much it costs to support himself. Well, he'll find out. The weakness of my position is that I can't let him starve, or be really up against it, and here is the house, ready for him if he goes down and out. If I weren't reasonably well off, and if I lived in a two room apartment, he would be forced to face life, and no fooling. Which is no logical reason for my selling the house.

Well. Children!

Have ordered 25 copies of Nixon article and will send one as soon as received.

Must now read a hefty ms. about the Third Republic,* a subject about which I am ill-informed, and do not expect to become much better-informed, as this is the U. Press book.

Loads of love, dears,
Avis

MARCH [N.D.], 1958

J. angel:
Not my day for good typing evidently. Do you want the book?† It certainly is no earthly good to me and if you want it as a horrible curiosity I'll send it along.

Note your letter to Dorothy — a fine very dignified statement.‡ But Gawd, don't go too damn far toward canned and frozen simplicity.

Please put in roast chicken that is browned on top of stove and contains that delectable stuffing with chicken livers and cream cheese. Also, pork roast boned and marinated, and ditto small veal and lamb. These treatments unknown to all I have told about them and most stimulating and attractive. Think also for sure, the puff paste American, Simca's orange cake as well as the chocolate-almond, your braised endive and celery. And when I see you I'll probably have more musts for the book. After all I'm the lucky girl who's eaten most of these things. I want to snare

* The French Third Republic, beginning in 1870 after Napoleon III's defeat during the Franco-Prussian War and ending with the installation of the Vichy government after the German invasion in 1940.
† Julia's manuscript.
‡ Julia's response to Dorothy de Santillana's letter of March 21 rejecting the manuscript (see page 258).

these first readers firmly and whet their appetite for more, and more, and more.

I have to read a kiddie book now—very busy day. Wish I could be with you three dear ones right now, a Saturday which looks like being warm, clear and balmy.

All love to all,
Avis

MARCH 25, 1958

Dearest J:
I had to take three pills to get to sleep last night—mind going round and round in circles. I can imagine how you are feeling, though of the two voices, Paul's sounded most upset. I can understand HM's side of the question—the state of business and the increasing expense of publishing anything, plus the enormous hurdle of bucking the ordinary cookbook trend and really re-educating the American cook. When you send me the copy of Dorothy's letter I will study it for clues. My impulse was to rush in and talk to her but I'm not going to. Too emotional at the moment. And hard for her, too, as she is the only one in there who really understands the why of the book, but is forced also to bow to the publishing realities.

I am perfectly positive that this is not the time to take it up with Knopf. We have got all these commitments, including June Platt which Alfred is dead set against and thinks will lose money hand over fist—he is probably right. It's Pat Knopf's baby—Pat sensed my feelings about the book, which were pretty negative, and has taken me off the editorial job. If the book flops, there is going to be a really ghastly family row—Pat against his parents. The Donon* book is classic French and is an editorial mess which is going to cost a fortune before it sees print—even if it is burden money. Alfred wrote me some time back how he felt about cookbooks—he says the only one we ever published that justified itself economically was Mildred's.†

In spite of this I would talk to him about your book if Alfred had the slightest idea how cooking is really done. He is a gourmet and he

* Joseph Donon, the last French chef to train under Escoffier and the author of *The Classic French Cuisine*, which Knopf would publish in 1959.
† Mildred O. Knopf, the author of *The Perfect Hostess Cook Book*, published in 1950. Mildred was married to Alfred's brother, Edwin.

has a palate, but he must think good cooks are born, not made, and the whole business is some kind of magic. We cook like this, we never have any difficulty with that recipe—which turns out to be Alphonse, or Blanche's cook. It is very like Blanche's reaction to the Mapie book—it was the best French cookbook she had ever seen in her life, because it was so simple. So easy. That's when we had our row and she still thinks I'm nutty.

This damned and unreasonable attitude, plus Alfred's intensely gloomy convictions about business these days, make the outlook at our place quite hopeless. Nevertheless, when I catch Alfred in an outgoing mood I am certainly going to consult him about the problem, purely as a publishing problem. I would have more of a chance of some kind of sympathetic understanding with either Herbert Weinstock or Bill Koshland, because they cook themselves. But they would be frightened off by the gamble, for a small firm, I think. Alfred said yesterday that we are publishing far too many books, and they are not selling. Meaning we only have one on the best-selling list, the Humphrey* novel, which has sold a modest 15,000, on which we do not get rich. I was left feeling why on earth is he employing me, to find books that are almost bound to be turned down, however good they are, as bad financial risks. I still don't know. Seems quite likely to me, if business goes on bad, that Alfred will let me go in September. Not that he said so, or indicated so, but why go on paying my salary?

I've changed my mind, overnight, about approaching Oxford† right now. I'd rather get an opinion from Raney first. I think he, too, will be frightened by the magnitude of the publishing job—Rinehart is not a big firm—but he knows a hundred times as much about publishing from the inside as I do, and I think he can give you far better advice than I can. He is not as emotionally involved as I am, and he will be realistic. If Bill advises cutting down in size, I think you'll have to face up to that.

If only you or I were a really good writer! Because what is needed, I keep thinking, is a carefully graduated series of articles—published God knows where—breaking the news of a new concept of cooking. Easy steps for little feet. Don't frighten them off. Perhaps even a first book like that—not too big, not too long, authentic French recipes but the easier

* William Humphrey, the author of *Home from the Hill,* a novel about life in rural Texas, published in 1957.

† Avis had considered approaching Oxford University Press about publishing Julia's book.

and quicker ones. A few soups, omelettes, half a dozen chickens including the Suprêmes which God knows are easy, a handful of meats and vegs. and a few desserts including that chocolate cake. I think a book like this would probably lead cooks on to want more and to overcome their fears. But God, I just don't know.

If Bill wants to take the ms. let him—all depends on whether he spends any real time with you in Washington, assuming he comes. It is hopeless, it seems to me, to try to deal with any publisher unless he knows and cares something about food. I will try to find out the ones who do, and who have a reasonably open mind about the techniques involved, as Alfred and Blanche have not. I don't know about John Brett-Smith* at Oxford—since he's English, I would rather doubt it. Next time I see him I will find out, but I don't know when that will be. And if Raney is of no real help, I would like to have the ms. long enough to show it to Charlie Morton† of the *Atlantic*. Charlie is passionate about cooking and suffers the tortures of the damned over it because he is greedy, has horrible allergies, and is going to kill himself eating and drinking someday. Of course he's on the magazine, but knows a good deal about publishing. And is a promoter at heart. It would take time for him to go over the ms. because he is a putter-offer and badly organized.

All this really means I think we shouldn't panic, but sit back and take a good look around and decide what's to be done before going off half-cocked. I am just sick for Simca—it is going to be such a shock and disappointment to her, away from home, and all full of high hopes. Wish I could be there to lend a hand with the general gloom, though I don't see how I could do anything but add to it. But we must roll with the punches, and see what is to be done next.

Life just seems to me less and less satisfactory and I suppose it is going to get more so. Two blows for you in rapid succession—trip called off‡ and then this. Well, all I know is this—nothing you ever learn is really wasted, and will sometime be used. You have come nearer to mastering a good many aspects of cooking than anyone except a handful of great chefs, and some day it will pay off. I know it will. You will just have to go on working, and teaching, and getting around, and spreading the gospel until it does. The alternative, that Americans do not give a damn

---

* The president of Oxford University Press in the United States.

† Charles W. Morton, an editor at the *Atlantic Monthly.*

‡ Paul Child's official trip to arrange exhibits in South America was canceled.

about fine food and refuse to learn how to make it, is one I simply refuse to face.

I'll write again after I hear from you and see Dorothy's letter—feel bitterly that I have not been of any help to you, but had to play it the way I saw it, and probably would again. Just have to keep trying, and slugging away, is all.

All my love,
Avis

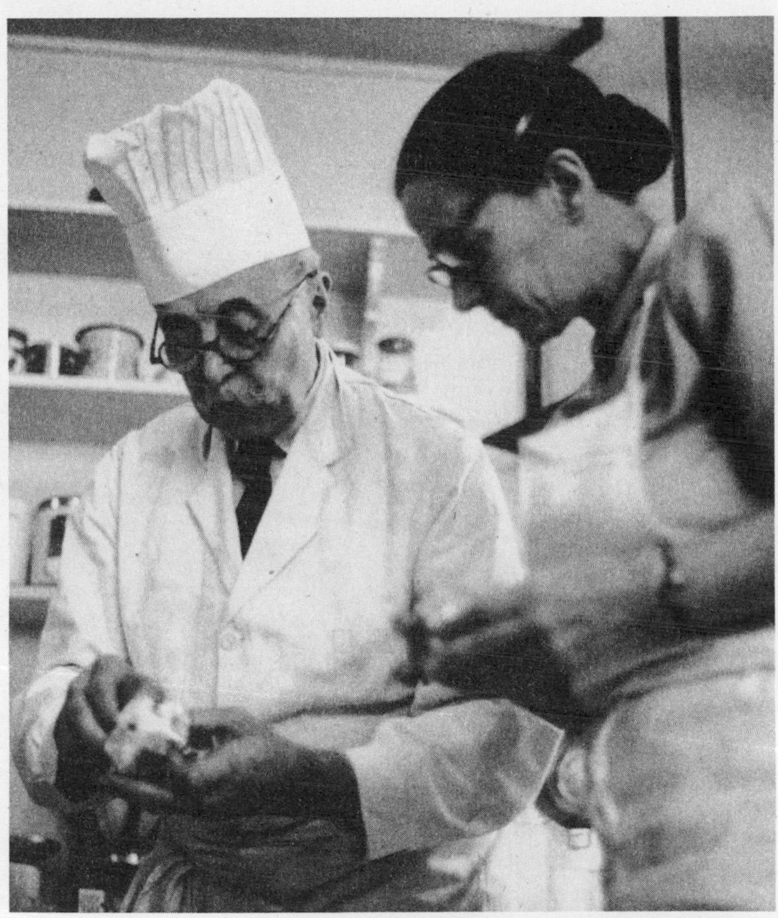

When Avis visited Paris in 1956, she observed a class taught by Chef Bugnard at L'Ecole des Trois Gourmandes.

AUGUST 7, 1958

Dearest Avis:

Well, things have finally jelled, and Paul has accepted the post of Cultural Affairs Officer to Norway (Oslo). He will receive six months of language training, and we should be leaving some time in March. HOORAY. I think this is wonderful, and we are getting excited. And thank heaven, I really didn't like the idea of our drifting around, and Paul really feels a calling for government service and wants to do something useful and valid. We just learned this yesterday, so don't know much. But how wonderful that he will get language, and we will have time to study up all about Norway, and learn all the geography, history, culture, literature, and everything possible. Sounds like a fascinating place, and think of all them Limfhord oysters, and them lobsters (Norway lobsters???), and Norwegians. We are also going right away up to Maine, as soon as Paul has found out more about things. Perhaps we shall leave even by Saturday or Monday, and we can stop in and have a day with you. Anyway, we shall let you know, probably call you up or something.

We have decided that possibly one reason for the Exhibits debacle, along with the fact that he inherited a division which had gone down hill and had a bad reputation, and probably anyone, whether genius or boob, would have had the thing collapse in his face no matter what kind of a job he had done . . . the one reason is that we have been trying to live a normal life in the framework of the bureaucracy. The having and eating of cake syndrome. So on this post, we are going to become more bureaucratic. We shall entertain people. Paul will make sure that he advertises himself and his wares, so that people know about him. (This he has never done, and we have always laughed at people who have done so . . . but there must be discreet but effective ways which can achieve the result with dignity.) And I, by god, shall be a diplomatic wife. I blame myself for much of this, as I have been but a cook and a cook book all the time we have been in the service. We are both a bit diffident about entertaining people we don't know very well, but this time we shall turn several new leaves. We don't intend to turn too many, but enough. Another thing I like about Oslo, from what I have heard, is that the social life is not very intense, and that the living is simple. So I shall be able to cook, but we shall, I suppose, be able to hire people to come in when we entertain.

Well, we feel better. Paul is still very much disappointed that he cannot go on with Exhibits. But he has always wanted to get into the Cultural Affairs end, too. And he will be out from under his most rabbit-like and ineffective superior. The Ambassador to Oslo is a woman, Frances Willis, who is supposed to be very good. What his immediate boss will be like, the Public Affairs officer, we have no idea, but shall hope for the best.

Much, much love, and I hope all is well with you. We shall telephone you as soon as we know what our plans are.

Julia

PS: Lucky we don't leave until March, so I'll have time to finish up the GD book!

NOVEMBER 30, 1958

Dearest J:

I don't know why I always think I can get some work done during holidays. I brought my books home from school and I haven't opened one of them.* Well, perhaps a breather is needed. I am not going to take any cruise or anything else, not in the foreseeable future. I am going to attend school all through January and who knows, part of February. I can take dictation fairly easily up to seventy words a minute if the material is not difficult, but I bog down when the dictation is full of 4 dollar words, and I really need 90 for bursts of speed. And I've got to lick that filing. Boy am I stupid at filing. I can't seem to remember the alphabet accurately, let alone all those little rules about names of married women and the like. Well, it will come.

Mattie was here all vacation, plus her brother some of the time, plus Mark two nights. Plus no end of others in and out. We had seven for Thanksgiving dinner. And a 17 pound turkey, dressed weight. This due to the fact that some demoralization hit the First National†—it closed Saturday night, damn it. Not space enough to expand, and the rent went up. So though I'd ordered a 12 pound fresh turkey, the fresh ones didn't come in, and I wouldn't take a frozen one. Sages provided me with the 17 pounder, and it was really very good, but, as always with those big ones, it was done before the time schedule. Went in the oven at 11:15 and was perfectly done at 3:30. I think we may regard this as constant. The 20 minute a pound rule for birds over 12 pounds doesn't work. However, with all the young things in and out, we're just finishing the last of it today.

* After Avis left Knopf in September, she took a class to brush up on her secretarial skills, having accepted a job as the house secretary at Lowell House, a dormitory for Harvard undergraduates.

† A supermarket chain.

Pleased no end to get the cakes, and one of these days I'm going to make a couple of them. Just let me get over turkey and mince pie. I have read the ms. with care and since I do not choose to send the copy back, I'll write you my piffling little corrections. In the first page discussions, what do you think about mentioning the fact that there is no baking powder in these cakes, leavening depending on egg whites? American cooks expect baking powder. I remember my surprise that French cakes didn't rise higher, and my unbelieving bliss at their delicacy when I ate them. I feel that if you mention no baking powder, cooks will take more care in the folding in of the egg whites, and also whip them more carefully. Same page you mention they may be frozen—want to add anything about thawing?

Page 2 directions—your cake tin, your ingredients, your spatula—and occasionally throughout, as on page 4. But mostly you use the. Want to make it all the? It sticks out of the page a bit, the your.

Page 6—typo—for about a week, you've got or week.

Page 10. softened, unsalted (butter) is missing.

Page 10. off heat, beat in butter.

Page 14. I don't understand this business of standing the baking sheet on edge to knock off excess sugar. With the batter on? Wouldn't the lady fingers run, or fall off?

That's all from your fussy friend. It's a most alluring chapter, guaranteed to drive the cook to the kitchen to try one.

Are you really going to end the book right there, with no concluding remarks? Seems a little abrupt.

. . .

Did you have fun at Lumberville? How was the ham when you cut it? Wish to God we could get decent ham up here, though I must admit a whole ham is much too much for this family. Sounds easy to do, the way Mary Moore [Molony] does it.

I am going to make no plans for a trip anywhere until I have licked this shorthand, and no I can't take dictation on the radio because I will never have that much speed, or need it for that matter. I just have to keep slogging away. When are you coming up to see HM Co.? I am just longing to see you both, to talk to you. If you don't come up, I may come down after I finish at school, just for a couple of days. But you are bound to be increasingly busy, even when you finish the ms., with all the packing and arranging to do. I couldn't have you at Christmas, even if you wanted to come all this way, because Mattie and her brother would be here, and Mark home. Mattie and G. are living together—since I don't know when, perhaps late Spring. I think it's wonderful, but it makes problems since I don't want Mark to know as he is very indiscreet. Mattie says she is going to marry him, when she finishes college, and she wants him to go back to college and finish up and she will work and support him while he does. This may be the miracle. Must say she can get him to do almost anything. Isn't he lucky? But you can see the difficulties when she is here. It took some getting used to for me, but I am peaceful about it now. Wonder what B. would have said?

Well, things to talk about. But not right now because I am going to have my drink and get some supper. Then sink into a coma, after the frantic life in this house since Wednesday evening.

Much love to you both,
Avis

JANUARY 9, 1959

Dearest A:
Good to hear from you and to know that you are all right. Anne Doyle must have had that same bug. She came to have a cup of coffee with us one evening, and that night (no connection I hope) woke up with a raging fever and was terribly sick to her stomach, and felt she was about to die. But got over it in 3 days. Doctor told her it was bug going the rounds. Awfully unpleasant.

Papers for 103 Irving finally got signed, and the mortgage has been granted, and Mrs. Mason* is cursing at the bank because they didn't read the agreement and wanted to grant the mortage right away, etc. All very complicated, and thank god we have her managing things! As we see it, the rent will not quite take care of all the fees, but the rent from this house will cover the rest plus the payments for the mortage on this house. (I know mortgage has that other G.) Well, it is all too high finance for us, but evidently things will work out. I, too, wish we were moving right in. I keep thinking of us in it, and feel at home there already. *Eh bien,* some day.

Crystallized Ginger. Just looked it up (Anne Seranne, complete book of home preserving).† For ginger root preserve she suggests (and I am sure it would start out the same way):

* Marguerite Mason, the real estate agent for the Childs' new house.
† Ann Serrane, a former editor at *Gourmet* and the author of *The Complete Book of Home Preserving*.

Wash; soak in cold water 1 hour.
Cover with fresh water, bring to boil, boil 5 min.
Drain, and repeat.
Repeat several times, depending on whether a mild or strong-flavored ginger is desired.
Then boil ginger until tender and cool in cold water for about 1 hr.

I remember when we also were given fresh ginger, we found it was awfully strong unless it was cooked quite a bit, but it didn't occur to me to blanch it several times, which would have been the sensible thing to do.

I am feeling a little better. The other day we had nothing but rice, cooked 6 ways. Floods of it, but necessary for final check up. Mary [Moore Molony] took home several quarts, and last night I gave the remainder with mushrooms to Paul's Norwegian teacher and husband. Anyway, I am absolutely sure of the recipes, timing, etc. (After years of making risotto, I am now convinced that success depends on (finally convinced):

Sautéing rice slowly in butter for about 5 minutes until grains turn milky (but do not brown), as that cooks the outside flour coating and prevents it from getting sticky.
2 cups liquid per 1 cup rice.
Rice must cook in 18 to 20 minutes, so heat must be regulated so it boils slowly and is completely evaporated in that time. Otherwise sticky, mushy, and tastes bad.

Paul said to me, what are you doing, you've cooked rice for years, and good rice. But when the chips are down, is one really sure of everything? Hah. Vegetables will be done by Sunday. Then Meat, which is a great virgin chapter, irregardless of how much meat I have done. Fish is short. Cold Buffet is short. So God willing I'll get them done before we go. Then, all that will be left will be poultry, soups, sauces, eggs, which is just a re-editing job. I think I will just send up to HM what we have done (6 chapters, I hope), and they can wait for the rest. At least they will know what they are getting. Phoo. Mary Moore is being wonderful. Just a perfect MS typist/editor, though can't be hurried. We just about keep up with each other. She has done quite a few recipes, including a Charlotte Malakoff, Choc. Almond Cake, Crepes, and seems to find they cook up as they should. This is very encouraging as she is a somewhat timid cook, and entirely self-taught, but interested. Quite typical, I should think, of the kind of an audience we would get.

Paul has switched Norwegian teachers and now has an extremely nice girl who is the wife of the Norwegian Press Attaché, and finally things are beginning to roll and he is very happy with her, and is also in the best of health.

Isn't it exciting that Arthur Schelsigner's book* has had such wonderful reviews in the *Times,* and also in the *Wash. Post.* I should have sent that one to you, by Justice [William O.] Douglas (I think, or was it Sen. Douglas),† anyway, it was extremely puffy. I am most anxious to see what *Time* shall say, and imagine it did not come out this week because they are thinking up horrid little digs to put into it. He must be mightily pleased. We are even thinking of joining the Book of Mo. Club just to get it. We have asked them if we couldn't get a *Webster's Unabridged Dictionary* as a membership beginning prize, and have heard no reply as yet.

Most anxious to hear what you turn up at Harvard, and please write immediately you know. And *Bon Courage, ma chère!* Please give our fondest to Mark, and thank him for his good Markian letter. Do hope the exams turn out well! And also hope he comes down here.

Much love,

J.

P. trouble OK—strained muscle. Thanks.

APRIL 15, 1959

Dearest J:

Ms. arrived Saturday, and I spent a couple of blissful hours leafing through it—I haven't had time to read straight through, and don't know when I will, but will ration it somehow as a special treat. I agree, it looks extremely handsome, and I bow to Mary Moore [Molony], who is a better typist than this professional. I will be very surprised if there are any real errors.

. . .

While I was gloating over the manuscript, Charlie Morton appeared to borrow books. If you have forgotten, he is assistant ed. of the *Atlantic Monthly,* a glutton and a gourmet and quite a bit of a cook. He also has

* *The Coming of the New Deal: 1933–1935* (1958).

† Senator Paul Douglas of Illinois.

considerable to do with Atlantic–Little Brown, and in fact persuaded James Thurber to do the book on Ross* of *The New Yorker* which is just about out. It started as an article for the magazine and grew. Well, I showed Charlie a recipe or two, knowing he would be fascinated, and he said if HM for any reason doesn't, why not for us, meaning Atlantic–Little Brown. And I said I had understood that LB were not doing so well with cookbooks, and not apt to be interested in new ones, and he said Nonsense, we're hot on cookbooks and do very well. So it's a possibility, if——. Mind you, I think HM have every intention of doing this book, and would be damned fools if they didn't, but there always remains the fact that only Dorothy knows about food at that shop. <u>If</u> HM reneges, I want to send this carbon at once to Bill Koshland at Knopf's, for a quick once-over, as he is devoted to food and loves to cook. But God only knows what is happening at Knopf these days—upheaval all around. I suppose you read in the *Times,* or in *Time,* that Pat Knopf has left the firm to set up his own house, with Mike Bessie† of Harper's and Hyram Hayden‡ of Random, which leaves [Alfred] Knopf without an heir, and one can only guess at the family rows that preceded this step. The sales manager has also left, or been fired, and they have a new one. And I was bowled over to read in *Publisher's Weekly* that Herbert Weinstock, executive editor and my former boss, is also out—maybe I wrote you this. A positive earthquake of a turnover, and Alfred writes me that Herbert wanted a year's leave to do some writing, and he, Alfred, didn't think he should have it. So I gather he fired Herbert. Wow. Alfred is now trying to hire Sam [Seymour] Lawrence away from Little Brown, and I share a general dim view of Sam Lawrence as a person, though I admit he is a good sharp publisher. Charlie doesn't think Sam will take the job. I dunno. I only know that the house of Knopf is at the moment rather gutted, and that Alfred and Blanche are 66 or so, and can't live forever. It is the greatest pity. In view of all this, I don't know how vigorous they are going to be about long-term commitments.

Now after lunch, which I had with two fellow secretaries. My, I love secretaries. One of them is a 28-year-old gal from Oklahoma City, who spent several years of her youth in China with missionary parents, be-

---

* *The Years with Ross,* a memoir of the humorist's early days with Harold Ross, the founder of *The New Yorker.*

† Michael Bessie, an editor at Harper & Brothers publishers.

‡ Hiram Haydn, an editor at Random House.

came a professional violinist, conducted mid-western orchestras, went on tours in S. America and Mexico, now wants to be a writer, has written one novel good enough to interest a top flight agent, has almost finished a second, is blocking out a third, and meantime has been a full-time sec. here, audited some classes, and is entering Law School in Oklahoma in the fall. So when I see an employer treat a secretary like an anonymous piece of office furniture, I get sore.

. . .

Your time is so short. I can't bear it. Why do you choose to go on the United States line,* after your sour experience last time? Is Paul feeling better at last? I don't like the way this thing has hung on and hung on, but I don't think there's any doubt that he will be the better for it when he is finally fully recovered. Hate to think of his energy being cut down, especially right now.

I am going to try to get down to the Vineyard [Martha's Vineyard] this weekend, if Allie [Harkness]'s daughter is back from her cruise and can put me up, though I have been offered a bed at the Jack Saltonstalls,† and can catch a ride down with them Friday afternoon. Would however have to come back before they do, and it is a puzzling question of winter schedules by boat and train. Hope it works out as I feel the need of a slight break. Though I hasten to say that I feel fine, and am really getting on top of my job, and don't shake at all. We finished up the Accounting Book manuscript, four huge boxes of same, and sent it off air express Monday, a great load off the shoulders. Now I am doing for that dear Trautman‡ a plagiarism case—former student of his now teaching at Pittsburg, I think, who simply lifted large chunks from two articles published by another man. Cannot imagine how anybody could be so criminally stupid—how could he ever hope to get away with it? It seems inconceivable. I am typing all the stuff out, and underlining the lifted passages. Trautman deeply distressed, especially as this oaf is coming up for a doctorate here, but of course it has got to be proved. Wonder if he will be disbarred. Perhaps not, but at any rate he has blotted his copybook for fair, and will be second rate in the law from now on in.

Suddenly turned warm and lovely, practically for the first time today—though, ahem, we completely missed the snow you had. I've been

---

* United States Lines. Because Paul did not like to fly, the Childs traveled by ocean liner to Europe.
† John Saltonstall, a professor at Harvard's School of Government.
‡ Donald T. Trautman, a professor at Harvard Law School.

picking white hyacinths and scillas and daffodils, and bringing them into the house where I can enjoy them, as I have no time to fuss around outdoors.

Something very surprising has happened to me and I don't quite know how or why. Suddenly realized I have something over $9000 in the savings account! Got to adding up bank books. Well, that's too much in savings, so I have had the pleasure of seeing Charlie Curtis.

. . .

I'm sure I've missed saying lots of things, but do something for me— when you have that lunch with Simca at Rouen,* drink a little glass of something beautiful for me. I will be thinking about you on Apr. 29. Meanwhile much love, and must now read three whodunits quite fast so I can be free this weekend.

Fondest to both,
Avis

American Embassy
Oslo, Norway

MAY 22, 1959

Dearest A:
So much has gone on since I wrote you from the boat that I find difficulty in beginning. Anyway, it was wonderful to have your letter when we arrived, and you sound in good form. I can imagine you will be relieved to start in with Perkins† at last. A shame about Gordon not having a job yet. I keep thinking and hoping the miracle will happen for him . . . but as you say, it would be foolish to be optimistic. It is such a drag on you, but you seem to have learned to live with it. We feel for you, dear!

Well, Paris was heavenly, as usual. From some food articles in the *NYKer* by Liebling,‡ I had rather expected to see it Americanized. It ain't at all . . . quite a few signs saying "Snack Bar," but the sign is the only American thing about it. The rest remains wonderfully French, nicely down-at-heel, and just as it should be. The greatest change we observed was the tremendous, really tremendous, increase in automobile traffic and the almost complete disappearance of big US-type cars. Gasoline is so expensive, $10 for 15 gallons, or a tank full, that everyone uses little bugs, many of them a bright terra cotta color. Our Rambler, which seemed

* On their way to Oslo, the Childs planned to stop in France.
† Elliott Perkins, a lecturer in history at Harvard and the master of Lowell House.
‡ A. J. Liebling, a columnist for *The New Yorker*.

so small in US, appeared to be quite gigantic. And, of course, it was just wonderful to eat food prepared by people who knew what they were doing. We didn't do anything particularly fancy as we had so many people to see that our restaurant time was limited, and so were our stomachs (would we had brass-lined ones). I managed to get in three good cooking sessions with Simca, and we tried out the various ham recipes I had worked on . . . finding, as usual, quite a few major changes in technique, proportions, and so forth which were required to put the recipes into proper and perfect shape. She invited, for a morning session, a fine little old Périgourdine* cook who hires herself out for parties and is completely taken up with luncheons and dinners for months in advance. Very interesting to see her working methods as she has to get a big and fancy meal ready, with usually no help, in 3 hours. Also had a luncheon session at Louisette's with dear old Bugnard and some of Louisette's pupils. L. is as completely disorganized as ever, and always tries to have far too much for a cooking lesson . . . which drives Simca crazy as it always turns out that Simca has to do most of the work. But the lunch was delicious. Filets of Sole with mushrooms and one of his lovely sauces, a chicken dish, and a most delicious orange mousse done in the ice box tray. The Gourmettes had a luncheon in my honor at the very fancy apartment of the Vice Presidente, which Paul attended as the only male member. And we had a delicious dinner at Les Restaurant Des Artistes with their Soufflé de Turbot, and long family session afterwards with the Caillon family.† We also did quite a bit of poking about the old streets and sighing voluptuously and nostalgically. My, it is a satisfying and heavenly city, and I do love French people. *Eh bien*.

Simca is not in good health. She has an excess of uric acid in her blood [gout] . . . I don't know what that means, but it doesn't sound good. She is going on a strict diet, and to a place which gives "cures." Damn, I hope she will manage to put herself in order. It is really impossible for her to "relax" as she is not the type. But she seems to be taking this condition seriously, and to be determined to do something about it. Please write her and give her the biz.

After Paris we drove through Belgium to Plittersdorf on the Rhine, which looks just exactly the same. We did have a beautiful lunch on the Rhine under the Linden trees . . . where we went once. And we stayed

* A cook from Périgord, a former French province corresponding roughly to the current Dordogne.
† The owners of the restaurant.

with some very attractive friends, the Van der Grachts . . . don't think you met them. He's Dutch, she Belgian, and they are just charming. Then we drove to Amsterdam to spend 2 nights with Erica and Hector.* Hector has been sent by his NY bank for a year's study of international banking methods, all expenses paid, 4 months in Amster[dam], 4 months in Geneva, 4 months in Hamburg. Then three days to drive up to the tip of Denmark and take the night boat to Oslo. We were blessed with beautiful weather all the way. Got up at 5 AM as we sailed up the Oslo Fjord, and were greeted with wafts of pine-scented air, a delicious surprise. Then the heavens clouded over, and it became rather nippy.

Sunday, the day we arrived here, is May 17th, the great Norwegian national holiday, celebrating their deliverance from the domination of Denmark. On that day all the schoolchildren from over Norway parade in front of the courthouse or, in the case of Oslo, before the palace. The parade starts at 10, and goes on for a good 3 hours. We trotted up to the palace and found seats at the side, and had a good 1½ hours of parade. Couldn't see anything but the top hat of the king, waving from time to time from between the palace pillars. After that, teeth chattering, we drifted down into the town along the parade route. People cheering, waving, cafés full to brimming, and everyone having a fine old time. In the afternoon some Norwegian friends of friends took us on a drive about the city, a dear thing to do . . . as it gave us some idea of things. It is a great big old city of no distinction whatsoever, rather old and cozy though. There are low-lying hills surrounding it, on which half the population seems to live, getting a fine view of the fjord. We've been house hunting the last few days and feel we may have stumbled into a jewel, but will not know if we can have it until next week. It was built by a wealthy old widower, is set on one of the hillsides just below some friends of ours. He died last Feb., and they immediately made inquiries for us. So we are keeping crossed fingers. It will be just perfect, with a good kitchen, Norwegian style, and quite big, good living and dining rooms, adequate sleeping, studio, and guest room, fine garden, beautiful view. Please pray for us!

The Norwegian spring, which we are just in, is beautiful. There are loads of flowering fruit trees, great fat dandelions, and lush green meadows. Once we get settled I think we are going to like it fine. The people

* Erica and Hector Prud'homme, Paul Child's niece and her husband.

look healthy, and hardly a fat one is to be observed. The young are particularly fine with glowing cheeks and bright eyes. Life appears to be simple and outdoorsy. But I must say we miss home, and were quite ready to turn around and take the next boat back as soon as we arrived.

Public food, such as we have had, is really bad, but the worst I have ever had was at a ladies luncheon given by the Embassy wives every fourth Tuesday. Gawd. First and main course was a tower of pink frozen mayonnaise and whipped cream with rock-hard frozen strawberries, bananas, dates and peaches. Final and sole following course was a very thick slice of banana cake-mix with a thick pale frosting. A triumph of Norwegian/American *McCallism,** and the lettuce leaf supporting each tower of pink mousse was too small to hide the better part of one's uneaten mousse. How can people concoct such things? Ugh.

Thanks for enclosing Dorothy's letter. I haven't answered her yet but shall say what the hell, better to have too much and cut down until the MS is completed.

Much, much love—write soon,
Julia

NOVEMBER 11, 1959

Dearest J:
Time presses like anything. This is a holiday [Veterans' Day], but I must go to the office for a few hours to finish up some reports and stuff, and get back here and tidy up a bit because Mark is having the Harvard-Radcliffe Music Society for the evening. And Henry Reck and wife [Betty Allen] coming tomorrow from New York and I wish they weren't.

You will by now have had Paul Brooks' letter and I hope to God you are not in the depths of despair. I had a lovely letter from Dorothy yesterday—wish I had time to copy complete, but I am sending it down this morning to Bill Koshland at Knopf, where, assuming authority I realize I have not got, but supported somewhat by our correspondence about what next, and Paul's questions about Knopf in case, I asked HM to send the ms. direct. Bill has seen part of the ms. when he was sitting around here, he is a cook, he swings great weight at Knopf where he has been anchor man for twenty-five years, and every time I have seen him (and he was here last weekend) he has asked. And I don't think they have any

---

* A Julia-ism referring to *McCall's,* the American women's magazine.

commitments for a big cookbook right now. So if you want to take my head off, do, but let's try this before exploring further. Paul's idea about Doubleday may turn out to be the right one—they are rich. This may take time but you will get published yet—I know it, I know it. HM has no cookery minded people but Dorothy, does best with revolting cookbooks like the Helen Corbitt Texas gimmick one. I suppose their reaction was a foregone conclusion but I didn't know it when I steered the ms. there in '53 or whenever it was.

Dorothy says in part—

> I hate to think of her disappointment. The ms. is now in perfect form and absolutely complete. They had several meetings about it in the executive department and the decision was based on a very cut and dried equation: probable costs against possible sales. The production costs will be extremely high; the book, although greatly cut from the Topsy-like growth of the original project, is still 750 pages of ms. The possible sales were discussed from the point of view of actual existing competition in the bookstores combined with an analysis of possible purchasers. The elaborateness of procedural information overweighed the book's foolproof quality. All the men felt the book would seem formidable except to the professional cook, and that the average housewife would choose a competitor for the very reason that it was not so perfect. They feel she wants "shortcuts to something equivalent" instead of the perfect process to the absolute which Julia's book is. We did tell Julia all this when we told her the original project had run away, and she re-talked her book at that time in terms of "the American housewife-chauffeur." This present manuscript is a superb cookbook. It is better than any I know of. But I could not argue with the men as to its suitability for a housewife-chauffeur. It remains time consuming. You have been so deeply involved in this project, and you were so nice to send it our way, that I want you to know right away not only Park Street's decision but the actual reasons behind it. Maybe you will want to send it to Knopf; it beats their recent "French" cookbook by chef Donon all hollow. But sadly enough, I think Donon is more what the sales department would want. Paul's idea of Doubleday is good because of their amazing sales gimmicks with their own book clubs. I will do anything I can to help. I promise you I feel very badly to see the perfect flower of culinary love—and the solidly achieved work of so many years—go begging. And they're such nice authors!

(I left out the paragraphing.)

Her heart is in it, no doubt whatever. I am sending this down to Bill. No more time now — I will write over the weekend. Do not despair. We have only begun to fight.

Lashings of love,
Avis

American Embassy
Oslo, Norway

NOVEMBER [N.D.], 1959

Dearest Simca and Avis:
Black news on the cookbook front . . . The answer is NO, Neg, Non, Nein . . . too expensive to print, no prospects of a mass audience. Too bad. If it had been one of those quick books which tell few details, they probably would have taken it: "COQ AU VIN. Take cut up broilers and brown them in butter with onions, bacon, and mushrooms. Cover with red wine and bake for 2 hours." (Copy of letter enclosed.)

Simca, I am most upset for your sake, as you have been in on this bloody thing for so many years and years. You just managed to hook yourself up with the wrong collaborator! *Eh bien.*

Avis, do you have any ideas of to whom we should submit the ms. Don't know anybody at Doubleday. Could ask the advice of Bill Raney; in fact, a letter to him is enclosed, forward on to him if you would be so kind. Should not think Knopf would be interested in another cookbook of this type as they have just released the Donan book. We did meet a literary agent in Washington at a friend's house. His name is John Valentine Schaffner, New York (don't know his address). He handles both James Beard and the Browns,* but I know nothing more about him except he seemed a very nice chap.

We must accept the fact that this may well be a book unacceptable to any publisher, as it requires work on the part of the reader. NOBODY has ever wanted to publish ANY of our recipes in <u>any</u> publication whatsoever thus far. So that may well indicate something. In fact it does indicate that we're not presenting things in a popular manner. I am frankly not interested in the chauffeur–den mother type of cooking, as we have enough of it.

---

* Evans Brown who wrote the famous *Westcoast Cookbook* and other cookbooks with her husband.

Well, I have little more to say, except that it is too bad. I certainly did allow my hopes to rise considerably after Dorothy's letter.

Any ideas you may have PDQ will be most welcome.

Julia

So now what? Maybe this is an unpublishable book. Wish I knew more what we did about these things. Maybe should just wrap it up and let it do the rounds of the publishers, but just sending it to them? Perhaps it is not necessary to know anyone? Pls. advise.

PS: Do I return the $250.00 advance to HM? I would just as soon—although I feel we have been strung along a bit by them—even more than a bit. But I would really rather return it. What you think?

NOVEMBER 14, 1959

Dearest darling woe-begone J:

Oh dear your hasty note after hearing the news made me feel so sad, because you were in the depths. Don't for God's sake apologize to Simca for letting her down because she is so damn lucky to have you for a collaborator, and all is not lost, and it is a fantastically good book, and there is too a market for it, and all we have to do is find it, which we will proceed to do. Peace, peace, perturbed spirit. I was so hoping my letter would reach you same time Paul Brooks's did but the holiday licked that as it wasn't collected. You have it now. And I hope your spirits have risen somewhat, and your dander too, and Julia is her old fighting self again.

Brief note from Koshland after my phone call to him Tuesday evening, but ms. hadn't arrived yet, or of course my Wednesday letter with the attached correspondence. But I don't think he will keep me dangling long because he is a dear thoughtful man and he is very fond of me and he knows how my heart is involved in this business. It will just take time for him to try some things, let alone read the ms., and then somebody else will have to read it so in all it will take time, but you must be resigned to that.

No for God's sake do not send that $250 back. Don't be ridiculous. It would only embarrass them, considering how long this thing has been going on. Kindly remember that every publisher has thousands and thousands of bucks in outstanding option money that they will never see again, and with far, far less justification than this token payment of $250. Don't think of it again. And don't let your leaning-over-backwards feeling get the better of you, please. It is perfect nonsense—especially considering what an enormous investment of time and money you and Simca have put into the book. Flea-bite.

That is really all I am writing to say. I will have hopes for Knopf, strong hopes, though he may demand considerable cutting, and if he takes the book you had better listen hard to him about cutting, because he knows what he's about. And if he doesn't decide to take the gamble, I will pump Alfred and Bill Koshland dry for the best possible advice. One step at a time, is all.

I thought I might write a real letter today, Saturday, but what a hope. Mattie is here. Henry Reck is here, attending meetings but in and out. Paul Cubeta* coming tomorrow to talk over Bread Loaf plans. Out to dinner and evening last night with A. [Arthur] Maass, who has become my most faithful escort, every week to Symphony and Dinner, so last night I took him to Sanders Theatre to hear that remarkable young actor Hal Holbrook in his one man show, *Mark Twain Tonight,* in which he is Mark Twain at 70 on the lecture program. Extraordinary. How I wish B. could have seen and heard it. There was a reception afterward, and I met him and of course he read everything B. wrote about M.T. when he was working up this show, so we are having lunch together Monday and I am going to give him a copy of that Mark Twain in the nude I am so fond of.† Then Arthur and I went into the Ritz and drank and I didn't get home until two, and tonight I'm going with him and an entrancing pair, Sue [Susanne] and Lloyd Rudolph, both of whom are in the Govt. Dept. [at Harvard] and give a course on India, to eat at a darling little new French restaurant and go to Symphony which tonight will have all the visiting Russians, Kabalevsky, Shostakovich,‡ etc. All gay but quite exhausting. And yesterday one hell of a day anyway, fourth anniversary of B's death which depressed me, Friday the 13th which seemed to put a hex on everything I touched, and turned out I forgot completely to send out the student invitations for High table§ a week in advance, and it is day after tomorrow, Monday, and I like to have died, and had to write them all out and send them around and then telephone students to apologize and see if they were coming. E.P. [Elliott Perkins] took it with great aplomb

---

* The assistant director of the Bread Loaf writing program at Middlebury College, in Vermont. The DeVotos were regulars of the program.

† The reference is unclear.

‡ Dmitry Kabalevsky and Dmitry Shostakovich, two of a small group of Russian composers who visited the United States in 1959.

§ A formal dining occasion at Lowell House.

and I think sympathy, but Mary P.* was horrified and showed it. Well. Not my day.

So here I go to the second page anyway, and shouldn't. I am about to write Dorothy thanking her for her so understanding and generous letter. I sent it down to Koshland.

Darling, must rely on you to copy out and send what I say about the book to Simca as I just plain have not got time to write to her, though I owe her a letter. And when there will be time I do not know as Thanksgiving coming up, for which I have one whole day off, whee! And Mark has asked three friends, and Mattie will be here and maybe her brother from Yale. They do all help, but you know perfectly well where the burden rests. I love it, but I get tired. Horrible stuff at the office about budgets and the like, plus the February room assignments† which I will have to do by myself and about which I know absolutely nothing, and during the Christmas slack period (ha!) with boys gone home; absolutely must master enough of it to get me through. I don't know why I took this job.

No more right now. Do please write as often as you feel like it or can, saying everything, and I will write when humanly possible, as I find I really don't care deeply about any correspondent I have except you.

All love,
Avis

NOVEMBER 15, [1959]

Dearest Avis:

Your letter arrived in the morning mail yesterday, and I immediately dashed off a copy of it to our poor old Simca. I haven't heard yet from her, but expect I will in the morning's mail tomorrow. I hate to think of her desolation after the years and years she had put in on this book. Your letter has naturally raised our (P's and my) spirits quite considerably, and of course we approve of your actions as we always consider it is your book too. What a good friend you are, but I shan't even attempt to elaborate upon that.

Your package of books arrived, and I have been pouring over the Mildred Knopf and the Donan. It seems to me that in print, Knopf recipes are not much longer than ours. It is a nice book and I should think would do very well, and I should think could well put people at ease because

* Mary Perkins, Elliott's wife.
† Students were allowed to change rooms during February.

of her chatty pieces at the beginning of each chapter. I think the Donan is very good and perfectly good classic French cooking with quite a few Americanisms put in, which is quite natural. However I don't think the book contributes much that is new as it is a book written from the point of view of a chef who is in the kitchen rather than the chef-hostess who is not. It is <u>most</u> attractively presented. Looking at these two books, I can't see that ours is so much more complicated. In fact, from the point of view of anyone who does any cooking and who does try out new and different recipes, it would seem to me that ours is easier because it tells a bit more.

For instance, having now started to re-arrange my life in case there is to be no published book and as I still have so many hundreds of parts of cookery which I have not yet gone into . . . Patisserie is the large realm . . . I have just been working on some petits fours. My first was on "Tuiles," which are almond, sugar, and egg white cookies which are laid on a rolling pin as soon as they come out of the oven and crisp into a curved shape, like tiles . . hence the name, "Tuiles." Terribly simple batter to prepare, but in none of the French recipes have I found quite the exact explanation of what is what, so I must start out blind. "Beat the egg whites and sugar for 2 minutes." Until what . . . what is it supposed to look like? Does it "make the ribbon." Does the sugar dissolve? What? Or is it just supposed to be thoroughly mixed? Then you get to the temperature of the oven. Some say *bon four,** others say *four chaud,*† others say 5 minutes, others 10. Etc. The extremely critical point is that they must be done only 9 at a time, and the pastry placque‡ must remain in the oven with the door open so the cookies will be very hot and soft as they are removed from placque to roller or the cookies break in two. Etc. Etc. In other words, the recipe should be written so one knows the [potential] problems faced. If various pitfalls are not discussed the poor old cook will feel he/she is stupid and that only a genius can cook.

Well, the fact that Koshland is a cook will make a great difference.

I am quite willing to do considerable re-writing, and am also quite willing to cut down on a number of pages. I also think the introduction could be written to aim the book at the serious cook who considers cooking as one of the fine arts, and as a true creative hobby.

How very nice of Dorothy to take the trouble to write you, and to be

---

* "Medium oven."
† "Hot oven."
‡ Baking sheet.

so concerned. I have not replied to Paul Brooks's letter. What would you suggest?

We have 18 Wegians [Norwegians] coming in after dinner tomorrow for drinks and eats and talk, and 25 Embassy wives coming Tues. for lunch (monthly ladies' lunch, and I took the rap this time). Have been in the kitchen all day, and must now study 25 Wegian irregular verbs. Their verbs are the worst and most irregular of any I have ever seen.

Much, much, much love, and hoping all is fine on the Harvard front,

J.

While the Childs were in Oslo, Julia taught cooking classes to small groups of her "Wegian" friends and diplomatic wives.

NOVEMBER 17, 1959

Dearest J:

Before I get into my electric blanket—

I wrote to Dorothy Sat. or Sun., and got a handwritten one back today as follows, which I won't send you the original of, because I may send it to Bill Koshland for what it is worth—haven't heard from him yet but it's early times.

> Dear Avis: Thanks for your very nice letter—the book is absolutely perfect for what it is—and what it is is a completely authoritative, fool-proof cookbook on French cooking. And it involves appreciation of fine food, interest in preparing it, and a fair degree of time, patience, love. It is not a "get rich quick" book—in other words throw the garlic in the casserole and think it's "continental" school. This latter school abounds. And Donlon (sic!) in the Knopf book had avocado recipes—when there aren't even avocados in France.
>
> Of course Julia is discouraged. But this book should go to a house which is solidly behind it. Our sales dept. wants only cookbooks with a ready made outlet. We sold nearly 70,000 of Helen Corbitt (poppy seed and sugar on salad!) because she had Neiman Marcus behind her. We did not have to create a demand. I have really begged off from reading cookbooks now because it doesn't make any difference what is in them. So it's too frustrating.
>
> I don't know anyone who cooks at Doubleday, but LeBaron Barker, their exec-editor, is Dean Briggs's* nephew, just terribly nice, and one hell of a bright publisher. At least you could talk to him and he would know what you were saying. Most don't. Love, in haste, Dorothy.

All emphasis hers. This ought to warm you up a bit. I'll probably send it to Bill, ask for it back, make a note about LeBaron Barker, and send it to you for your memory book. If Knopf falls through, will probably make time somehow and go down and talk to LeBaron B. in person. Or send him ms. first and then go down. We are fighting.

Paperback from you arrived today—will try to read soon and let you know what I think.

Forgot to tell you Charlie-Freddie stopped in couple weeks ago, with live lobsters, which we cooked and ate in the kitchen. Guess I did too

* LeBaron Russell Briggs, the dean of the Faculty of Arts and Sciences at Harvard, and a friend of Bernard DeVoto.

tell you. Too many things happening around here for me to keep up. House exactly like Grand Central this last weekend—Henry Reck arrived Thursday afternoon late, Mollie [Brazier] and he to supper of chicken livers. Friday the *Mark Twain Tonight* show with A. Maass (and I had lunch with Hal Holbrook who does it all yesterday, and he is quite a young man, and going off on a European tour sometime, and if it gets to Norway, better see it) and Symphony with the Russians Saturday, and callers dropping in all Sat. afternoon and Sun. same, let alone Mattie was here. And Mark lectured on Satie* at the Signet Society Sunday at a supper meeting, introduced by Wat Edmonds, who called me last night to say M. was just wonderful and god-father most proud of him. Mark talked, played, played records, and apparently charmed an audience of 35. Well. And Herb Scheinberg at lunch at Square today, and feel I have seen enough old friends to last a while. And Lowell House is to entertain Moss Hart† for four days in March, so chairman of House Committee has pulled off a coup, and much excitement today. Well, wish life were quieter, wish you were here, wish we could sit down and just talk a bit. Wish I really understood the nature of veal. Ordered, over the phone this morning, a thin cutlet weighing a pound or so, which awaited me when I got home from work, and which I trimmed, dipped in egg and water, crumbled in home made crumbs, chilled, and sautéed in clarified butter. For about fifteen minutes, slowly. Well, just never ate anything so good, so juicy and succulent and tender. But had done precisely the same thing couple of weeks ago, and it wasn't very good, dried out, non-tender, no juice. Don't get it. I always cook with loving care, never hurrying a sauté. Quality couldn't have varied all that much. Veal my favorite meat, practically, but baffling. I've asked you before, I ask you again—what is this thing about you cook veal just a bit too long and it gets tough, like liver? Think maybe two weeks ago I cooked it 20 minutes instead of fifteen.

Good night sweetie—me for the hay,
Avis

* Erik Satie, a French composer and pianist.
† An American playwright and musical director.

DECEMBER 5, 1959

Dears:

There is just no TIME. Must compress like anything. Lunch at Dorothy's last Sunday, and no talk about the cookbook because the Murdocks [Kenneth and Eleanor] were there, and another couple, and we then all went on to Castle Hill to the Thompson reception, where there were 500 people and a string ensemble and lashings of champagne and all the works. Quite fun.

. . .

Hasty note from Koshland, who is back and swamped, saying he has turned the ms. over to Judith Jones (while he was away) and she is experimenting with the recipes at home. She is a very good girl down at that office, an editor now, and knows quite a bit about food, and her husband [Evan] more, according to Bill, and he says he hopes to get at the ms. this week. So let us wait.

Very, very busy at office this week, but somewhat more at ease there, and certainly no dull moments—Perk had to deal the other day with a student, ex-Marine, who is being slapped with a paternity suit, by a lady-Marine—this is one I spotted in September, when the boy (my he is a handsome one) asked me if there was any way he could keep his name off University lists. Quite impossible of course, and I wondered then if it was alimony or whatever. Now we know. We are also dealing with a Long Island polo playing Hitchcock* who turned on the sprinkler system in a local night club and did a thousand bucks worth of damage, and I guess he will walk the plank, and he should. I don't deal with any of this, but the hoorah goes on about me.

Last night dinner and symphony as usual, and in an hour I leave for a party at the Gilmore's† (Myron will probably follow Perk as master of Lowell House, when Perk retires, and gossip about that date varies, and I don't ask) and fortunately I love him, and like his wife enormously—she is the granddaughter of Whitehead‡ who entertains herself by cooking lunches at the Commonwealth School in Boston—I think I must have mentioned her to you. Thanks for recipes—I sent them to Koshland.

Avis

---

* Tommy Hitchcock was a famous polo player from Long Island's "Gold Coast" aristocracy in the 1920s and 1930s.

† Myron P. Gilmore, a history professor at Harvard, and his wife, Sheila.

‡ Alfred North Whitehead, an English mathematician and philosopher.

DECEMBER 27, 1959

Dearest Julia, and Paul, and Simca-Jean, if this catches up on time: Happy New Year. Much, much love to one and all and I wish I were with you.* We had the quietest Christmas ever, just the two boys, as it turned out at the last moment, since all of Mark's invitees decided to go home, so we had a fifteen pound turkey anyway and I am quite sick of it.

Thank you J and P for the delicious soft little shawl, which is just the thing to put over my aging shoulders as I sit on my sofa, behind which the one opening pane in the storm window simply will not stay closed, so that a nasty little draft trickles in. It is light as a feather but just the right weight, and pretty as well, and I love it. Notes from the young enclosed. They were, naturally, delighted to be remembered.

I have been out every night for a week and am going out again tonight—such a lot of pleasant parties and such wonderful food at them. I was only supposed to have a day and a half off from work, but took two extras to get all my medical and dental things done, and am in the usual state of rugged health. I look fine, too, part of which, alas, is about six pounds around my middle which maddens me, but which is undeniably becoming to the face. You can't win. I find, somewhat to my surprise, that when I am poking around home for several days I feel very much at loose ends, and actually missing the office. Must be getting to be a career woman.

Return enclosed clipping, which I must send elsewhere. This was the most terrible blow to me—among other reasons, because I had so looked forward to having Charlie† know you, and you him. Quite selfishly, I feel terribly bereft because he looked after me and all my little affairs, and kept an eye on the boys, and I never felt anything could go really wrong when I could turn to him for advice. I had just sent a mess of papers dealing with renewal of copyright on *Mark Twain's America*‡ down to Stonington for him to deal with, and I suppose they went up in smoke. [His wife] Frances has been immobilized with arthritis for about 20 years, and when I saw her a year ago was enormously fat, but she had recently lost 75 pounds and was feeling well enough so that they were going to Mexico in January. The fire started in Charlie's bedroom study, where he had some photographic equipment, and a log rolled out of the fireplace and the photo stuff, I can't think what, blew up in his face. He got Frances out

* The Childs were visiting Simca and her husband, Jean, at their home in Plascassier, in Provence.

† Charles Curtis had died in a fire that destroyed his home.

‡ A book by Bernard DeVoto, published by Little, Brown in 1932.

on the porch roof and went back for his manuscript. He would have been blind if he had lived. She is up here in the apartment on Mt. Vernon Street and I have talked to her and will go in tomorrow. It breaks my heart that you never knew him; he was the most exciting man, bubbling over with ideas and gusto and eagerness. Hell and damnation. Remember always to use a fire screen, and if there is a fire never go back for anything.

I feel fine about Rockefeller* bowing out, except I wish he were a Dem. But I think now we can elect just about any Democrat, and holding my nose firmly I shall vote for [John F.] Kennedy or whoever. Shocking, that the Repubs. are so sworn to an extreme right-winger as Nixon is. There is so much distrust of him all over the country that I do not think he can possibly win. I don't even think it will be close.

Now good news, only we mustn't count chickens. Bill Koshland called me up this morning and we had a fine long talk about this and that, I carefully not alluding to the cookbook. When he got round to it, it was to say that he is impressed to death with it. He left the ms. with Judith Jones when he went to San Francisco, and she took it home and started cooking out of it, with great success, and is trying quenelles today, he said. When he got back, he took the second half of the ms. home and started cooking out of that, with several smash hits. He and Judith are now switching halves, and Angus Cameron, who is their new editor and I haven't got time to go in to his checquered career but he used to be with Little Brown and was a famous Communist sympathizer but is now reformed, is also taking a hand as he, too, cooks. Well. Bill says what Dorothy says—best damn cookbook he ever ran into, and unique. He did not think highly of Donon and does not use it at all. He loves the arrangement, says it makes all kinds of sense and is easy to use. He is going to consult with Sidney Jacobs about production costs, and wanted to know how many more drawings there are, and I didn't know. Wanted to know if there were photographs, and I said I thought not. Sidney Jacobs is production manager, has been there practically from the beginning of the firm, and next to Alfred and Blanche is the power in the firm. Bill says the firm has <u>no</u> commitments for any new cookbooks.

Now of course Alfred has not yet seen the manuscript, and God knows what he will say, do, or think. Alfred was quite sold on Donon, but Bill says he also had some reservations about it. When I wrote to Al-

* The liberal governor of New York, Nelson Rockefeller, had been a Republican candidate for president. Vice President Richard Nixon was now the front-runner.

fred, after sending the ms. to Bill, Alfred sent my letter in to Bill with "Doesn't she know about Donon?" scribbled on it. So I am just not going to get my hopes up too much until Bill and Judith and Angus have stopped playing with the ms. and sent in their report. As Bill said, three enthusiasts can swing a lot of weight.

I really should not have written you this, since it is so very iffy, but could not resist, and have to tell my pen pal everything, and only hope I have not raised up hopes to be dashed to bits later. But hope is important, and I am hoping. How excruciating all this is — the waiting and wondering. So much worse for you and Simca and Paul and Jean. Well, I will inform you instantly if anything happens. Will cable if news is good.

So I'm back to work tomorrow, anxiously anticipating the great rooming crisis which breaks on me in February and March. Had planned to get some expert help on it during the Christmas recess, which is quiet, from Evelyn Spidell, who is secretary for Houses and who fortunately likes me, but dammit, she has gone to Canada. By the time she gets back, the students will be back too, but somehow I hope she can give me an hour a day for a few days until I get the principles bashed into my head.

Must now get this into the mail, hoping it will reach you at Bramafam, and not just miss you. Oceans of love to all of you, and tell Simca that someday I will write, but I don't know when, but anyway I think about her every day and miss her. And I miss you too, dears.

Love,
Avis

PART FOUR — 1960–1961

Great Gratitude to the House of Knopf

Publishing *Mastering the Art of French Cooking,* a Group Effort

The Childs' newly remodeled kitchen at 103 Irving Street in Cambridge, Massachusetts.
PREVIOUS PAGE: Julia and Paul pore over the proofs of *Mastering* in their Oslo home.

As you know, the enthusiasts around here are absolutely convinced that this book is revolutionary and we intend to prove it and to make it a classic. There's a big job ahead, particularly along promotion lines, and we don't propose publishing before Fall '61 but the sooner we get going the better. I am so grateful for Angus [Cameron] because he has had so much experience in this area (he was in on the launching of THE JOY OF COOKING years ago at Bobbs-Merrill) and together we are hatching up all kinds of schemes.

Judith Jones to Avis DeVoto, May 6, 1960

The Childs ushered in 1960 with a visit to Simca and Jean Fischbacher at their home in Provence as part of a holiday vacation that started in England and included stops in Paris and Marseille. In the meantime, the recipes from *French Home Cooking,* as the book was tentatively titled, were energetically being tested by Judith Jones and Knopf's vice president, William Koshland, at Knopf in New York.

On her return to Oslo, Julia tried to put thoughts about a possible Knopf acceptance of the book out of her mind by giving cooking lessons during the season of short days and long nights, which the Norwegians call *voersyk.* She also conducted classes in her students' homes, limiting the groups to about seven or eight women, who were, for the most part, Norwegians, friends, or embassy wives.

Avis's letters were encouraging, often indiscreet, confiding progress on the Knopf front. She had carefully laid the groundwork with Koshland, who referred to her "missionary work with me [on the book]" over the years, and he reported that Jones was increasingly impressed with the thoroughness of the research and testing. But the situation at Knopf was tenuous. Blanche Knopf was promoting *The Classic French Cuisine* by Joseph Donon. Alfred Knopf was in the midst of the company's merger with Random House and not interested in another French cookbook. So Koshland and Jones enlisted the aid of Angus Cameron, who was now a senior editor at Knopf and the former champion of the *Joy of Cooking* at

Bobbs-Merrill. They also priced out the book with Sidney Jacobs, the long-time production manager at Knopf, and then presented the Knopfs and the board with an irresistible number of reasons to publish the book.

On May 6, 1960, Jones sent Avis word that "our publication proposal for the Child, Beck, and Bertholle has just been approved. I can't remember when I've been as excited about a project and I know how much it means to you. Bill is off in Greece but I'm anxious to let the authors know of our decision (heaven knows they've waited long enough) and to start negotiations."

Avis immediately called the Childs in Oslo, and on May 10, Jones wrote to Julia:

> As you know we have spent months over the superb French cookbook by you and Simone Beck and Louisette Bertholle, studying it, cooking from it, estimating, and so on, and we have come to the conclusion that it is a unique book that we would be very proud to have on the Knopf list. Bill Koshland, who would ordinarily be handling preliminary negotiations, is on a vacation in Greece just now, but I have been authorized to make you an offer and start things in motion. I know you have been waiting a long time for our decision.

Julia kept her coauthors informed of all the details regarding the book. But the problems of collaboration surfaced with Simca's insistence that they avoid what she considered an emphasis on "*goût américain*" (American tastes). Difficulties with Louisette regarding the 10 percent royalty share that she had initially agreed to were more vexing. She rejected the title of consultant and insisted on being cited as one of the three authors, listed alphabetically. She also demanded 18 percent of the royalties of the book. (Much later, in 1968, she finally agreed to a buyout and no further partnership in volume 2.)

Simca and Julia did the lion's share of the editing. Julia especially went through the manuscript several times, and she worked out several of Jones's suggestions regarding portion sizes, the deletion and inclusion of recipes, and illustrations for the book. By the end of July, Julia had replaced hamburger recipes with *carbonnade à la flamande, pièce de boeuf, paupi-*

ettes de boeuf—Belgian beef stew, prime cut of beef, stuffed thinly sliced rolls of beef—and cassoulet. Paul wrote the dedication, but the book's title remained unsolved. Paul enlisted the advice of his twin brother, Charlie, and friends. Lists of possible titles were made and sent to Avis and Knopf, and in November Jones and the Knopf staff christened the book *Mastering the Art of French Cooking*.

With proofs arriving in Oslo, Julia's work schedule was demanding, and Paul's involvement was considerable. Julia confided to Avis:

> We think we shall get out of the Foreign Service at the end of our two-year tour next year, and come home for good. It is just not the kind of work which Paul enjoys as he has nothing along the creative lines he likes—all administration—possibly a lecture or two, but that's all. And no time at all to do his own life . . . I don't know when we shall be able to get into our house in Cambridge, as they have a lease until Sept. 1961, and 3 to 4 months notice. But if we tell them around Christmas time, or later, that we shall give notice on Sept. 1st, they might then find another house by that time. We can only hope.

From the inauguration of President John F. Kennedy in January 1961 until the Childs' arrival in New York City on June 5, Julia reviewed galleys, compiled lists, and checked details. Avis welcomed them into her home and became part of the promotion of the book, planning a cocktail party in Cambridge to launch it and writing letters destined to be sent with review copies to people she considered influential.

Julia and Paul moved into 103 Irving Street at the beginning of August and slowly began to remodel their home. A month later, Julia wrote to Simca:

> Avis telephoned last Friday to say she had just had a call from Wm. Koshland of Knopf, saying he had the book in his hands at that moment, that it was beautiful, that he was so grateful to Avis for having steered it their way, and that they were planning to do a full page advertisement in the NY Times Book Review Section. (Comes out every

> Sunday, and is read all over the country.) He said Avis and I were each to be receiving our copies right away. As yet, nothing has arrived. I have asked J. Jones to send you one copy airmail. PS: I must say I think this book of ours is MARVELOUS. It tells so much, and has so many wonderful recipes. I am struck all over again by what a remarkable book I think it is. And I think of all those years of struggle you have had ALL ALONE when collaborating with L. [Louisette] on your first gigantic MS. But those years were not wasted, because look at all the wonderful things you discovered, and the great recipes. My, I do hope others will feel the same way we do about it . . . The timing is perfect, and we couldn't be in a better place—and we are now not UNKNOWN WRITERS.

The months following the publication of *Mastering the Art of French Cooking* on October 16, 1961, were nothing short of a triumph. The *New York Times'* influential food critic Craig Claiborne's review proclaimed it "probably the most comprehensive, laudable, and monumental work on the subject, [written] without compromise or condescension." The writer and cooking teacher Michael Field said that it surpassed every other American book on French cooking already in print. Simca arrived in New York on October 22 to plunge into a schedule of meetings and lunches arranged by Julia and Jones. The editor of *Vogue* met with them at the Cosmopolitan Club, a private women's club. The next day, they had lunch with José Wilson, the food and features editor of *House & Garden* magazine, to plan a six-page article, and they were interviewed on the *Martha Deane* radio show, hosted by longtime radio broadcaster Mary Margaret McBride. They also dined with James Beard at the elegant Four Seasons, and they met Dione Lucas, who offered to give them a party at her restaurant, the Egg Basket.

Julia exulted, "We are in. Hooray!" She had arranged a book tour that took them from New York to Boston, Detroit, San Francisco, Los Angeles, and Chicago. Long before Simca arrived, Julia had prepared their props—white aprons with the L'École des Trois Gourmandes' monogram, French omelet pans, utensils, and sharp knives carried in what was to be the first of the many "sacred black bags," as the Childs called them. Julia also had made all the reservations for trains and planes. Because the trip was strictly do-it-yourself, not subsidized by Knopf, they stayed with friends and Julia's relatives. Visiting prominent department stores, they started preparing Roquefort quiche, sole in white wine sauce, and chocolate madeleines at 10:30 A.M. and continued nonstop all day. They baked Julia's favorite Reine de Saba, a creamy chocolate almond cake, in the test kitchens of big-city newspapers as part of their interviews. Sometimes the facilities were so unsuitable for cooking demonstrations that their pots and pans had to be washed in the ladies' room. But Simca, Julia, and Paul, who accompanied them everywhere, had hit their stride.

Before leaving New York, Simca and Julia accepted the *Today* host John Chancellor's invitation to appear on the show. They both agreed that Simca should demonstrate how to make an omelet. "With panache, Simca," Julia coaxed. "Toss that omelet! Ham it up, and make it visual fun for the audience." Not only was it a foreshadowing of things to come, but Simca's TV appearance also established the link between the book and the cook in action. That wasn't a new concept to Julia, who had learned that teaching and cookbook writing were interconnected much earlier at L'École des Trois Gourmandes.

In seven weeks, Julia and Simca saw their book successfully launched in a sequence of difficult and exhilarating events, and Julia's first letter to Simca after Simca's return to Paris in mid-December glowed:

> How much we accomplished. I think it was a good thing, and thank heaven you came. Without you, I would have been quite useless as a promoter, and it would have been too bad. And the fact that we are now accepted by the big food people is perfectly wonderful—we are now *quelqu'un* [somebody]—Hooray.

Oslo

JANUARY 17, 1960

Dearest Avis:
We arrived home last night to find Oslo cold but bare of snow, which was a great surprise and disappointment. The whole of Europe as we came through from Rome, was covered with it and I just don't understand what has happened around here. It was a beautiful bright day, temp around 28. Today it has clouded slightly and a faint powdering of white has been produced but no more. Well. Meself I love piles of snow, and especially around here as it makes it seem especially Norwegian and just right.

It was a great pleasure to find your letter waiting for us among the piled up newspapers and advertisements. What a weird thing about G's lung.* We've never heard of such a thing, but are grateful to hear that all is going well and that it is a heard-of thing. Weird. You sound all right about the job, too. Paul knows the work of Stuempfig† and likes it immensely. Marya Mannes Clarkson, as a matter of fact, has one in her apartment and we have always admired it immensely. Will be most interested to hear what a man of his talents does with a portrait to make it interesting. If they take a picture of it, do send us a copy (which we will return immediately).

Was also happy to hear you had a real Channel sole in Cambridge! We met a Mr. Prince, a Frenchman, once in Paris, who has a wholesale fish import biz. in NY and gets soles airmail . . . Wonder if this came from him. My, they are good . . . We had quite a few lately. Have had only one here which came from Denmark; they seem to be rare. But it was fine and fresh tasting, and we did it in one of those nice buttery white wine sauces . . . and the sauce was so good just because of the sole fish fumet which seems to be more elegant and tasty than anything else.

We were horrified to hear about Charlie Curtis . . . What a really stupid kind of accident, just as stupid as Camus' death at the hands of that bloody French driver.‡ He sounded like such a nifty fellow, and we know how much he meant to you. Now what on earth will become of his wife? So sad.

---

* In a previous letter, Avis had reported that Gordon's lung had collapsed for no apparent reason. After a week in the hospital, he was well again.

† Walter Stuempfig, a Pennsylvania artist of the mid-twentieth century, whose expressive paintings were done in the style of the old masters. While Avis was employed at Lowell House, Elliott Perkins had his portrait painted by Stuempfig.

‡ The French author Albert Camus was killed in an accident in 1960, when the driver, his publisher, Michel Gallimard, lost control of his sports car. Gallimard died a few days later.

Well, we had a lovely trip. Took a Swedish boat from Götheborg (pronounced sort of—Yúh-tah-bore), Sweden, about 5 hrs. from here by train, and aimed for Tilbury, which is the Port of London. Because of very big high waves, we could only go 3 knots an hour most of the time, so we were 12 hours late. But we didn't mind at all as it was a most comfortable boat with splendid food and quite a gay group of passengers . . . Took us two full days instead of 1½ days, which cut our time in London a bit short. However we did manage to see everybody we wanted to see including Nancy Balfour who invited us to a splendid oyster and sole lunch. We also met, at a dinner, a Mr. Peter Wilson who turned out to be one of the managers of Sotheby's. In London we met Erica and Hector Prud'homme, our niece and husband, and took the train up to Cambridge for Christmas. All Peter Bicknells in fine form and madly busy because their eldest daughter was about to be married shortly after the New Year . . . only 19! And the next one who is barely 17 is now engaged. Seems awfully young to us . . . and probably to the PB'S . . . but what to do in such cases. Cambridge was lovely, as always, and we did a bit of walking around and admiring as ever.

The Sunday after Christmas we took that fine day train, the *Blue Arrow,* from London to Paris, arriving at the Pont Royal around 10 PM. We immediately trotted up to Lipp's* for a nightcap and a breath of the Paris that is Paris. What a city, as we said to each other every other minute. With only 3 days there, we didn't do nearly as much walking about as we would like, but we did see the main people we should have. Poor old Bugnard is getting so lame he can barely move. We asked him to lunch with us at Lapérouse but he said he just couldn't manage it, so he invited us for coffee and liqueurs at his house way in a little *banlieu* outside of Paris. We went there in an old taxi with a chauffeur who had never been out to Parc St. Maur and kept moaning loudly, "*Ah les banlieux, c'est affreux, c'est abominable, on ne sait pas où on est.*"† We finally found the Bugnard residence, and he was just as sweet as ever, and looked well, but has terrible difficulty moving around. It's sad, particularly as I think his little old wifey is rather a bore for him . . . a nice little body, but with none of his class.

Mme. Perrier, our dear former landlady who is now 86, is hardly

* Brasserie Lipp, a Left Bank favorite of Parisian intellectuals.

† "Ah, the suburbs, they're awful, they're dreadful, it's impossible to know where you are."

functioning mentally, can't remember what anyone said and keeps asking the same questions over and over again, but is as darling as ever. Hélène Baltrus[aitis]' mother had a stroke 3 days after she received the Legion of Honor from Gen. Charlie de Gaulle's* own hands with a private luncheon afterwards, and that very day her apartment building started in to be demolished. So Hélène has been having a hideous time, and finally managed to get her mother in a clinic. Heaven knows how long she will last, but she is a tough old thing (83) and may go on for months or years paralyzed on one side. Luckily, according to Hélène, she does have enough to keep herself cared for in a clinic . . . but that will probably eat up any inheritance Hélène might have received, I suppose.

So, from the side of old acquaintances, Paris was a bit sad. It is always hard to realize that life and people will not go on forever, including one's selves. Woe.

The last day we were there we trotted up to Les Halles and bought a 1½ lb. fresh foie gras (raw), and 2 fresh truffles about 2½ inches in circumferences, and took them down in the train with us to Cannes . . . That smell of fresh truffles in the compartment was almost suffocating. While at Simca's we marinated truffles and foie gras in Madeira and Cognac, then poached the truffles in the marinade, cut them into fat sticks, stuck them in the foie gras, wrapped the foie gras in a *crépinette* (filament cobwebbed with little lines of fat . . . pig's cawl), and poached it in a casserole. *Quel parfum*. Then, after it had cooled and chilled, we all ate it with a spoon . . . 8 of us, and we ate every bit of it. An unforgettable gastronomic height.

Bramafam† is charming. It's an old Provençal house, the kind they call a *mas,* set on a hillside facing a little medieval village on another hillside. All around are grape vines, and rose cultivation for the perfumeries of Grasse. The weather was lovely, and smelt already of spring (though since we were there it has snowed) . . . Roses were out, little beginnings of mimosa, and that lovely smell of Provence, which includes wild herbs, and whatever they use there to burn in their fireplaces which gives a particular Provençal perfume. Heavenly.

Simca seems quite well but ought to lose some weight. She says her blood condition is much better and she is following her regime of diet . . . except over the holidays. But she is just not the type of person who

* Charles de Gaulle, the president of France.
† Jean Fischbacher's family's land was called Bramafam, "the cry of hunger."

can relax. I think also that with her old mother, and his old father, both around 85 (people seem to live long in France), she has a great deal to do and little time to relax. I just hope she is all right.

After 3 days with the Fisch[bachers], we took our little rented Fiat . . . a bit bigger than a Volkswagon and started down the Riviera to Italy. A few weeks ago we read somewhere about a man who complained to the Fiat salesman that the foot pedals of his car were so close together one foot covered them all. The salesman said if he expected to drive an Italian car he should wear Italian shoes . . . all too true. Otherwise it is a fine little car and just the right size for medieval streets. Our first 4 days were perfect weather, but by the time we got to Siena things had gotten a bit gray and foggy, so though it was all fascinating and fun, P. was unable to take any great photos and was mad. We were the only tourists anywhere, which is always nice, and loved poking about old streets. But next time we shall do Italy in a bit warmer weather . . . We kept thinking, when we bedded down in our cold hotel rooms, how nice and warm we would be if we were in Norway! Possibly because of our Norwegian, our Italian seemed much better this time . . . a lovely language.

Simca and I were naturally very much interested in your report on Koshland and the cooking they are doing out of our MS. That's wonderful, as it is just what the damned thing is for . . . to cook by. Also delighted that the Fairbanks [Ben and Ginnie] are using it. Well, we shall patiently wait. There aren't any formal photos, only the ones Paul has taken for our illustrator to work with. We will happily provide any amount of technical drawings, and indicated only a necessary minimum in the MS. Our illustrator, John L. Moore, is now living in Albany and working for [Governor Nelson] Rockefeller, so is easily available. (He is not a Republican.)

Must stop, and get to Christmas letters, and the pile of bills and so forth which has accumulated.

Write as soon as possible, and keep well and happy, and
much, much, much love from us both,
J.

FEBRUARY 18, 1960

My pets:
I have been hanging by my eyelashes long enough and I can't do it any more or let you go on doing it. You may have noticed that I can't write you proper letters; well, I've had enough of that. I promised Koshland

that I wouldn't let you know what was going on until it was wound up, but much as I love Bill I love you considerably more and it kills me not to be able to write and tell all.

Even so I will make this short. What is happening is that Bill called me up—nearly three weeks ago I guess, though I haven't time to look up the letter I wrote him in my hysteria—and said he is mad about the book, so is Judith Jones, so is Angus Cameron, all of whom have been cooking out of it at length. Just like I've been droning on all these years, they say it is unique, nothing like it or ever will be again, a real break-through, perfect. It's really French and it works. They all—three weeks ago—turned in glowing reports. Neither Blanche nor Alfred has seen it. Blanche said "I don't think we need this, do we?" Bill talked to production, sold them completely. Bill, plus production, says they can do a beautiful book for $7.50, no cutting, maybe they even want more material, double columns and all. Bill says they think they can sell 15,000 to 20,000 the first year. The first year. And God damn it all to hell, I have been waiting ever since, my stomach tied in knots, for the final word. It just takes time. Bill says they are so sold, they are completely behind it, and once the final figures from production come through they will ram it through the board no matter what. And said then, wait until it is final and then you can have the pleasure of telling the ladies. So I've been waiting. He just don't know how hard it all is on this old lady. Which is why I wrote for your telephone number. Oh Gawd. Well I am waiting, but I can't keep you dangling one minute longer. Whether you can keep Simca dangling one minute longer is up to you. Probably would be better but I will leave it up to you.

Now I will feel guilty, jumping the gun on Bill. You can't win. But I can't stand this any longer and three weeks is too much. I truly don't think it will fall through. But it was pretty ghastly having dinner with Alfred and the elder Schlesingers and not having A. say one word about the book. He [Alfred] was being difficult and shut up Elizabeth [Schlesinger] and did a monologue. I didn't talk at all and didn't bring up the book. Coward. Seems weeks ago at that.

Bill talked to me for over an hour on the telephone so you can see how sold he is. They are now getting figures—from HM which evidently worked some out, so that you could guess at page length and book length. And working out their own. And Bill says they are not keen on the drawings, but mad keen on getting Paul's working photographs. Says he thinks it's the kind of book that should have those photos. How they can do this at $7.50 I just wouldn't know. So Paul better start getting those photos together, when the word comes. I said they might decide to do both photos and drawings, because maybe some things need drawings more than photos, don't you think? Also, told him about [Paul] Sheeline's part in all this, and what do you know, he knows Sheeline well and said something about Sheeline being in and out of his house when he was growing up. I don't know if he has got hold of Sheeline. I have tried hard not to bother him or nag him, which is probably why he just doesn't know how I am suffering, not being able to write you freely.

This is where it is now and maybe I might even get the call tonight. I even stay in nights waiting for that call. I am dying anyway, quite apart from this cliff-hanger, because I am in the middle of the room re-engagement, tired to death, having insomnia, falling apart. But Bill doesn't know how hard it is, with our special relationship, and even if God forbid the whole thing falls apart, I have to tell you, and will you please, please just talk to each other about it and no one else, except Simca if you think she has been pushed as far as she can go?

All my love,
Avis

Oslo

FEBRUARY 22, 1960

Dearest Avis:

Your letter in this morning just before I took off *pour le ski,* and as it is a holiday, Paul was home and we poured over it together. Very interesting indeed. I refuse to let more than a coal of hope glow quietly, but it is certainly encouraging. In *fact*, has made me think that I should not spend so much time on *le ski*. (All ski and no work makes a healthy pin-head . . . And how true, as there is no time for anything else, and I can easily see why athletes must be meat balls. In a manner of speaking, one can be one or the other, but never both.) What most impresses me is that these chaps are really cooking from the book, and it is not hot air.

As to illustrations, I am air-mailing you today a parcel of Moore's working sketches, and some of the working photos. Both of us feel that if photos are to be successful in a technical book they must be perfectly reproduced . . . which seems to happen rarely except in Swiss printings. Nothing can be indistinct or fuzzy. The backgrounds must be perfectly plain and non-obtrusive, and everything concentrated on what is being shown. From the point of view of illustrating how to do something, we both feel the drawing is usually more successful as in a drawing one can eliminate all non-essentials, and point up the parts which are to be emphasized. Paul also feels that drawings go better with type. In any case, if photos were to be considered, none of the ones enclosed are suitable as they are purely working photos and no attention is given to background, etc. A new series would have to be done.

A great many of these for Moore were blown up to 8 by 10-inch size. He has them all . . . which were taken for the first book. He also made a lay-out with type specifications, etc. If anyone would like to get in touch with him he can be reached:

John L. Moore
6 Woodridge Road
Delmar, New York

(He is working with Nelson Rockefeller in Albany.)

Well, I shall say no more and think no more, but am happy to report that the old stove feels as though it were being kindled again. I have really gotten myself quite out of the cooking rhythm as I have become so interested in trying to get on top of Norwegian, and now the snow has come have been so mad about nature and exercise. It is so lovely here now and so deeply Norwegian. Every two or three days we have about 3 inches of snow, and a mist and snow clings to the trees so that every prickle and knobble is outlined in white. We can put on our skis and start right out from the front door. Or we can go up into the hills above the city and join ski classes which take place on some great gentle hills. Or we can start out from there and slide through the woods on innumerable trails which lead up into a vast wooded preserve called the Nordmarka. Sunday the local electric train, *trikk,* from the center of town to the ski slopes registered 65,000 passengers! It's so nice to see women of 65 and over with little lunch packs on their backs skiing along a trail. Little babies of 4 dressed in blue and red on tiny skis plow down the hills with happy ruddy faces. Little boys of 7 and 8 are learning how to jump. Fathers and Mothers are instructing the young, while families are on a day's outing, which they call *løype* and will go 15 or 20 miles into the woods. A friend of ours and his 64-year-old father were on a 25-mile run yesterday, leaving at 8 and returning at 4. What a life indeed.

Do hope you have that GD room arrangement business finished, and are less tired, and are sleeping. You were in such fine form the last time you wrote! But one does have inexplicable ups and downs. My, it would be nice to see you if even for an evening. Had two frozen Denmark soles for dinner with a lovely Bugnard sauce and thought of you. Funny they don't have fresh sole here, but they don't.

Shall probably report the sense of your letter to Simca, but you need not worry that it will fire her hope too much as we have both become fatalistic.

Much, much love,

J.

Oslo

MARCH 25, 1960

Dearest Avis:

This is really to be short. Yours of March 13 in yesterday, and we hope to god you are less tired and are getting rid of those bloody house assignment things . . . Sounds just the worst kind of dog work, but as though you had managed it very well indeed. One can see why they need a superior type of person for those house sec. jobs. Not only does one have to be a good dog-work type secretary, but administrator and mother. Lowell sounds like a very good house, I must say. Can't wait to see you in your office after having heard so much about it. (I'm behind one letter . . . am now at yours of March 20 . . . That's the one which came in yesterday.) Must say the electric typewriter does a beautiful and even job, and your typing looks more professional than ever. Wonder if you managed to see Moss Hart! And your New York plans sound good, even if a few things have fallen through. We always found it was the greatest vacation just to hole up in a hotel and poke around. It's such a fascinating city, and one feels one is at the hub of the world.

One is certainly at a far corner of the world here in Oslo. I keep forgetting that the whole damned country has only a bit over 3,500,000 people . . . or is about the size of Chicago. And Oslo itself has only about 350,000. Imagine trying to be a whole country with army, navy, TV, radio, king, and so forth with only that amount of population. I kept wanting the Norwegians to win everything at the Olympics, but had to keep remembering that there are only that number of people to choose from. Oslo itself is really a one-street town, and I am sure if you lived here long enough you would know everyone in your particular field and all the related ones. There is no really snappy hotel or restaurant in the NY sense . . . It is really quite a bit like New England but minus the Ritz in Boston and SS Pierce. There is quite a considerable amount of money in the ship-owner class and, I suppose, in a few of the big businesses. But as it is a labor government, taxes are high and it is only when you get really into the big money that you can take advantage of deductions and other fancy deals. The nice thing is that no one seems desperately poor. Our cleaning woman, whose husband is a worker in the building trade, lives in a very nice house with a garden, central heating, ice box, clothes washing machine. The educational system is truly democratic. Except for about 1 private Catholic school in each large town (of which there are about 8), all schools are government run. Everyone has an equal chance to get into the universities. And there is no Ivy League university . . . so no social significance. And as far as we can make out, the education is extremely good. By the time they have finished our equivalent of high school and are ready for the University, they can qualify for junior year in our universities. And besides all the good education, they get wonderful outdoor exercise. All in all, if I had children I would love to bring them up here, and then send them to the US for university.

Paul is furious at the Washington office who insist on sending him a stream of lecturers and harpsichord players in Easter week . . . which is the time of year when there are no Norwegians at all in Oslo. Everyone, almost, takes off the Sat. before Easter and spends the whole week skiing up in the mountains. If you don't book a room in a mountain hotel before Dec. you don't get in, it is that crowded. He keeps sending Washington telegrams to this effect but it is quite like spitting down a well. They also don't realize back there that when May comes around, there is no one in town after 3:30 in the afternoon . . . They are all out sailing or hiking or gardening. And of course everyone wants to take a trip to Scandinavia and lecture in the summertime . . . but they will lecture to empty houses.

Next A.M. I see I shall have to finish this later, but will send along this much anyway. I have to be at my Norwegian class armed with a translation of archeological finds in the land of Cush.* Hoh.

First cooking class was held yesterday and seemed to go off very well, with just enough to do. What with each of 8 people with own bowl making pastry dough it took quite a bit of time, but was most useful to all as one could learn from the mistakes or successes of the others. And I also found my short paste recipe was not quite as descriptive as it could be, so have made up another one for the MS.

* One of the earliest civilizations on the Nile, south of Egypt in what is now Sudan.

Have enclosed

1) New recipe for short paste (copy for the DeVoto/Fairbank MS, and 1 for original with Knopf)
2) New recipe for choc/almond cake (DeVoto/Fairbank and copy has already been sent to Ben)
3) Schedule and recipes for cooking school . . . in case they might be of interest to Knopf, and to show flag is waving

Much, much love. Keep healthy and happy! Will continue soon as poss . . . but have to make up recipes for next week first. We've bought ditto machine! Great fun, and most useful. Electric typewriter would be a help for this, but ain't going to spend no more money.

Love again,

J.

APRIL 9, 1960

My dears:
I have been to the big city. I am back, late Wednesday night. This the first minute I've had to report to you. And we are still cliff-hanging, because I got a brief note from Koshland this morning which says, "I don't think you should write Julia until I say that I've got the initial on the important sheet of paper."

Briefly, situation is about the same, but Bill is beginning to move, and when he does he will probably move fast. We must be patient. But we can be quite hopeful. I had about a half hour's meeting with Bill, Judith Jones, Angus Cameron (all of whom have cooked with the book) and Sidney Jacobs who handles production. Don't know what earthly good I did them all, as most of the talk was way over my head, about production costs. Coated stock and all that sort of thing. I think they don't like coated stock. Sidney is going to get some sort of rough estimate, and Bill is going to draw up a formal production proposal for the Knopfs. All awaits on their decision. Bill is also writing at once to Moore, in upper NY State, to ask for the rest of the drawings and working photographs. Nobody likes the drawings—says they aren't good enough. Everybody likes the photos, just as they are. I put in a strong plus for a good number of drawings, said they would turn out to be absolutely necessary in some places. Some agreement there. But I doubt if they will use Moore. However, please sit on this like anything, and certainly don't write Moore until we are set to go and you know what the decision is. If they take the book, it's bound to be quite a long time before everything is ironed out, and I would be very surprised if they managed publication before a year or more. I have to keep reminding myself that they haven't signed on the dotted line yet.

Most enthusiastic talk from all four about the quality of the book, and its complete uniqueness. After I went back to Bill's office (meeting having taken place in the library, where in the middle of things Alfred walked in and I gather was dumfounded at seeing me. Had a note from him today, saying, "I really was very embarrassed to walk in on you the way I did the other morning, and even more embarrassed not to have had any chance to talk to you while you were here. The explanation is that I never knew you were coming or that you had arrived. Why Bill didn't inform me once the appointment with you had been made is the sort of thing I never attempt to understand or explain"). Pure Alfred, and he probably had a tantrum with Bill, but he frequently does. They are all used to it. I went round to his office when I was leaving, to pay my respects, but he and Blanche were closeted in her office, and her secretary didn't want to break in.

Talked longest to Judith, who has done most cooking from the book. She is a complete convert—says she goes home for lunch and blanches the vegs. the way she learned from you, then finishes the cooking at night. She could dispense with the hamburger recipes, but wants cassoulet, and maybe some other recipes. None of them want more desserts. All of them convinced that amounts—that is, this serves four, or six, are too small for American habits of eating. We eat more of the main dish, because not in the habit of beginning with a substantial first course. This something only you can do, and you might be thinking about it.

Well I started "after I went back to Bill's office"—anyway, he had to zip off to a meeting, and left me with the reports, and I took some quick notes, by no means complete, but if I can decipher my crazy shorthand, here they are. Its shorthand plus condensation, but will give you the idea I hope.

Judith Jones:

> I've tried this cookbook for two months, think it is not only first rate but unique. No other like it. Because the authors emphasize technique. It's like taking a basic course at the Cordon Bleu. Better than that because the authors' whole focus is on how to translate the tricks learned to the problems that confront you at home—different meat cuts, utensils, materials, etc. For the cook who wants to improve, to take the giant step from fair-to-good accomplishment to that subtle

> perfection that makes French cooking an art, I swear that I learned something from every few pages.

Followed by a description of what the ms. contains.

> The authors are perfectionists, opinionated, and culinary snobs in the best sense—they will approve of a frozen short cut, when time demands it, but they tell you how to add some tastiness to the packaged goods.

More description I hadn't time to take down.

> I would be reluctant to cut the book in any way, even though it is going to be a big one. The authors certainly need to have their names built up, and the book must have a first rate title that tells what it is.

(Here we go again, but perhaps they will come up with something—I told them we had sweated over the title, and you didn't want anything cute.)

> Could become a classic. But it is a big job and only worth doing if we do it well.

Angus Cameron:

> A remarkable manuscript. The <u>best working French cookbook</u> I have ever looked at. Does for French cooking what Rombauer did for standard cooking, makes it as nearly fool proof as a good cookbook can be. The first really useable cookbook of fine French cookery I have seen. I would use it to the exclusion of all others. What these authors have done has never been done before, namely to organize a step by step collection of fine recipes, and in the process, to make the climate and theory of French cooking graspable and useable. I am convinced that a path will be beaten to the door of the book and I do not think the high praise which will be involved will hurt it.

(And more like this.)

> I agree with Koshland and Judith that the 4–6 quantities is wrong simply because so few American households do the full course meals even on special occasions. I feel sure that some like me would leave some of those tables hungry. This can be fixed, either by changing quantities in some cases or by precise instructions in introductory sections or individual recipes about increasing quantities.
>
> I think there is a solid sales appeal in the best and only <u>working</u> French cookbook. Of course, it would stand alone as the only one

> needed, but we need not injure other French cookbooks by making that point exclusively. This manuscript is an astonishing achievement and there is simply nothing like it. Cooks will know this by word of mouth very soon, I'm sure. I think we should have this confidence, and venture it with the knowledge that others will have to look to their laurels when this one is available.

Then there was a very short note from Julian McKee who works with Sidney on production that the book looked terrific and he thought they could count on 20,000 sales the first year.

So that's that and now we must wait. I feel very, very encouraged, and you can fan your small coal of hope a bit. With this strong support, I don't see how Alfred or Blanche can do anything but go along. If they don't, will never speak to either again. Well, not really, but my heart will be broken. Transmit this to Simca if you want to—damn, why didn't I think to make a carbon? But remember, all of you, that Bill told me not to write you, and I did anyway, and I feel horrible about it, but I had to tell you, so you must protect me and never, never let slip that I have told all.

Avis

APRIL 23, 1960

Dears:

Not a letter, just to send on a little assorted reading matter of possible interest to you. One hundred niggling little chores to be accomplished this weekend. One of these days if the House office ever calms down I simply have got to take a couple days off and catch up with my own house, which grows slowly shabbier and shabbier. Can't be helped, as I am not here to watch over it and Mae has no initiative at all. If I only had Mary back I would have no domestic worries at all. When I think what Mae has done to ruin my fine floors, by sloshing water wax on top of good paste wax, I could weep. Gray to white in places, impossible to budge even with steel wool. When windows and doors can be opened for hours, have got to have in a strong man to clean them up with benzene, and perhaps even sand them and start all over again. Mae stubbornly says it's because we have too much heat in the house, which we haven't. A real dope, though a nice woman and willing.

Hasty letter from Koshland, saying he will call me this weekend—hope to God about the book, but probably just to say things are progressing, and to tell me about the merger. Hope he catches me, as I will be out both tonight and tomorrow night.

Never thanked you for Oslo branch lesson three, which is extremely interesting, and which I will send down to Judith. She may as well see what you are doing, and will undoubtedly take the recipes home and do them anyhow, and perhaps want them for the book. So keep sending. And I see no earthly reason why Cassoulet shouldn't be as long as you like, and of course include the doing of preserved goose. Why not.

Well you know I do think that except for me the book would be dead, perhaps not as dead as a codfish. Because I think you and S. would just have given up, convinced the times weren't ripe for it. But I did, do, and will continue to do so, and if by any wild fluke Knopf turns this down, I will simply start pummeling another publisher. But the more time goes by, the more I am sure that Knopf will do it, only I do not say this aloud.

If and when K. starts, Judith will undoubtedly be your editor. She wants Cassoulet so do it properly. The time of preparation doesn't matter, I think. Types like Judith (and Bill K.) are perfectly willing to spend three evenings after work messing around with a really fine dish. Judith is a quietly distinguished looking blonde, about thirty I would guess though it is hard to know, very nice but rather cool—not hard. Very bright and a very hard worker as any woman would have to be to make good at Knopf. I think she has been there four or five years, and has her own office and is depended on.

My office is modified hell these days, as Perkins labors over picking 148 freshmen to come in to the House next fall. It is the most elaborate damn process you ever saw in your life. The freshmen name three Houses in order of preference, applications come to the House, plus second choices on pink paper, third on green. This year Leverett (because the Lev. Towers—a great skyscraper thing—are opening in the fall), and Quincy which is very modern and opened this year, were top choices, plus Adams (evidently because it has its own kitchen and thus better food, and the sole House swimming pool) were top choices. (We are on the central kitchen, which supplies four Houses, and the food leaves much to be desired.) We were third from the bottom in choices, trailed by Kirkland and Winthrop. Much heartburning. Rivalry is intense. I am untouched by this, as I feel strongly that Freshmen are almost unformed, do not know what they want, make choices for odd reasons, and are unknown quantities themselves. So EP took all but two of his first choices, and then the second choices came around, having been turned down by house of first choice. And next week third choices, having been passed on by two houses. All of these choices have to be weighed according to the quota—which is rank, list standing, schools—private, Andover-Exeter, HS, and so on—rents they can pay, and what have you. We all have to have a cross section. Very, very difficult to fit it all in. I do none of this—EP labors at odd hours, shut up in the file room, quietly going mad. When third choices are taken, the remaining men go to the Pool—poor things, nobody wants them. Then we all pick from the Pool. And a lot of this depends on intangibles, at that. Perk pounced on a boy the other day, who had been declined by two Houses, because he wrote on his application that his interests were world affairs and girls. All men have to be picked by a certain date, and then I will have to compile a great gawdawful list with all details on it, and draw up statistics on the lot, and I must admit I am terrible at statistics, and am always counting wrong, and missing a man because he turns out to be a special student or something. Most fortunately my new assistant Eleanor Hess really is an assistant, bright as a button, calm, capable, and a real darling. Am I ever glad Jane left—she was a brute, as I look back on her. Eleanor makes me feel that I am quite bright after all, and that all will be well. Bless and bless her. And while all this is going on, Commencement is drawing closer by the minute, and there are no end of letters to be mimeographed by the hundreds, and envelopes typed, and things to be kept track of.

. . .

This was to be no letter, so after two pages I will stop. I want to get my reviews done so that I can catch *The Mouse that Roared** at the University Theater tomorrow afternoon, with any luck.

I love you both,
Avis

Oslo

MAY 3,† 1960

Dearest A:

Well, you did it—did it—remarkable girl. Seems very hard to believe, and shall feel it is really true when that contract comes. I've just spoken to Simca, finding Norwegian woven into my French to such an extent that I could barely talk, and she was thrilled as you can imagine. Told her

* A movie based on the best-selling novel by Leonard Wibberley satirizing the cold war.

† This letter, though dated May 3, was clearly written after Avis's call to Julia and Paul on May 9 (see Avis's letter of May 9).

to hold mum until that official paper arrived. Well, is all I can say. Now what. From all I have heard, the dog work now really begins. That was a most interesting article on book publishing, and I sent it along to Simca and now have it back, but gives quite an idea. Paul is just delighted as he was never terribly enthusiastic about HM as he didn't like their type setting, and has always been an extreme admirer of Knopf's work. And the fact that they are really enthusiastic is wonderful. Well again. It is certainly going to be some problem to build us up. Only Louisette is fairly photogenic, and no one has ever wanted to publish any of our recipes in anything. (Wonder if they know that?)

How good it was to talk to you. First time we've talked to America from Norway and are just amazed how clear as a bell your voice was. We felt just like leaping right over there. Some day, of course, it will be easy!

The last 4 days have been really spring, with that actual warmth in the air, and all the leaf buds which have been slowly swelling are slowly unfolding. It is one of the first times I have really sat and watched spring come. We have an espaliered plum tree (beautifully trimmed this year by Paul) which is at the side of the house. Last Saturday morning there were little firm buds just sitting there on the branches, tight and unmoving. We went out in the afternoon and there were a full dozen branches which had leafed, just in that short space of time. We have bought 2 bales of peat moss, fertilizer, and have put it to the berry patch and the fruit trees, and are hoping we've done right. Paul has also clipped and tied the climbing roses and they look quite healthy, and most professionally cared for. But we are lacking rain, and "they" say if we don't get some in the next few days our flowering fruit trees will not be at their best. And suddenly the long light evenings have started. It is now 9:10, the sun has just gone down, and we shall have a long twilight until after 10. And at night it doesn't get pitch dark anymore. (We are on daylight saving—but the sun rises before 5 AM.) Our lovely view of the fjord is back again, and the wonderfully varied cloud formations have started up. It is really a heavenly place to live.

From Wednesday to Friday last week Gay Bradley* and I took a little trip to Hardanger Fjord, on the West Coast going inland from Bergen. We were too early by 2 weeks, and the weather was not bright, but it was fun all the same. We flew to Bergen, over high snowy mountains and plains, and then bussed up the fjord to a little hotel. The next day we

* A friend of Julia's from her childhood in Pasadena.

bussed again and landed at a splendid new hotel at Ulvik. The place had been destroyed by the Germans, was splendidly rebuilt and re-opened in 1952. Really one of the most attractive places I've ever stayed, so comfortable, light, airy, and full of great views down the length of the fjord. Then we bussed a bit more and caught the train back to Oslo—a long but interesting trip. Great fun to be again with Gay and we talked and talked and never finished what we had to say. But what a different life she leads than we. Her husband is a lawyer in San Francisco, and they live in Burlingame, and must have so much mazuma that it is quite beyond our ken. They live an upper level San Francisco type of social life and, I guess, know everybody who is anybody including all the visiting big wigs who come out there. Go to the dinner for Khrushchev* and sit facing the table of honor, and all that. It takes her 1½ hours to get dressed in the morning, even when just here with us, for instance. And then there is the endless shopping, matching of colors, taking things back, re-fixing of hair, finger nails, face oil, bathings. Most interesting to observe for one who can easily get dressed for a big evening in about 7 minutes. For a bit I felt like a smelly old frump, but luckily reverted to my usual what the hell, as one couldn't compete or compare. Most interesting to think about, however.

Finally had our dinner for 10 last week, and the Cassoulet. Very good it was, but a bit too much fresh garlic, and it is indeed delicious but indigestible. Had a most typically French response to my Cassoulet research from dear Simca—I honestly wonder how anything scientific comes out of *la Belle France,* as this is so typical. None of my sources (which include the most reputable) are even considered. The true story is this which comes from some friends of her parents who live near Toulouse, that nothing is a Cassoulet which does not contain preserved goose, and that Toulouse was the originator of the Cassoulet. Etc. *Eh bien*. I think that her remarks must just come under the list of dogmatic instances. And her parents' friends say that it is only dry white wine which should be drunk with a Cassoulet. Hah indeed. Well, this can easily be straightened out by a more explanatory letter from me, but we were both highly amused.

Much love as ever, and C*O*N*G*R*A*T*U*L*A*T*I*O*N*S!

J.

YES * YES * AND YES! You are a wonderful friend; & we appreciate the persistence, the never-failing enthusiasm & the skillful diplomatic

* Nikita Khrushchev, the Soviet premier from 1958 to 1964, visited the United States in 1959.

pressures which you used at just the right times & places. With just the right people. My god, am I delighted! Julia has worked so hard, so long, and so carefully—and so has Simca, though she cannot truly be said to be careful; but she's a character & persistent worker—that I love to see her (them!) get this psychological gold star. There's no doubt that external recognition—even when you know you're good—is a treasure & a boon. In spite of all the work still to be done, the fact of those knowledgeable people at Knopf's being generous enough to show the kind of enthusiasm they did show means everything to me vicariously.

Love and thank you for the telephone call. It was a blessing, a solace, a wonderful & a happy surprise.

Paul

MAY 9, 1960

My loves:
This about an hour after I talked to you on the telephone, I find I am somewhat limp and stunned, and not much to say except deep, deep pleasure and great gratitude to House of Knopf. Wow.

Only to send on Judith's letter, and I have taken two copies of same, and sending one to Simca with a brief note. Simply will not have time for a real letter until maybe weekend, as I will probably be working quite late at office Wednesday and Thursday, when crashing deadlines have to be met including final lists of Freshmen taken, and letters to same (140 of the little monsters) which are mimeoed but have to have room numbers and things written in. Well with this shot in the arm, I will undoubtedly make it, no matter what the interruptions and crises.

Have written Judith, incoherently (and also sent her a night letter so she can sail ahead and write to you) and sent along the last lesson, and the Cassoulet research, and the titles. Just wish I could cable Bill Koshland, who is presumably by now among the isles of Greece on his cousin's yacht. Can't manage this bloody typewriter either.

It was lovely hearing your two calm voices and I just wish we were all together so we could dance around the maypole, emitting loud cries of joy and relief. It all very nearly makes me forget briefly how boobishly dismal our CIA plus State Dept. is looking at the moment.* Being an American is not always easy, is it?

---

* On May 1, a U.S. spy plane piloted by Francis Gary Powers was shot down over the Soviet Union, resulting in an embarrassing diplomatic incident.

I must now fall into bed as it is all I am good for, that is after I have written Simca note.

Love, hugs, and kisses, in unlimited quantities,
Avis

JULY 13, 1960

Dearest Avis:
We do hope you are feeling a bit better after your last hot and tired letter! Lobsters and the 4th of July weekend must have done you quite a bit of good. Here we have had nothing but rain for the last 3 weeks, and after complaining about the drought, we are complaining about the wet. Our strawberries have not the flavor they should have for lack of sun, I think, but are pretty good anyway and fun just because we grew them.

Well, our contract finally came, very handsome on white heavy paper with a beautiful blue ribbon and red seal, and a check. So we are really in. But I wish Judith Jones would write me the final word about the MS—now I am at a standstill waiting, and it is supposed to be in their hands by Aug. 15th. Oh well. And, of course, we are being flooded with tourists. I think we made a mistake in saying how wonderful Norway is and that everybody should come, because that is just what they are doing. In the midst of all this, the new Fulbright students have arrived, and we have to have a reception for about 50 on July 26th—but luckily we don't have to pay for it, so I am having sandwiches made, and a man and a maid. And a very dull school friend and husband and 3 children are arriving in 10 days, so we have to have 14 for supper. A nice thing about entertaining here is that you can call a meal an *aftens,* and just have a buffet table with cold meat and maybe some scalloped pots; so it can be very simple indeed. That is what I am now going in for!

We have finally come to a decision which is to be repeated to NOBODY (we have told ChaFred* only), which is that we think we shall get out of the Foreign Service at the end of our two-year tour next year, and come home for good. It is just not the kind of work which Paul enjoys as he has nothing along the creative lines he likes—all administration—possibly a lecture or two, but that's all. And no time at all to do his own life. And, as we realize, he will be 60 in 1962, and looking at things coldly, there are at most 20 to 25 productive and active years left—probably only about 15, but one never knows! I don't know when we shall be

* Charles and Freddie Child.

able to get into our house in Cambridge, as they have a lease until Sept. 1961, and 3 to 4 months notice. But if we tell them around Christmas time, or later, that we shall give notice on Sept. 1st, they might then find another house by that time. We can only hope. Well, we shall be living quite modestly. But I figure if I can give 2 cooking lessons a week at about $40 a throw, that will bring in a tidy sum. Paul will need a year or so to get back into his art rhythm, as it gets rusty. Anyway, we love the idea of thinking about living at home for good.

It is terrible not to be in on the conventional doings,* though our Norwegian papers are full of it—but it is not the same. It is now Wed. morning, so we don't know nothing, and won't until tomorrow AM. Seems as though Kennedy were in though.

One question is: people (Republicans mostly) say Nixon has had so much more experience. But I don't see that it is such an enormous amount. He has had 2 to 3 trips to abroad, and a TV talk with Khrushchev. He has been VP for 8 years, and has been pretty much in the center of things physically. But he has run nothing, has been in no position to make decisions, or to administrate anything, or to initiate anything. Would most like to know how the Dem. big-brains are going to answer the youth and inexperience attacks which will come, and will appreciate any remarks you may care to make on this subject, please.

. . .

Love to you all, and write soon as poss.,

J.

Oslo

SEPTEMBER 11, 1960

Dearest Avis:

I'm finally back from my trip to the fjords, and in one piece, and the family are off and away. I think their visit was a success. Pop, who will be 80 in October, is a bit unsteady on his feet, but cuts a fine figure nevertheless. Mentally he has solidly crystallized into the model of an extreme right-wing Republican. Right is right and everybody knows it and there is no discussion, as no other point of view is conceivable; what Ike and the good republicans say is THE TRUTH. Needless to say, we had no discussions whatsoever on the subject and all in all things were calm

* The Democratic National Convention was under way in Los Angeles. Senator John F. Kennedy would be nominated as the party's presidential candidate, running against Republican Richard Nixon.

and pleasant. He really dislikes travel, distrusts foreigners, and wishes he were comfortably at home the whole time. He is interested in hospitals, and not at all in sight-seeing. How Phila* got him to come on this trip, heaven knows. They are now off to Sweden, Germany, Spain, Portugal, and home and I hope the poor dear survives it all. I found the best way to act was to talk about nothing but personalities, such as how is old so and so whom I haven't seen for 30 years. Well, it is good to get back to normal life.

Your fine letter describing the doings up to Bread Loaf was here on my arrival, having been thoroughly enjoyed by Paul. It sounded like one of the best sessions ever. And we were most interested to hear about Allen Drury.† We did indeed read that book, and were absolutely fascinated from beginning to end. Wonderful that he has made such a success of it, but I can't imagine how they will do it as a play. Would be terribly difficult, I should think. A shame about the Frost-Morrison mess.‡ Do you suppose you will be able to take those letters and put them under seal? It would be a shame to have them destroyed.

Thought you would enjoy that letter from Louisette. The last agreement we signed gave her 18%, which I think is crazy when she hasn't done one single sentence of work on either of the two huge manuscripts we have prepared single handed. I have written her another letter stating this fact, and asking her to consult her family and friends and come up with what she and they would consider fair and just under these circumstances. We are quite willing to offer her 10% to shut her up, then have no further collaboration, and retain exclusive right to "Les 3 Gourmandes" as a trademark. Haven't heard from her yet. I just don't think she has any conception of what work is.

Poor Simca writes that she has some kind of a liver attack and is in bed. She just isn't very well, dammit. I think that having her old mother, who is quite dotty, and Jean's old father, who is not dotty, but 85, is quite a strain on her. Also she is not one to take anything in a relaxed manner.

A second letter in from Judith Jones with comments on the balance of the MS, so I am about to incarcerate myself for the next two weeks. I

---

* Phila McWilliams, Julia's father's second wife.

† The author of the Pulitzer Prize–winning novel *Advise and Consent,* who had given a well-received political speech at Bread Loaf.

‡ Kay Morrison, the wife of Theodore Morrison, a poet, novelist, and Harvard English professor, was Robert Frost's secretary and adviser and was said to have had an affair with him. Letters between William Sloane, Frost's publisher, and Bernard DeVoto referring to the relationship were discovered by Frost's official biographer, Lawrance Thompson, in the DeVoto collection at Stanford.

am now ⅓ the way through my 3rd proof-reading and feel like re-writing the whole MS. It's really a nightmare business. She is keeping the ribbon copy,* and I am making clear changes on your copy (which she will get), and sending along replacement or insert sheets for the ribbon copy. So she will have a bit of work to do. We have re-done the omelettes. First put in 3 recipes, to see how she liked them, but she thought that was too much—as we do. So we shall have two, the rolled and the scrambled. The scrambled is certainly far easier. When I get the final typed up, I will send it to you and would be most interested to see if you can do it from the directions. (I really think they must be shown, however.) So now I must get back to the MS. We shall be off for three weeks starting Sept. 26:

> Sept. 26 to 30, c/o Simca, Bramafam, par Plascassier, Grasse (AM)
> Oct. 2 to 14, Hotel Pont Royal, 7 Rue Montalembert, Paris 7

It would be glorious if you came over here next year. We hope to finish up with Paul's work here by the 1st of June, then will take 2 or 3 weeks touring around Norway, then 2 or 3 weeks on the continent, and home probably August 1st. But all is fluid, as so much depends on when my work on the book will be done. Anyway, we have told Mrs. Mason we would like to get into our house on Sept. 1st. We plan to live in the little apartment until we make whatever changes are necessary in the house proper.

Anyway, let us keep the thought. When could you get away? Just the thought of that gives us a warm, rich feeling (not financially, of course, just spiritually).

Much love. Will write later as a relaxing relief from MS work,

J.

OCTOBER 29, 1960

Dearest J:

Returning herewith the omelette recipes, which should have done long ago. Ms. contains B. Fairbank's scribbled commentary, some of which I cannot read. Ginnie had success after muffing one; I did also. But I haven't done it since, not being for some reason in an omelette eating mood. I know it just takes practice and practice, and you should say so. No other omelette recipe ever says it's a knack, and should. Mark hasn't tried it—he

* The original typewritten copy.

is almost never at home except late at night—but then he makes fine omelettes without any difficulty. I never stood over him and watched him do it. He does it your tilt and jerk method, having watched you. Also he has the requisite height and strength, which I have not.

My God I can see that you simply cannot go on nit-picking on that ms. Comes the time when one has got to let it alone, and send it forth, all imperfect. It will still be light years better than any cookbook ever published in this country.

. . .

I have been hit, God knows why, by the bread making urge, and wrote Freddie for the recipe for that fine, coarse, rough, dark bread with the wheat kernels in it. So I made it twice last weekend, with great success and many customers in the house, and have a batch rising now, and will do it again tomorrow. I do love making bread. Of course, it means I eat some, with gobs of butter, and this cannot go on forever or I will get fat. But my goodness, how the boys do love it, and my students, and my assistant at the office. Queer, this urge all of a sudden. I really feel that it is a proof that I have some energy left over after the week's work, and marks an advance in self confidence, which I certainly do feel. I am not as tired when I come home, and I don't grind my teeth in the middle of the night, and I don't experience those horrible moments of doubt. Of course there will be some, but I am feeling pretty good, and pretty competent. May be famous last words, of course.

I hope by now you have got the Kennedy-Nixon book* I sent you, and read it. I have circulated six copies, and made five converts. I think it is a brilliant job. I'm going to the Schlesingers on election night, I think, and it will be quite a change if we win, as I am feeling very sure we will (as of now), because I was there the last two election nights, and it was awful. Arthur is just back from Minnesota, where he reports a massive Lutheran revolt against Kennedy's Catholicism, and says that even Humphrey† is in danger as a result. And tomorrow is Reformation Sunday, and the Bible belt is going to let fly with all it's got against the Papists, and I just hope it backfires, is all. The anti-Catholic vote is the joker in this campaign—nobody knows how deep it goes. Concealed Republican money is being poured into absolutely virulent pamphlets which are very widely circulated—I have not seen one, but have heard about them and

* *Kennedy or Nixon: Does It Make Any Difference?* by Arthur Schlesinger Jr.

† Hubert Humphrey, the Democratic senator from Minnesota.

read about them. As of now, it is thought that we may lose the South, but take the big industrial states, New York for sure, Pennsylvania, California, and so on. As [James] Reston said a few days ago, the moment has arrived when "the bees are settling on Kennedy's hive," and we Dems. feel victory in the air, so much so that some of us are scared stiff. I'm not, but I know plenty who are. I am just quite sure that we will win, maybe by a landslide.

. . .

Goodness I'm glad you are seeing the debates. Fascinating indeed. I think they are going to do to Nixon what they did to McCarthy, at any rate, I hope so. He looks more like an Armenian rug peddler every time. Shifty. And the debates have made Kennedy, who was not known to the public before them.

Louisette. Ten percent of what? What royalty percentage is Knopf giving you, I have forgotten. I feel sorry for Louisette too, though relieved that she has finally split from that spurious man,* but hell, the time comes when you have got to fish or cut bait, get a job and put your back into it. Oh dear, oh dear. I hate to see her getting a big chunk out of a book she didn't do a lick on. Get papers firmly signed, for heavens' sake. She must have no handle whatever on the next book.

. . .

Well must now knead the bread down, cook myself a steak for lunch, and tackle a great many more letters. Much, much love to both of you, and how wonderful to know that in less than a year you will be here, for life, and I just hope and pray that Cambridge turns out to be all you expect of it. I know damn well it will be better than Washington, New York, or California. With all its drawbacks, it is for me the hub of the universe. The drawback for most people is housing, and thank the Lord you've got a nice one, the greatest stroke of luck. You can teach cooking forever in Cambridge with no lack of pupils. Paul will be welcomed into the art world with open arms. You can live on what you've got, for money. I know, I do it, and I have more dependents than you do. The climate isn't bad at all. Politics, in the state, is lousy. So is it other places. And you'll have the pick of the best and most exciting people in the USA (including, ahem, me).

Well speed the day is all I say,
Avis

---

* Louisette's marriage ended in divorce, and her financial situation was difficult.

NOVEMBER 16, 1960

Dearest Avis:

So good to get your letter and hear about your election day. The Schlesinger affair sounded perfectly wonderful. Aren't you lucky to be in with all those big wigs! Always so much fun to feel you are right there in the bull's eye. Now I can't remember if we have written about our election day or not, but I think not. Anyway, down in the Embassy cafeteria our press officer had a set up with charts, radios, and the works. Some hardy types arrived at 2 AM Wednesday morning. We had to go out to a dinner, so got there at 5, armed with 3 bottles of champagne and all ready to celebrate. Of course, nothing was possible. And the big Kennedy lead kept dwindling, though it seemed clear he would get it. By the time Nixon's "I'm going to bed" speech came around 9, I went home and turned on the short wave radio every half hour. It was not until 7 PM our time that Nixon conceded. What a day! It was wonderful to have that little get-together at the Embassy, with 50 people off and on, all huddled to the radio, keeping scores, swilling down coffee, tea, and sandwiches. It made us all feel we were part of the big event. Next one, by gum, I shall be home. Having the TV must be perfectly wonderful. Actually the Nixon little half-concession speech was the best thing I heard him say. He sounded like a human being, and it was a decent generous speech, and his voice did not sound like McCarthy on that record you sent us (of the McCarthy in heaven business, which is so frightening). Well, thank god is all. You wonder if Ike had got going sooner* it would have made any effect. And one thing which always amazes me is the way the Republicans really believe their own propaganda, such as that there shall be vast inflation, wild spending, etc. etc., and socialism, and labor dictators. I suppose they scared quite a few people. I just feel sure that once Kennedy gets started that people will like him very much indeed. Certainly wonder who Sec/State will be. And what will happen to Adlai, etc. etc.

Your house sounds good, with all the new painting. But what a mess it all is when you have to tear it up. All that dusting of books, etc. Hope you get it back into shape quicker than you expect—that does happen. And so glad you are feeling fit! That is the most important thing.

We are both fine. I was worried that Paul might begin to wilt, as his boss is away on 3-months' home leave, and he has the whole bloody thing to run by himself. But he seems OK, and has a good rosy color, and is ex-

* Eisenhower did not endorse Nixon until late in the campaign.

cerising (can't spell that today) with his iron dumbbells every morning. I shall be certainly glad when June 1st comes and he will be out of it for good. But I think the fact that he doesn't see at all what his future life will be makes him feel a bit uneasy. It will take a good year for him to re-think and re-do; but he will not be under this awful pressure which harasses him here, and should eventually do just what he really wants to do, and have joy in it.

In many ways I hate to leave Oslo, as we are, after a bit over a year and a half, really beginning to get acquainted and feel at home. I really love it here. The city is just the right size, there are some nifty people about, and having free open spaces and beauty all around are all lovely. We have made quite a few good friends here among the Norwegians — not any, so to speak, among the American government types, but that is normal for us, it seems. And I can just feel more good friendships around the corner. Ah well, at least this will be the last time we shall spend 2 years making a life, and then have to leave it all again.

There is lots of theater here, and great interest in it. In Oslo alone there are 8 or 9. The National Theater gives [Henrik] Ibsen and Bjørnson,* and modern ones. I saw a fine performance of *Hedda Gabler* there 2 weeks ago, and see it has just started up in New York with what seems to be a fine performance. We also saw Françoise Sagan's *Château en Suède,*† which we had seen in Paris and wanted to see here in Norwegian just to compare the two performances. The one here was excellent, quite different from the Paris version, but extremely well acted and presented. It is a thoroughly decadent and original comedy, and needs a light and sophisticated touch, which they gave it. I'm amazed at Sagan, only 25 years old — born the year after I graduated from college — and look at her. She has real talent and originality, and it will be fascinating to see what she does with it as she matures. Paul has had to give up his Norwegian as it takes too much time away from his work, and he is too pooped to go on with it when he gets home. I haven't been able to fit in the University courses this year, but read an hour a day and can now read with ease. But the practice is hard to get as the Norwegians all speak perfect English. The only practice I get is with my darling cleaning woman, and out shopping. Too bad. I can now get ⅔ to ¾ of a theater performance, which is not too bad. Of-

---

* Bjørnstjerne Bjørnson, a Norwegian writer and Nobel laureate.

† *Château in Sweden,* a sexy romantic comedy about a group of jet-setters who cavort around a château dressed in 1750s costumes.

ten get everything just up to but not including the crucial point, though!

And I now have two cooking classes going, which makes about 1 a week. 8 people in each. We go to various houses, never come here. I could probably start 2 or 3 more, but shall not, as it takes too much time. My number 2 class started up last Monday, with pie crust dough, quiches, etc. It did not go quite as well as I had wished, as the refrigerator didn't work and the dough didn't harden as it should; the oven was poor, etc. But now I know I must bring some ready-made dough so we don't have to wait for it to harden after each pupil has made her own. Mushrooms should be mostly all washed except for a handful, etc. etc. And there was one French girl who was there as a guest, who had lived 2 years in the US but doesn't speak a word of English, or a word of Norwegian (typical old froggie)—and she kept mumbling to her companion "*Oh, moi, je ne la fait pas comme ça—jamais de la vie!*"* Put me off, darnit. But it is good practice psychologically, and I should be able to dominate such types. Yes—but how.

Yes, that is one thing which really got my goat in France this time. That insane one-up-man-ship, which is so often based on nothing factual at all. Simca has it to quite a degree, as in that ridiculous Cassoulet research. It makes no difference how much research one does, using native French sources, no difference at all. If the French person you are discussing something with has his own ideas, that's that. Just make a great dogmatic statement and that is the undeniable fact. Fah. Where this myth of the "logical French" ever grew up I would certainly like to know. From my 12 years of dealing with the French in the most intimate circumstances I think they are the most illogical people in the world. But they are certainly fun, gay, affectionate, inventive, quarrelsome, sticky, talented, and thoroughly French. AND THE FOOD!

Don't know what's happening to our book. J. Jones said the chief copy chap might send the whole thing back here. That will probably happen at exactly Christmastime, if I know such things. No title as yet. I keep sending in lists of suggestions, but haven't quite hit on any thing yet. LA BONNE CUISINE FRANÇAISE (with subtitle) is the one I now like best. But Paul thinks not enough people in the US know French. I say that anyone who wanted to learn French cooking would know those 4 fundamental words.

* "Oh, I've never made it like that—never in my life!"

Now must get to work on me new class recipes for next week.

Love and love, and write every day,

J.

PS: I am running out of those fine white towels. Could you order me 2 dozen?

NOVEMBER 26, 1960

Dearest J:

A note only, which will skip around like a grasshopper, as I seem to be in a very addled condition, due no doubt to late evening and much too much food and drink with Bill Koshland last night, after fairly grueling Thanksgiving week, with G's new girl visiting still (she really helps) and Mark bringing three classmates. Turkey was just fine, a 14 lb. dressed weight fresh one. But since I had only the one day off, no time really to cook, so ordered a couple of pies from the Window Shop,* which were very expensive, and not good at all. Ah well.

I have been no end entertained by your translations from the Norwegian, and have showed it to lots of people, who laughed like anything. Including Koshland last night. I did not get too much out of him about the cookbook, except that he says Judith is nearly finished with it. The four of them grow more enthusiastic by the minute, he thinks it is going to sell well, he thinks they've about licked the design problem. I asked him to send me a copy of the galleys when they are ready, so I can at least pick away at them. All depends on when they come. If during my spring rooming crisis, I will probably not be able to do much; or at Commencement time. But I do want them. Another eye is a help. I asked if Alfred had said anything about it; he said, only vague reference from time to time, as "that book Avis is interested in," and he hasn't looked at it, and is in a frightful temper about many things, and probably not very happy about the merger. And Blanche will have nothing whatever to do with the book and knows nothing about it. They are indeed a queer pair. But do not fret, you have the strength of the firm behind you, and Alfred will be pleased enough when he gets round to looking at the book. Really in his present mood I would be just as happy if he did not look at it now, as he might be tempted to take a hand, and they are managing very well without him. Right now he is being the terrible tempered Mr. Bang in a big way. I scarcely know what to write to him, in reply to the note I

* A Viennese restaurant and bakery in Cambridge.

had before he went to France, which was in September I think. Couldn't write him before election, because I think he voted for Nixon. His attitude is a pox on both their houses. Very odd man, and getting more choleric with age.

Towels. Well indeed I do wish I could order some, and knew where to get them. I rather wonder if they are still available. Not, certainly, in the East. I need some myself, and don't know where to turn. The ones we both have were bought at a J. C. Penny store somewhere out West, I forget where, by Benny's niece Dee Boicourt, who is now dead. I think you would find these towels perhaps at a chain like Penny, but only in the middle west or far west, and how to get there, or how even to write? If you do any driving around California, after you return, and get into any small town where there would be a Penny store, do by all means inquire. And if you find them, buy a gross, and let me have half. I paid 29 cents each for these, but it was a good eight years ago. I am longing for some more, being down to the last dozen. I did, last winter, send a [letter] to Vrest Orton who owns the Vermont Country Store in Weston, Vermont, and specializes in fine old-fashioned merchandise, and asked him to stock them. But he never replied. Knowing him, I know he would think they were wonderful, but perhaps he never could find them. I bought from him, incidentally, two (one for me, one for you when you get home) rock maple holders for paper towels, which will never wear out in the world. He publishes a wonderful catalogue and I always buy odd things from him — stone ground meal of various sorts, samp, odd teas, kitchen aprons and the like. I do hope one or the other of us can locate those towels, if they still exist. But now that all flour comes in paper bags, maybe they have vanished.

. . .

Cambridge an absolute hotbed of political rumor and speculation, as it is a center of Kennedy brain-trusters. McGeorge Bundy, my neighbor and Dean of the College,* is the hottest candidate locally — gossip says he will be Sec. State, Under-Sec., Chief of the Security Council, or Ambassador to St. James's. I certainly think he will get something, as he looks like the cat that swallowed the canary. He only just left the Republican Party, as he could not tolerate Nixon. As I went from cocktail party to cocktail party after the Yale game a week ago, the rumors got wilder and wilder, and I really doubt if anybody knows anything for sure. Headline

* The dean of the Faculty of Arts and Sciences at Harvard.

on a Boston afternoon paper yesterday—Jack, Jackie, John, and Joe All Jake*—somebody had some fun. This of course after the baby's birth.

I really doubt if I will go abroad next summer—would be fun seeing Simca, but frankly, without you, I don't want to stay with her more than a few days, and not mad to go to France anyway. Am definitely not going to travel around with A. Maass, so don't worry. Am not rich enough, this year, to travel quietly around England as I would enjoy doing. So I will probably wind up at Bread Loaf, which after all is great fun. Your cooking classes sound great fun, and I would personally have strangled that Frenchwoman—you are so very right—they know the answers to everything. That was what drove us absolutely mad about Jean Guislain.† Well dearie must now have a cup of tea, very strengthening, and get pretty to go to a cocktail party and then immediately to the *Pirates*‡—the Loeb Theater shows begin at 8:00, so I shall have a bite when it is all over. I like *La Bonne Cuisine Française* and do not agree with Paul that not enough here know enough French. Koshland is very worried about the title. I forgot to ask him about that title last night—we talked mostly politics—and after dining sumptuously at Locke-Ober,§ were too pooped to talk sense about anything.

Well, no more, except lashings of love to both,
Avis

Oslo

DECEMBER 29, 1960

Dearest Avis:
Here is hoping you are feeling fine again after your doleful sounding cold—and that you have managed to keep off smoking. Congratulations, anyway; and would your example could be transferred to me. Keep feeling we should give it up. What, as you say, do you do with those wandering little desires which smoking seemed to quell? Or do the desires gradually subside? Please keep us informed, and keep up the good work.

We ourselves are feeling like a couple of old crocks. Paul's shoulder began acting up a few days ago, so he is hanging himself. Then the other night he got up and whacked his middle toe on the leg of the desk and it

* Jack, the president's nickname; Jackie, his wife; John, their newborn son; Joe, the president's father, Joseph P. Kennedy. "All jake" is slang for "fine."

† Jean-Marie Guislain, a French painter and poet, was married to Bernard DeVoto's cousin Rose.

‡ Gilbert and Sullivan's comic opera *The Pirates of Penzance*; Mark DeVoto, a Harvard senior, was the music director.

§ One of the oldest and finest restaurants in Boston.

swelled up and turned blue. And he has picked up pink eye to boot; so is feeling pretty seedy. Says this always happens just before his birthday. Yesterday I woke up with a cold and was dying, but managed to spend most of the day in bed and am now over the worst.

We had lovely feathery snow for Christmas, but now it has mostly melted off. Presumably a bit more will turn up today, and I hope so. When it is winter, I want snow all the time. Christmas went off very nicely. We had a friend up from Rome, and that livened things up. Fisher and Debbie Howe,* who live just up the hill from us, and their two children, came for dinner. And we had goose. Before drinks, and as they arrived at the house, everyone had to go through a spiderweb hunt to get their presents, and that was rather nice. Paul and Bob, our guest, did all the decorations and they looked splendid, including a mobile of Christmas ornaments. First time we have ever had Christmas in our own house!

The great thing around here for Christmastime is the Julebord, or groaning Christmas table, filled with cold meats, pickled herrings, etc. etc. We have been to numerous ones, and even had one ourselves. I got a half of a smoked pig's head, which you cook just like ham, and then glaze. That was fun. Not much meat on it, but style. In fact, we have eaten and drunk so much we are feeling a bit bloated.

If you do go to Europe, when will you go? August? Let us know. We shall be arriving in New York on July 18th, and don't seem to have gotten beyond that point. Wonder when we can get into our Cambridge house? That will depend on when Mrs. Mason can get the tenants another house. We are hoping to get in at least by Sept. 1st. We shall have to take a quick trip out to Calif. at some point; and we'd like to get a week up in Maine. I also haven't the slightest idea what kind of work will be needed for the cookbook, or when. Presumably I am to hear about the final corrections at some point, but so far nothing. So all is fluid.

Our first menu in the French magazine† is to appear in the February number, and I shall most certainly send you a copy. It is a tremendous amount of work, but also good, as it puts new recipes into the till. Simca is still on the vague side as to measures, etc, but that is all right. We work perfectly together, as she feeds the recipes and I get them into shape. She is going to give a lecture demonstration to 200 American SHAPE‡ wives

---

* Debby Howe was a college friend of Julia's sister, and the Childs had met Fisher in India in the OSS. He was now deputy chief of the embassy in Oslo.

† *Cuisine et Vins de France* had run an article by Julia and Simca.

‡ Supreme Headquarters Allied Powers Europe.

in January; and the French magazine is talking about starting some cooking classes with Simca, Bugnard, and possibly Thillmont. That will be fine; and I hope she will able to keep up her forces. This is just exactly what we need, so that by the time the book comes out at least one of the authors will be a "recognized authority." But I keep wishing I knew what in Hell was going on at Knopf; but suppose I will know in due time.

I keep, also, being more and more enthusiastic about Kennedy, even with Bobby [Kennedy] as Atty. Gen.—but I understand from Krock* and [Joseph] Alsop that that is OK. Can't wait for them to take over, though. I must say I am amused at Eisenhower saying Dillon† should have had a written statement that "sound financial policies" would be followed. How can they speak of sound financial policies when they dump the results of theirs in his lap? As you say, Republicans are the limit.

You asked what we would do to the Cambridge house. Don't know yet, as we can barely remember what it is like, having seen it only once. Kitchen will have to be fixed, as that was pretty barren. Public rooms will need fixing. As I remember, there were very few bookcases in the house, which seemed odd. Paul will use one of the big bedrooms as his studio; and I may use the other as mine. These two rooms and our bedrooms (unless I use my studio as a bedroom also) will be simple. Can't remember how the bath and running water situation is upstairs. Apartment will have to be nice but not too elegant. Really don't know; and feel things will have to be gone at somewhat gradually. We shall certainly need help with workmen, etc. In Georgetown, we had a contracting chap who furnished all the plumbers, electricians, etc, and that worked out very well. We'd also like to have a garage, and wonder if they come pre-fabricated. Wish we were moving in right now, is all I can say. This is not much of a letter!

Much love, anyway, and hope you are fine,
Julia

PS: Thanks so much for the K. Clark book.‡ Paul is reading it, so I have been able to have only a glance or two, but find it very interesting. I, who am not a visual type, need instructions at how to look at pictures, so I shall find it very useful indeed.

PS: Happy New Year!

---

* Arthur Krock, the Washington bureau chief of the *New York Times*.

† C. Douglas Dillon, Kennedy's Treasury secretary.

‡ Kenneth Clark, *Looking at Pictures*.

Oslo

APRIL 24, 1961

Dearest Avis:

This is a long W*A*I*L from the far northland. How I wish you were here to help me with these galleys! I have now received two batches which takes us to the last page of the sauces. But I am presumably to wait for the illustrations to be sent so they can be keyed into the galleys before I send things back to Knopf. They were supposed to have been sent on the 18th, and in the mean time J. Jones says everything must be hurried as they are behind schedule, etc. etc. But the instructions are really not quite adequate so I know what I am supposed to do, but I have done what I decided to do, which is to answer every query no matter whom it seems to be addressed to, and some seem to be addressed to the printer. Ah well. I didn't know I was also supposed to be a typographer, but it says in the instructions "Answer every query." However I have decided I have to participate in the typography because there is a passel of illogic in headings. It all looks splendid indeed when you look at it casually, but then when your nose is in you find that headings are not consistent, pages farther on has another, etc. etc. ETC. ETC. J. Jones says I must be warned that I am fully responsible for proof reading—does that also mean type? I must assume so. At least, if I do not point out errors, they may well be missed. But I cannot believe any serious publisher would leave all that kind of stuff up to an amateur! So, going over the setting copy, I have made a long list of all the type faces and symbols, etc. etc. and have just marked up the galleys the way I think they should be done.

It would be much easier for authors if fuller instructions were sent, and if a type guide were also included. Certainly I know how I will do my next book. 1st send the working draft to a good copyeditor, then have it fully and neatly typed, and make marks of ORDER OF IMPORTANCE, so all this mess could be avoided.

Well, I wish I were home, is all.

I am also having some ridiculous trouble with dear old Simca, who is inclined to rush at things without half reading them, to forget what she has pronounced previously, and to act, in general, enragingly French. We are working up some recipe for almond cookies, *Tuiles,* for *C&V de F.* So I send her the recipe which has as a basis egg white, sugar, and powdered almonds (having tried the cookies with and without powdered almonds and concluded those with almonds were the best). So I get back the following from her, "If you add the powdered almonds to the batter the cookies are no longer *les Tuiles Françaises* . . . They MUST NOT HAVE THE TASTE OF COOKED ALMONDS IN THEM, but only toasted almonds on top, nothing more." And then a long go around about *goût américain,* etc. etc. It happens that this recipe is one that she gave me, typewritten, and dated July 29, 1958, and titled TUILES CLASSIQUES, and with almonds. It also appears in all our FRENCH BOOKS. She has also received a bit of the galleys and taking a quick glance over them runs into a passage (which she has not fully read) about sprinkling the surface of the soup with beef drippings (it is not soup, but bread rounds). So we go into a long go-around about that—*Pas du tout française, ce beef drippings me chocque beaucoup**—and more about *goût américain*. This recipe for croûtes has been in the MS since 1950; she has read and approved it for each version of the MS; and it is probably her recipe anyway. *Eh bien*. I had intended to send her the typescript of the MS so she could see all the final corrections, when I am through with the galleys for each corresponding section. But now I wonder if it would not be a foolish thing to do, as she will rush into it, make wild decisions I am afraid, and want to change everything and conclude it is not *goût français*. So I am in somewhat of a quandary. I am mulling over a calm letter, however. I think Simca does too much, gets tired, is not in awfully good health, and is an instinctive anyway. (No amount of research when presented to her will ever make her change her mind about anything, for instance.) But without her, this book would be nothing at all, as she has had the major suggestions for recipes, and the real French touch, and she works like a Trojan.

And Algeria† and Cuba.‡ Everything is too horrible. Just a shame about the Cuba business, and I wonder how they did, in Washington, have their pants down so far. I feel so sorry for Kennedy that this should happen! Was most interested to see in our morning's paper that he had been to see Ike, had conferred with Truman, and was about to visit with Rockefeller. Smart boy. And what about Algeria—looks hopeless of settlement—or can we hope that this resistance is the battle of the bulge

* "Not at all French, these beef drippings greatly offend me."

† In the eight-year Algerian war for independence from France, the right-wing French–colonialist sympathizers and the Algerian liberation movements vied for power. President de Gaulle advocated self-determination, but on April 22, 1961, French sympathizers defied him and staged a coup in Algeria that was also intended to overthrow the president. The coup collapsed after four days.

‡ On April 17, CIA-trained Cuban exiles with the covert support of U.S. troops invaded southern Cuba in an attempt to overthrow the Communist government of Fidel Castro. The so-called Bay of Pigs incident, named for one of the invasion sites, further destabilized U.S. relations with the Soviet Union and embarrassed the Kennedy administration.

and that a settlement will now be forced. Hélène Baltrusaitis's son, Jean, is in the army in Algeria at this moment, and I can just imagine how she must feel.

P. and I were discussing at breakfast that perhaps we just should come right home the end of May. I couldn't bear to visit France in the midst of troubles like these. Also, from the point of view of bookery, it would be better to be home for the page proofing. Well, we shall think about it. I am really dying to get home, and have no mad desire, actually, to do any traveling around.

Well, must back to work and get a calm, nice, but somehow pull-your-pants-up-for-god's-sake letter to S.

J.

Oslo

APRIL 28, 1961

Dearest Avis:

I am just going WOOOOBIS with our dear old friend, Simca, and I have to talk to somebody or I'll burst even more. (Paul is in Trondheim [Norway].) After my burst to you the other day, I get a letter from Simca yesterday afternoon, commenting on a partial set of galleys I asked J. Jones please to send her so she could see that things really were in print. So I get a long letter of comments and corrections, in effect—change everything. How did you ever put this in? This is not correct, etc. etc. The big boob has had all of this stuff for years, she has OK'd everything, again and again and again, and again and again. I took the whole MS down to her last October, and she presumably went over everything. Triple phoooooo. Section on the wines at the beginning of the book, for instance. This not right, that not right. And she doesn't even know what she's talking about, what's more. Spell it Château Yquem, NOT Château d'Yquem (am sending her a label from a bottle, Château d'Yquem). NOT CORRECT: Champagne has sugar added to make the bubbles. She doesn't even know how their own damned wine is made, even. So I have to copy her out an article by J. R. Roger, *de L'Académie des vins de France* explaining the whole process. (Sugar is added to make the bubbles; she is talking about the second dose which may or may not be added to a Brut.) WHY ADD A SLICE OF BREAD TO THE PISTOU SOUP, she says. This was HER SUGGESTION, god dammit. Etc. Etc. Etc. (My god, I pity J. F.

Kennedy trying to make any headway with de Gaulle.* He's even worse than Simca, from all I've heard.) And now she and Jean are going to trot around Paris: *"Oh ce livre, c'est plein d'erreurs. Ce n'est pas la faute de Simca. Ça—non. Ce livre était hors de ses mains complètement, c'est goût américain. Ce n'est pas français, pas de tout."*† I am not going to make one correction on the stuff she has changed her mind on, not one. I will correct any downright errors, certainly, but not until I have consulted proper French sources, which luckily I have.

I really got so mad last night I was trembling inside—never have done that before, and can see that if one is of that temperament (which I didn't believe I was until now!) that ulcers, heart trouble, and all kinds of things could be a most natural result. If I'd had any Miltowns [sedatives], I'd have taken a handful! Then to top it all I had to go to an utterly ghastly performance of *The Importance of Being Earnest* by the Oslo Dramatic Society, a collection of English, American, and Norwegian amateur players. And whom should I be sitting next to but a FRENCH WOMAN. I all but threw up.

Well, so I shall have to write another letter, calm, nice, as though to a child. "*Ma Chérie,* etc." Urging her please to get out her file copy of the MS typescript, and insert all the many pages which I have sent her as replacements for old ones. Get out also all her old correspondence with all the OK's for each page, etc. etc. UNFORTUNATELY just the other day, having a free moment, I decided to toss out all the old copies and copies of copies of old MS I have had hanging about for years—so I threw everything away, including her comments and OK's. *Eh bien et vel-vel.*‡ I don't think it will come to anything serious, but I wish I had that ammunition just in case. I shall, of course, pretend I have!

I don't know, really, what you can do with such people. Reason has no effect, they have no memory of what they have done or said before, everything just comes out of the stomach and the top of the head at the moment—and with tremendous dogmatism. That is certainly one thing I have learned from living and working with the French—listen to what they say, but do not take anything as a fact, ever. (But then, how much can one rely on French written documents?) Indeed. But with people like

* The two leaders were engaged in talks to work against the Soviets in Berlin and on the role of France in NATO.

† "Oh, the book, it's full of errors. It's not Simca's fault—no. The book is completely out of my hands—it's now for American tastes. It's not at all French."

‡ "Well, well" (Norwegian).

that, however, the written printed word sometimes has some effect. But also they are just as likely to say, when you cite Escoffier, for instance, "*Oh, bien, Escoffier, vous savez** —hmmph."

End of section. There shall not be a break between us, as I am far too fond of the old goat, but it will take some delicate maneuvering to try and make her realize some of the truths of publishing, and of the truths (if such they be) that she herself has written and/or approved.

We have decided that we shall come right home, and have confirmed reservations on the OSLOFJORD sailing from here May 27, and arriving in New York June 5th. And are we delighted. I am just aching to get home, and so is Paul. He hates to face the hot weather, though—feels about it just the way Benny did—it is a nightmare to him. But even so, just the thought of being home one day less than a month today is unbelievably exciting. We were figuring out the other day that we have hardly spent 4 years in the USA since 1943! The reason we decided to speed up our return is that it is evident that those page proofs will be ready just about the time we were to be cavorting in the fjords, or eating our way about France. At this point, things have just got to go through on schedule. So now we shall just stay near enough to J. Jones so we can be on call at any time. We have therefore made no plans other than arriving in New York June 5th. And I doubt very much if we shall go out to California in the summer, what with the house to do, and all. This will also give us time to consult with Mrs. Mason, and painters, and so forth on what to do with the house; and when Aug. 1st comes, things can start right in. So you will be seeing us at Berkeley Street quite a bit sooner than planned! And maybe we can also arrange to have that weekend at Bread Loaf, which would be such fun.

Next weekend all the bigwigs arrive for the NATO conference here in Oslo, and the town will be stuffed to the gills. They have persuaded one of the transatlantic ships to stay in port so they can have a few extra beds for the press! Sec. State [Dean] Rusk will be staying with our Ambassador, and heaven knows where all the rest will be. Embassy wives are to be on tap night and day, in case they are needed, and the US delegation is taking over just about the whole Embassy. I am afraid we, at our level, shall not be seeing much of the bigwiggery—too bad, as I am a great rubbernecker, and would just love to see Rusk. (I am also sure that a big batch of galleys will arrive just at that moment, and that the final batch

* "Oh, well, Escoffier, you know."

will arrive just as we are trying to pack up the house—such is life. I have now had nothing since the first batch. But I did, at last, get a letter from J. Jones with some specific instructions. I am not, thank heaven, ultimately responsible for the typography!)

Perfectly lovely spring weather here, and Paul can't keep himself out of the garden, even if he knows his work will be for naught. He has trimmed and fertilized the roses to a fare-thee-well; weeded the strawberries; tied up the raspberries; seeded the lawn. He is now, I am sure, about to tackle the driveway. He just loves to be out doing exercise and making things neat and shipshape. And he is just hanging on until May 19th, his last day of official life. I shall be so glad when he is out of it! Best decision we ever made.

Now I must to work composing the perfect (I hope) and persuasive letter to that old archi-*française,* S. Beck Fischbacher.

Best love,
Julia

APRIL 30, 1961

Dearest J:
My poor lamb. Would it indeed be possible for us to be together to do the galleys. Knopf didn't even send me a set, though I asked for one, some time ago, hoping that I'd have a little time to look at them. Which I haven't, and looks as if I won't have. There won't be a breathing spell until after Commencement, June 14 or is it 15? Then a couple of weeks of very busy clearing things up, and the Perkinses take off for the West by car and I will be able to let down considerably. Mary has never been West at all, Perk not since he spent a year in Montana after two years of law school, when he decided to teach English History instead.

Actually I don't think you can go too far wrong, and I don't think you have to take it as hard as you are. Isn't there someone connected with the Embassy who has corrected proof? Hard to believe there is not. I never use anything for correcting proof except the single page in any dictionary, of proofreader's symbols. I certainly do not think you have to do anything about the type—never heard of such a thing. Surely those queries are for the printer. I'd know in a minute if I looked at a page. I don't see why you have to key the illustrations in either, and wish you had cleared that up with Judith. Well, hell, it will really all turn out about as well as one could hope for, and do remember, there just are no books at all, bar the Bible and *Webster's Dictionary,* that are free of howlers, and I'll bet there are a few subtle ones in those vols. Horrid truth is that as editorial help gets more and more expensive, there is less and less time even for a publisher like Knopf to turn out an admirable book. No you won't send the working draft of your next to a good copyeditor—unless you propose to pay same yourself. You will just have to go right along with Knopf, as to how they want it done, but next time, you'll be here and I can lend a hand. Dearie, don't fret so.

Must say your description of Simca's tangles with the galleys curdle the blood. I should if I were you simply ignore anything she says, unless you agree completely. Hope you can keep out of arguing with her at all. *Goût américain* indeed.

Well it would certainly be nice if you would come home end of May, and more than welcome to come here, as well you know, and you will have two rooms to spread out in. But the French [Algerian] thing has cleared up somewhat, and what about Charlie and Freddie? Anything will suit me. Mark I suppose will stay right over at Eliot House until Commencement, as he has a right to—we clear all the other classes out of the Houses early June, to clear rooms for reunion classes. And he says he is going to sleep in the library this summer, as he insisted on doing last summer, on a hospital cot. Gives him quite a bit of privacy, and he likes to sit up late, have friends in, write music and so on. Except this summer he has got a $300 a month job at the Chem. Lab, doing some kind of steroid research for a friend of his. So he will have to keep reasonable hours, should think.

Not yet announced, but to be announced late May, so don't spread it around, is that Mark has won a very nice fat prize from Broadcast Music Incorporated,* whatever that is—$750. Think of it. He submitted his Toccata† and a thing whose name I never can remember as it is Old French, and which I missed hearing when it was performed, because it was December and I was sick. I am not sure what Broadcast Music Inc. is except that it gives away a flock of prizes to young composers every year. I couldn't be pleaseder, naturally. So too is Mark. What with $1250 from Princeton, eight or nine hundred that he will make at Mallinkrodt‡ this summer, and this $750, he will have very nearly enough for his first year in graduate school, and Ma will only have to scrape up a few hundred,

---

* BMI was formed in 1939 as an advocacy group for songwriters. It began awarding cash prizes to student composers in the early 1950s.

† A rapid musical composition for a keyboard or stringed instrument.

‡ One of the chemistry laboratories at Harvard, where Mark would be working in the lab.

which can probably do out of income, with any luck. I think also I will give him for his graduation present a set of tails — he needs them, if he is going to do any conducting, and should have them made, so they will fit properly for flailing about. And Lord, Lord, at Princeton Graduate center they dine in gowns, but I will let him find out the lowdown on that. Perhaps they just wear fragmentary ones as at Oxford, token gowns.

. . .

Marguerite Mason telephoned a week or so ago about your request for a contractor, and I told her to send me a carbon of her letter to you, and she did. I must say I think I agree with her that you do not need a contractor, but perhaps that is because we never had one, and I don't know one, and we managed very well indeed with far greater changes in the house, structural changes and all, with no contractor. I think you will just have to give up hopes of having all this done reasonably and tidily and at once. If you can get John Malcolm for your plumber your troubles are half over. He is a miracle and a very nice man, and he will work beautifully with the electricians and carpenters, whoever they may turn out to be. His father did all the plumbing in this house when we bought it, supervised the whole heating job which was a big one, and took care of us until he died. John is even better. We bought the dishwasher and pig with his guidance, and he installed them, and he services them when anything goes wrong which it seldom does. I would take his advice on anything in the world.

I feel the same way about my carpenter — Haleys.

Marguerite is right about waiting on the garage, too. I hadn't realized that the floors in the house were soft wood. You may want to replace these, a few at a time. And if I were doing it, today, I would find out about the new flooring materials, and not necessarily order up the usual hardwood floors. There are all sorts of tile and composition floorings now that you don't have to wax and fuss over. And they are beautiful too. I have marvelous hardwood floors in this house but occasionally they are a real headache, as no cleaning woman knows what to do about them, and refinishing and waxing is difficult and heavy work.

But there's no reason why you shouldn't move into the house and have a lot of this stuff going on while you are in it. Just don't fret, is all. Both Marguerite and I can give you very good leads on workmen. And I see no reason whatever why Paul [Kennedy] and John Malcolm and whatever carpenter you get can't plan a kitchen that will knock your eye out. You get in a professional kitchen planner and he's going to charge you $5000 for shedding the light of his countenance, believe me.

Well. Quick drink and quick dinner. Lemon chicken again. It is so delicious and so easy that I will bet I have eaten it forty times in the last two months, and one of these days will probably not be able to look it in the breast again. And fresh asparagus which is wonderful right now. Who could ask for anything more?

I cannot discuss Cuba and all that. It is too dreary, too awful, too depressing. I only know the world limps on, and occasionally I get a bit of a laugh.

Best love to both and hurry home so we can
get a bit of a laugh together,
Avis

MAY 2, 1961

Dear girl:
Just got your letter dated 28 April, plus accompanying carbon of letter to our dear Simca. My first reaction—good old selfish ego—feeling of profound relief that I did not carry through and go visit at Bramafam this summer. Second reaction, sooner you two wind things up and come home, the better. Though I'd be hard put to be absolutely rational about that last.

Well am quite sure it will all come out in the wash, the book will appear, in better shape than you think, Simca will be quite pleased, and so on. But oh Lord how hard on you right now. Am sure that by now, having got it off your chest in a great burst, you are feeling more relaxed. Hope to God so anyhow.

Leave us face it. No relationship is flawless. And a relationship like yours with Simca is in many aspects like a marriage. Very good ups and very bad downs. But has been a working relationship and on the whole, good and productive. And the child you have produced is going to have flaws too, but will be on the whole, good. We must settle for what we can get.

I am extremely glad that you are coming home, and hope that you come straight here, or shortly thereafter, even if only to check in. June 5th. H'm. Let me know, soon as you can, whether you will come here, or stay in NY for a bit. I will be in orbit, but do not mind. There will be beds, rooms, food, drink. Even though I will be out and away long hard hours. It will be just plain heaven to have someone my own age-like here occasionally, to have a drink with, to have a dinner with. Let alone while you are here you can lend a hand with my marketing which is driving me mad. I am too tired at the end of the day, want to trudge right home. Sage's has taken to not opening until nine A.M., so that I cannot stop in at 8:30 as always and personally pick things over. Ordering by telephone very unsatisfactory, as today when I ordered three quarters lb. calf's liver all in one piece, unsliced—so of course they sliced it, much too thick. One cannot win, and I know it is unimportant, but to come home and look at those slabs, just too thick to re-slice thinly, made me swear like anything. I love coming home bushed and fooling around in the kitchen for a bit before my drink, but not to do the marketing at that end of the day also. Well, just another case of life not being perfectible.

Tomorrow I will see if I can order a couple of commencement tickets for you two, assuming and hoping that you will be here to watch Mark commence, June 15. Great day. And even if he were not commencing, it is quite a spectacle, and you will enjoy it. Cannot promise to get the tickets, but will try, and if you cannot promise to be here the 15th, *n'importe,** they can always be used, but it would be fun. Not that I can be with you much, as have to be at L. Hse [Lowell House] at 8:30 for start of academic procession, which with us includes a piper in a kilt! And have to be at University Hall at 9 to check diplomas, and then free to watch things until after commencement proceedings, and then this year, and this year alone, over to Eliot House for lunch, leaving Eleanor Hess to preside at L. Hse and help give out diplomas, which is done House by House—but I will be at Eliot watching my child.

Well kiddo, come home. We are probably going to have no spring at all. It is ice cold, with no really balmy days yet. This pattern will break one of these days and very likely plunge us straight into blistering weather. So if too hard on old Paul, you can get him a portable air-conditioner and move it to your new house. We will cope somehow. As you are coping now. Your letter to Simca is merely wonderful. If anything could calm her down, bring her around, reduce her to reason of a kind, this is it. I can't wait to see you both and talk it all out.

Fondest love to both, and all will come round right in the end,

Avis

* "It doesn't matter."

Epilogue — 1962-1988

I've finally finished my TV recipes for the already taped shows, and can now do a little bit of correspondence, and answer some Christmas letters, and think about the shows to come, and life. We feel that a new life is opening, without all this terrible bookery hanging on the shoulders and weighing down the mind, so that there was no freedom from guilt and no hiding place. Praises Be!

JULIA CHILD to AVIS DEVOTO, January 5, 1971

On the set at WGBH, Julia's crew, out of camera range, waits to hand up ingredients and finished dishes to the star for her first show, *The French Chef*. PREVIOUS PAGE: The first program of *The French Chef* television series aired in February 1963 and catapulted Julia into the nation's view as the avatar of French cuisine.

From October 1961 to August 1962, 100,000 copies of *Mastering the Art of French Cooking* were sold, and the book went into a fifth printing. Julia was thrilled with the success and wrote to Bill Koshland at Knopf, "We are absolutely amazed at that 'somewhat in excess of 35,000' Book of the Month orders. Can it be true? After you have digested and confirmed it a bit more, tell us again. Avis says she's not surprised! (Our greatest promoter in the area.)" But neither Knopf nor Julia could imagine the sales that her forthcoming TV series *The French Chef* would generate for the book.

A friend from the Childs' Paris years, Beatrice Braude, worked at Boston's WGBH TV as a researcher, and she mentioned that Julia should contact the station about a possible appearance on Professor Albert Duhamel's book review program. Armed with a copper bowl, a whisk, an apron, knives, mushrooms, eggs, and a hot plate, Julia appeared on the show to talk about her cookbook and to demonstrate how to cut and chop, beat egg whites, decoratively "turn" a mushroom cap, and make an omelet. Usually a ten-minute segment, the *I've Been Reading* program ran for thirty minutes that day. Letters from viewers prompted a series of three pilot programs on omelets, coq au vin, and onion soup, and their success led to a follow-up series called *The French Chef.*

On February 12, 1963, following the first broadcast of the show, which had aired the night before, Julia wrote to Judith Jones:

> We have gotten quite a few calls, etc, and people seem to like it. It will be shown in NYC over Channel 13 every Wednesday afternoon from 3 to 3:30—too bad it won't be at night so you can look at it. The first show, on Boeuf Bourguignon, was, of course, our first taping. It went quite well, I thought, though a bit rough and hurried in spots. We have now finished 11 of the tapes, and I think some of the performer's bloopers have been cleaned up, such as too hurried, looking at wrong camera, whipping things out of view of the camera before camera has had time to focus, etc. It is certainly fun to do, but, my heaven, what a lot of preparation. Every spoon, dish, onion and piece of fish has to be on hand and in its right place. Luckily a very nice girl, the associate director, is able to work all day with me rehearsing before we tape, otherwise all would be chaos.

Between 1963 and 1966, 119 black-and-white shows aired. Simultaneously, the logo of the larger-than-life Julia, jauntily carrying her iconic larger-than-practical wooden spoon over her shoulder like a garden shovel, radiated across the country. For all her lightheartedness and occasional clowning on the air, she single-mindedly kept her mission to teach front and center, and Americans not only learned to cook but had fun doing it.

Succeeding in a medium that had eluded her predecessors James Beard and Dione Lucas, Julia inspired viewers with her can-do approach to French cuisine. Her use of gourmet and ethnic ingredients propelled the opening of specialty food shops nationwide. Entrepreneurs such as Charles E. "Chuck" Williams imported and sold cookware and tabletop items. Williams, who started selling such items in his hardware store in Sonoma, California, later expanded to San Francisco and then throughout the country, making Williams-Sonoma a household name. Kitchen design, including colored appliances, became an architectural focus, as families and guests moved out of the family room and into the kitchen.

The WGBH studios resounded with Julia's laughter. She was increasingly in demand to give demonstrations on the charity and Smith alumnae circuits. Her TV recipes were published in newspapers and magazines and collected into a book called *The French Chef Cookbook*. And she was featured as the symbol of America's "food revolution" on the cover of the November 25, 1966, issue of *Time* magazine. The proceeds from the continuing sale of *Mastering the Art of French Cooking* permitted Julia and Paul to build a modest home, called La Pitchoune, on Simca and Jean Fischbacher's property in Plascassier. During those years, Julia also acquired a dedicated staff, including Avis, who had left her position in the office of the dean of students at Radcliffe in 1969 to concentrate on Julia's fan mail and to keep her informed about events in Cambridge and the rest of the country when she was in France.

Very soon after the successful publication of *Mastering the Art of French Cooking,* a second volume was in the offing. But with her busy TV schedule, Julia was unable to commit herself fully to it. The completion of La Pitchoune in 1966 enabled Julia and Paul to live there from December to June so that Julia and Simca could work together on the new book. They followed almost the same pattern established for their first book. Simca supplied most of the recipes, and Julia refined, tested, and produced the book. During a holiday visit in 1966, Avis watched them in action and wrote about the experience:

> I was realizing again what a great technician Simca is—Julia in this instance was following her lead. This is certainly one of the great collaborations in history. They are absolutely necessary to each other and it is a happy miracle that they met and began to work together. It is the <u>combination</u> which makes their work so revolutionary, and for my money they are benefactors of the human race.
>
> Simca is a creative genius, as Julia wrote me, "a great fountain of ideas." She is also inaccurate, illogical, hard to pin down, and stubborn as a mule. Julia is also very creative and is becoming more so. But the two women think differently. Julia is deeply logical, orderly, accurate, painstaking, patient, determined to get all this knowledge clearly on paper. And she can be just as stubborn as Simca is, and will plug away trying to convince Simca until suddenly Simca changes her position, and from then on she will talk as if it were her own idea all along. I have seen her do this on a number of occasions. Like the great *cassoulet* argument, mostly conducted by mail.

> But I am sure that deep down Simca just has never been convinced that all that accuracy and measurement is anything but a crazy American idea.

Knopf asked Avis to help proofread and do the index for volume 2 of *Mastering the Art of French Cooking,* which was published to high praise in 1970 and became famous for Julia's nineteen-page recipe for French bread. Although volume 2 was the end of Julia and Simca's cookbook collaboration, they continued to work together to keep L'École des Trois Gourmandes a well-known and successful venture, and they continued to publish individually.

Throughout the filming of Julia's first five TV series and the publication of her first six cookbooks, Avis remained Julia's friend, enthusiastic fellow cook, and commentator on the national political scene and the local Cambridge community. It was Avis who kept Julia in contact with her audiences and her inner self. Julia called Avis her "pen pal," but both knew that Avis was much more. First and foremost, as a part of the publishing community, Avis was responsible for securing the publication of Julia's masterpiece. As her neighbor, Avis introduced Julia to the intellectual set that she and Paul valued, both in Cambridge and at the Bread Loaf Writers' Conference. Avis also sympathized with Julia when Paul's mental confusion progressed following coronary bypass surgery and a series of small strokes in 1974. As she herself had done after Bernard's death, Avis encouraged Julia to continue to throw herself into her work after Paul's health sharply declined in 1989, and Julia had to place him in a nursing home. When Avis died of pancreatic cancer on March 7, 1989, Julia lost her favorite correspondent, her sounding board, her confidante, and one of her best friends.

AIR MAIL
AVION
REPUBLIQUE FRANÇAISE
1,40
GORGES DE L'ARDECHE
Mrs Paul Child
103 Irving Street
CAMBRIDGE, Massachusetts 02138
U.S.A.
PAR AVION
VIA AIR MAIL

Par Avion
REPUBLIQUE FRANÇAISE
18F
8F
MARSEILLE
25F BROUAGE CHARENTE MARITIME LES REMPARTS
40F
12F
QUIMPER
Mrs. B. DeVoto
8 Berkeley Street
Cambridge 38
Mass.
USA.

Acknowledgments

Were it not for the accomplishments of the following women who have made things happen across America's culinary landscape, this book would not be possible. Julia Child, of course, inspired a huge number of people to enter the culinary profession, but she also made the home kitchen accessible to the women and men who adopted her can-do spirit and became serious about defining cooking as both technique and fun. Avis DeVoto was almost iconic in her multitasking roles as wife, secretary to her husband, mother, career woman, community organizer, political activist, enthusiastic cook, friend, facilitator, and so much more. Dorothy de Santillana, Julia's first editor, outvoted by her peers at Houghton Mifflin, encouraged Julia, Simca Beck, and Louisette Bertholle in their efforts to publish *Mastering the Art of French Cooking*. And Judith Jones, Julia's editor at Knopf, excelled in a publishing world that focused on the cooking celebrities of the moment by taking a chance on three unknown authors who knew how to communicate the techniques of French cooking "to the servantless American cook" rather than simply collecting and reproducing the classic recipes of French *grande cuisine* that up to then had been shrouded in mystery.

The efforts of the following people also made *As Always, Julia* possible. William Truslow, Julia Child's lawyer and a board member of the Julia Child Foundation, and Mark DeVoto, the younger son and trustee of Avis DeVoto's estate, generously granted permission to select and edit these letters. I am grateful to both for their trust in me. I also owe a considerable debt of gratitude to Kathy Jacobs, curator of manuscripts; Ellen M. Shea, head of public services; Diana Carey, reference librarian for visual resources; and the rest of the generous staff of the Schlesinger Library, where both the Julia McWilliams Child Papers and the Avis MacVicar DeVoto Papers are archived.

My special thanks go to my agent of more than twenty-five years, Doe Coover, a consummate professional who championed this project and made it happen. Rux Martin, my editor at Houghton Mifflin Harcourt, believed in this book early on. Her judgment calls were always on target, and her enthusiasm was boundless. I'm grateful to Barbara Jatkola for her painstaking copyediting, as well as to the staff at Houghton Mifflin Harcourt, including Rux's assistant, Tim Mudie; the cover designer, Martha Kennedy; and the interior designer, Melissa Lotfy, who made this book a handsome reality.

My friends and past editors Darra Goldstein, Sheila Levine, and Linda Rosenberg read the proposal and encouraged me to publish these insightful letters. My husband, the historian John J. Reardon, immediately recognized their value and supported the project with his constant interest. I gratefully acknowledge the assistance of Hilary Thorsen, who copied many of the originals at the library; Jacinta Monniere, who retyped the letters; and Frances Kennedy at the Doe Coover Agency, who helped with the research.

At no time did I ever feel that *As Always, Julia* was my book. It belongs to Julia and Avis. I simply fell in love with their letters and thought that others would, too. I am grateful for their legacy of words and for the insights into food, friendship, and the making of a masterpiece that they have given us.

Index